D0079301

NOTHING LESS THAN VICTORY

NOTHING LESS THAN VICTORY

DECISIVE WARS AND
THE LESSONS OF HISTORY

JOHN DAVID LEWIS

PRINCETON UNIVERSITY PRESS PRINCETON AND OXFORD

Published by Princeton University Press, 41 William Street, Princeton, New Jersey 08540
In the United Kingdom: Princeton University Press, 6 Oxford Street, Woodstock, Oxfordshire OX20 1TW

Library of Congress Cataloging-in-Publication Data
Lewis, John, 1955–
Nothing less than victory : decisive wars and the lessons of history / John David Lewis.
p. cm.
Includes bibliographical references and index.
ISBN 978-0-691-13518-2 (hardcover : alk. paper) 1. War—Termination—History. 2. War—Moral and ethical aspects—History. 3. Peace—History. I. Title.
U21.L524 2010
355.02—dc22 2009024513

British Library Cataloging-in-Publication Data is available

This book has been composed in Goudy Old Style
Printed on acid-free paper. ∞
press.princeton.edu

Printed in the United States of America

1 3 5 7 9 10 8 6 4 2

CONTENTS

MAPS AND ILLUSTRATIONS

ACKNOWLEDGMENTS

Students in my university class, "Warfare: Ancient and Modern," confronted me with compelling questions and divergent conclusions about warfare across the ages and suggested this book. Sections have been presented to the Society for Military History, the 179th Squadron of the Ohio Air National Guard, the Social Philosophy and Policy Center, Rice University, and Ayn Rand Institute Conferences, and in many lectures in Europe and North America. Barry Strauss, Talbot and Barbara Manvel, Lin Zinser, Nick Provenzo, Tammi Sharp, and Alan Germani read sections and offered challenging comments. Special thanks go to John and Betty Allison, Sidney Gunst, Ed Locke, Brad and Sidney Thompson, Ted and Marilyn Gray, and others for reading or listening at various stages. The Anthem Foundation for Objectivist Scholarship and its founder John McCaskey provided a grant that allowed a semester's leave. Three articles based on chapters have appeared in *The Objective Standard*, whose editor, Craig Biddle, has taught me much about writing. I finished the book as a visiting scholar at the Social Philosophy and Policy Center, Bowling Green State University, whose directors and staff showered me with opportunity and encouragement. Final preparation was done at Duke University; thanks go to Michael Munger and everyone in the Department of Political Science. Princeton's ace editor Brigitta van Rheinberg and anonymous readers gave the best support of all, solid criticism. Most of all, my wife Casey demands shorter sentences, more pointed paragraphs, and death to all digressions. She is right, as usual.

Introduction

Victory and the Moral Will to Fight

Americans today have been told to expect years of military action overseas. Yet they are also being told that they should not expect victory; that a "definitive end to the conflict" is not possible; and that success will mean a level of violence that "does not define our daily lives."[1] A new administration is now bringing more troops into Afghanistan—where American troops have been operating for eight years—but without defining the terms of victory. The change in American military doctrine behind these developments occurred with astonishing speed; in 1939 American military planners still chose their objectives on the basis of the following understanding: "Decisive defeat in battle breaks the enemy's will to war and forces him to sue for peace which is the national aim."[2] But U.S. military doctrine since World War II has progressively devalued victory as the object of war. "Victory alone as an aim of war cannot be justified, since in itself victory does not always assure the realization of national objectives," is the claim in a Korean War–era manual.[3] The practical result has followed pitilessly: despite some hundred thousand dead, the United States has not achieved an unambiguous military victory since 1945.

Historically, however, this debasement of victory in military planning is radical. Aristotle knew that "victory is the end of generalship," and no Roman army fought for anything less.[4] The change in doctrine is not due primarily to the horrific destructiveness of modern war, for American leaders have adopted such aims even for conflicts that do not threaten to "go nuclear." We inhabit a moral climate in which any attempt by victors to impose cultural values onto others is roundly condemned. We have largely accepted that the pursuit of victory would necessarily create new grievances and guarantee an even more destructive conflict in the future. This is a moral issue.

But this idea should be questioned. This book presents six major wars in which a clear-cut victory did not lead to longer and bloodier war, but rather established the foundations of a long-term peace between former enemies. Each of these conflicts began with an act of military aggression. Each stagnated during years of carnage that ended when a powerful counteroffensive and an unambiguous victory reached deeply into the moral purposes behind the war, and forced one side to give up its cause and to renounce the fight. The result, in each case, was not "universal peace," but an understanding between former enemies that definitively ended the war and brought enormous benefits to thousands or millions of people. How and why these successes were achieved is the subject of this book.

The causes of war and peace run far deeper than the movements of armies and troops (strategy and tactics) into the reasons why armies form and move at all.[5] War is an exclusively human activity: animals eat each other, and clash over mates, dominance, and territory, but absent the capacity to pursue chosen goals and values with an organized commitment, they do not wage war. Contrary to Freud's conclusion that mankind has a universal "instinct for hatred and destruction," wars of a continental scale lasting years do not just happen, by chance, circumstance, or instinct.[6] The wellspring of every war is that which makes us human: our capacity to think abstractly, to conceive, and to create. It is our conceptual capacity that allows us to choose a nation's policy goals; to identify a moral purpose for good or for ill; to select allies and enemies; to make a political decision to fight; to manufacture the weapons, technologies, strategies, and tactics needed to sustain the decision over time; and to motivate whole populations into killing—or dissidents into protest. Both war and peace are the consequences of ideas—especially moral ideas—that can propel whole nations into bloody slaughter on behalf of a *Führer*, a tribe, or a deity, or into peaceful coexistence under governments that defend the rights and liberties of their citizens. The greatest value of the examples in this book is to show the importance of ideas—especially moral ideas—in matters of war and peace.[7]

Moral ideas as they relate to war must not be conflated with the rules associated with deontological just-war theory—for instance, of proportionality and absolute prohibitions against attacks on civilians.[8] Such rules divorce ends from means, and are often considered by their advocates to be absolute strictures apart from context and consequences. In this moral framework, the goals of each nation are granted no import in evaluating the conduct of the war, and those fighting to maintain a system of slavery become morally equal to those fighting for freedom. That such rules can become weapons in the hands of an enemy who is fighting for conquest, loot, or slavery is said to be irrelevant to the categorical commandment that each side follow those rules regardless of result. But surely we should question moral rules that exempt a belligerent from attack because he hides behind civilians whom he intends to enslave. The moral purpose of a war—the goal for which a population is fighting—sets the basic context for evaluating a conflict and determines the basic moral status of the belligerents. Those who wage war to enslave a continent—or to impose their dictatorship over a neighboring state—are seeking an end that is deeply immoral and must not be judged morally equal to those defending against such attacks. It is vital to evaluate the purposes of a war when evaluating both the means by which that purpose is being pursued, and the social support for those directing the war.

Because warfare is first and foremost a clash of moral purposes, acknowledging the place of moral ideas over physical capacities—expressed in Napoleon's dictum that "the moral is to the physical as three to one"—establishes a hierarchical relationship between the ideas fueling a war and the shifting details of terrain, tactics, and technology.[9] Certainly the tactics of Roman foot soldiers cannot be applied directly to tank divisions today, but the Romans might be able to tell us something about the motivations of a stateless enemy that is subverting a world power. This perspective on war leads to a certain conclusion about war's proper object, which is not the destruction of an enemy's army or industry. The goal of war is the subjugation of the hostile will, which echoes Carl von Clausewitz's identification that war is "an act of force to compel the enemy to do our will."[10] Clausewitz also wrote that "war is an act of human intercourse" and that "the essential difference" between war and the arts and sciences is that "war is not an exercise of the will directed at inanimate matter, as is the case with the mechanical arts, or at matter which is inanimate but passive and yielding, as is the case of the human mind and emotions in the fine arts. In war, the will is directed at an animate object that reacts."[11]

Clausewitz wanted to refute the idea, promoted by the Napoleonic officer Baron de Jomini, that military principles could be turned into charts and followed as rules.[12] Some theorists today, trying to understand warfare in scientific terms, are adopting the analytical methods proper to complex physical systems, using principles derived from chaos theory.[13] The idea here is to see every situation as a dynamic, interactive whole, to grasp an enemy's reactions as "feedback," to recognize that the complexity of the system does not allow us to grasp every detail of the action within it, and to understand the whole without being overloaded by its complexity.

But there is a limit to such analysis, for no mathematical methods will ever be able to quantify the primary factor that drives a war: the willingness of the leadership, soldiers, and civilians to butcher thousands in pursuit of an abstract moral purpose. Wars do have a fundamental cause: the moral purpose that motivates the decision and commitment to fight, which is expressed, in one form or another, from the highest levels of leadership down to every grunt with a rifle, a spear, or a club. The complexity deepens given the different ways such ideas are understood in different cultural contexts, the degrees of human ingenuity and technological sophistication available to pursue this purpose, and the commitment to achieve the moral purpose embodied in a nation's policy goals. The possibility of conflict increases to near certainty if a nation comes to accept that glory, loot, or moral sanctity can be achieved by the initiation of horrific force on behalf of a tribe, a ruler, a deity, or a collective.[14]

The will to war is the motivated decision and commitment to use military force to achieve a goal. How, then, can a commander use physical force to compel an enemy to change his mind and to reverse the decision and commitment to fight? Military planners have been consumed with this enormous question; debates among American strategists in the 1930s about the role of a new air force, for instance, grappled with the question of whether the destruction of an enemy nation's industry could break its will to fight. The selection of targets became the dominant issue in air strategy, and the solution required industrial experts to determine which factories had to be destroyed to break the will and the capacity of the enemy population to support the war.[15] This issue remains contextual—Germany was defeated by "boots on the ground," whereas Japan surrendered under air assault, just as the Romans had to smash their way through the Goths, while Palmyra surrendered when threatened by a siege—but commanders knew that the defeat of the hostile will was invariably the object of the war. In his 1934 testimony before the Federal Aviation Commission, Lt. Gen. Harold L. George stated that "the object of war is now and always has been, the overcoming of the hostile will to resist. The defeat of the enemy's armed forces is not the object of war; the occupation of his territory is not the object of war. Each of these is merely a means to an end; and the end is overcoming his will to resist. When that will is broken, when that will disintegrates, then capitulation follows."[16]

Clausewitz had seen certain principles behind this issue long before air power made its debut, in the nature of war as a duel akin to two wrestlers, each grappling for balance, albeit founded on a political decision that depends upon a network of political, economic, social, and military support.[17] The result, he concluded, is a certain "center of gravity" for each side, the point of greatest vulnerability, which is not necessarily its army. "If Paris had been taken in 1792, the war against the Revolution" would most certainly have been ended, Clausewitz explained. "In 1814, on the other hand, even the capture of Paris would not have ended matters," given Napoleon's sizable army.[18] Clausewitz elaborated: "For Alexander, Gustavus Adolphus, Charles XII, and Frederick the Great, the center of gravity was their army. If the army had been destroyed, they would all have gone down in history as failures. In countries subject to domestic strife, the center of gravity is generally the capital. In small countries that rely on large ones, it is usually the army of their protector. Among alliances, it lies in the community of interest, and in popular uprisings it is the personalities of the leaders and public opinion. It is against these that our energies should be directed."[19]

But Clausewitz's conclusion needs to be more solidly anchored in the social support for a war, which provides irreplaceable ideological, material, psychological, and moral resources for a nation's armies and its leaders. The "center" of a nation's strength, I maintain, is not a "center of gravity" as a point of balance, but rather the essential source of ideological and moral strength, which, if broken, makes it impossible to continue the war. A commander's most urgent task is to identify this central point for his enemy's overall war effort and to direct his forces against that center—be it economic, social, or military—with a view to collapsing the opponent's commitment to continue the war. To break the "will to fight" is to reverse not only the political decision to continue the war by inducing a decision to surrender, but also the commitment of the population to continue (or to restart) the war.

To force this reversal of the decision and commitment to fight, a commander must *know himself, his own people, his enemy, and his enemy's people—and, he must know his own moral objective as well as that of his enemy*. An effectively aimed, well-planned, and quickly executed counteroffense—or, even better, a credible show of strength that collapsed opposition with little killing, a feature of three of the examples here—was the climax of the six conflicts in this book. But each of these counteroffenses worked not only because of its physical success but because it forced that enemy, including the civilian population, to confront the enormity of what it had done, to recognize its hopeless inability to continue the fight, to lose heart as it lost support, and to give up the pretenses, misunderstandings, and delusions that had fueled the war from its start.

The chapters illustrate these effects on several levels. Strategically, each illustrates a major war, lasting more than a year and costing thousands of lives, which began with an attack against a political state that controlled a distinct territory, and whose survival was on the line.

I do not here deal with usurpations by competing Roman emperors, long-standing struggles for power between medieval families, or wars within a single territory, whether religious, civil, or revolutionary. The American Civil War warrants inclusion because the United States was geographically divided into two distinct territories, with opposing governments motivated by opposing moral ideals, and it is necessary to consider the two territories as distinct in order to grasp the nature of the conflict and the constitutional peace that has followed. In each of the examples, the war became a bloodbath without end until the side under attack launched an energetic offense against its opponent's social, economic, and political center. The military success was accompanied by a forthright claim to victory, so that there were no illusions about who had won the war.

But strategy alone does not explain the result; the tide of war turned when one side tasted defeat and its will to continue, rather than stiffening, collapsed. To understand this, it is vital to consider the cultural background of the opponents. For the Persians of the early fifth century BC, to take one illustration, the destruction of Xerxes' navy was an indirect projection of power against the ideological center of his rule, which was founded upon a culture in which people expected him to assert his magnificence through conquest. The defeat laid bare his fundamental weakness, which potentially undercut his position fatally. To protect his reign (and his neck), he had to abandon the goal of expansion that he had inherited. This became the policy of the king's court beyond his own lifetime. Despite differences in politics, terrain, technology, and tactics, similar results followed for each of these examples; in each case, the military failure reached deeply into the very identity of the regime itself and undercut the moral basis of the regime's actions as it destroyed material support for the war. The result was not only a change in the strategic balance of power, economic resources, or technology but a long-term change in *policy*.

In chapter 1, the Greek city-states faced decades of attacks by the Persian Empire. The Greeks ended the aggression—and discredited the ideology behind it—by ruining the enemy's army and threatening the king's position on his own soil. The Persians never again attacked the Greeks. In chapter 2, Sparta, home to the world's most feared infantry, mounted years of attacks against an alliance of farming communities headed by the city of Thebes. The conflict should have been no contest—but the Spartan mirage was shattered when those farmers marched into the Spartan heartland and made Sparta's defeat unmistakable. The Spartans never again attacked the Thebans. Chapter 3 pits Carthage against Rome, via a long indirect route leading to Hannibal's attack against Italy in the Second Punic War. After years of indecisive warfare, the Romans won quickly with a direct attack on Carthage's homeland. Carthage accepted its position and never again attacked Rome. In chapter 4, third-century AD Rome was rent by internal usurpations and external attacks that divided the empire into thirds. The emperor Aurelian reunited Rome in an energetic campaign that collapsed the threats posed by its eastern and western enemies. As in the previous cases, a display of overwhelming force exposed the physical and ideological bankruptcy of those advocating war, which led to an immediate collapse in the will to fight.

The ancients are worthy of this much of the text because they reveal the basic issues behind every war in terms that are stripped of their modern nationalistic, technological, and logistical embellishments. Ancient writers placed the human elements first, especially the motivations that fuel a war and that must be reversed if long-term peace is to follow. We are hard-pressed to find such lasting victories, either in our own day or in the ancient world—they are in a certain sense anomalous in history—but there are three modern examples in this book. In chapter 5, a long, deeply rooted war of rebellion within the United States ended quickly when one Union army pinned the southern forces in the North, while another marched through the South and destroyed the economic and social foundations of the rebellion. Once again, military defeat accompanied the psychological and ideological collapse of the will to fight.

Chapters 6 and 7 consider World War II—the worst slaughter in history, which killed millions until the defenders mounted an overwhelming offense against the capitals of Germany and Japan. Chapter 6 takes up a very specific aspect of this conflict: how certain moral ideas conditioned Britain's response to Hitler and became causal factors in the onset of war. This chapter differs from the others in that it deals little with fighting and not at all with the end of the war—its concern is with the prelude to war and the way in which a clear-cut aggressor was empowered to attack. Warfare studies are not only about strategy and tactics but also about why armies end up on the battlefield when and where they do. Chapter 7 turns to the Pacific war and considers the nature of the victory over the Japanese decision and commitment to fight.

A rich variety of details adorns these events. In each case, the leadership of the defeated side was discredited, emasculated, or demoralized by the visible evidence of defeat; neither Agesilaos of Sparta, Hannibal of Carthage, Zenobia of Palmyra, Jefferson Davis, nor Hirohito could rouse his people to further action. For others, cruel war came directly to their homes; they saw their economies destroyed and their former vassals rise against them—as the Spartans, the Carthaginians, and the Palmyrenes saw their support evaporate, and southerners in America saw former slaves set free by a Union army. In some cases their cities were surrounded and pulverized, as the Goths, the people of Atlanta, and the Japanese saw their towns burned. In others, civilians shook with fear, knowing that an enemy army was on the way; the bloodless campaign of the Theban leader Epaminondas parallels the campaigns of the Roman emperor Aurelian into the East as well as Sherman's march through the South. Reliance on divine providence—through Ahura Mazda in Persia, Baal-Shamim in Carthage, Sol in Rome, and the Chrysanthemum Throne in Japan—played a part in unifying the efforts of some, and in the victory or defeat that followed. In one case, the focus is on the prelude to a war: the defensive posture assumed by British leaders, who failed to challenge the moral claims of the Germans, left them unwilling to oppose Hitler while it was possible to do so. This is a powerful example in reverse of the effects of certain moral ideals on the policies of rational statesmen who genuinely wanted to avoid a new war.

Each of these examples has been studied in great depth, and there is a mountain of scholarship for each. No chapter-length essay should aspire to provide the details that specialists will crave, or to even attempt to exhaust the studies made of these events. Some readers may criticize the omission of many wars that could not be included. But all authors must be ruthlessly selective. B. H. Liddell Hart, in his important book *Strategy*, surveys history in order to present his thesis of indirect strategies. Yet he reduces the medieval period to eight pages (given "the drab stupidity of its military course"). Similarly John Lynn, in his important book *Battle: A History of Combat and Culture*, claims to take a global view of warfare in order to refute claims to historical continuity from the Greeks. Yet Lynn ends his discussion of Western antiquity with Greece circa 400 BC, skips past the Hellenistic and Roman periods, and pauses on ancient China before jumping ahead seventeen hundred years to the Hundred Years War, which he equates with the medieval period. No Alexander, Hannibal, Caesar, Justinian, et al. beyond scattered mentions—and neither were these necessary. Each of these authors illustrates his own thesis and invites readers to look further at the areas he chooses not to cover. I do the same.[20]

There is no single strategic pattern, no universal "theory of war," and no moral "rules" divorced from context or purpose to emerge from this book. The major point is to *take moral ideas seriously*. The lessons of history relate not to tactical or strategic rules but to the ideas that motivate people to fight and their consequences in action. An aggressive nation can be empowered far beyond its physical strength by a conclusion that its opponent does not have the will to fight—surely a factor in Xerxes' invasion of Greece and Japan's commitment to the "Asian War"—and then be demoralized and beaten by an offense that exposes the physical and moral bankruptcy of its position. Conversely, a powerful nation may give up if its people come to think that a war is unjust, their nation's position is morally untenable, or its goal unclear or simply not worth it. In either case, our recognition that war is the product of human ideas, ambitions, intelligence, and morality allows us to put the primary focus of warfare studies where it belongs: on human beings, who are the locus of the decision and commitment to fight, and the only agents capable of creating freedom and peace for themselves.

Chapter 1

"To Look without Flinching"

The Greco-Persian Wars, 547–446 BC

The Hostile Will

It was the summer of 480 BC, and the Great King Xerxes, ruler of the mighty Persian Empire, son of Darius and heir to the Achaemenid throne, King of Kings and beloved of the deity Ahura Mazda, stood at the head of his army, looking down on the object of his revenge: the Greeks. He had every reason to be pleased, and to anticipate swift victory over a ragtag enemy. For months, the largest military force ever seen had marched and rowed to his command, drinking the rivers dry as city after city sent tokens of tribute and submission. The last of his Greek enemies would soon be ground under his feet. Yet within weeks everything had changed: the king was in full retreat, his dream in ruins, his navy scattered, and his army facing annihilation. It was as if all the energy of empire, once pushing forward in an unstoppable juggernaut, had stopped and turned inward on itself. Greece would never submit to this king's will, and no Persian king would ever again invade Greece. Why this sudden turnaround? And, most important to the future of the Western world, why was it permanent?

The basic story is well known. A Persian naval attack ten years earlier had ended on the beach at Marathon, when ten thousand Athenians defeated some thirty thousand Persians. The honor in this victory was so high that the playwright Aeschylus is said to have inscribed on his gravestone, "I fought at Marathon."[1] The Persian commander retreated, but King Darius swore vengeance. His son Xerxes inherited the call to revenge, and in 480 he marched a gigantic force into Greece, rolling over the Greek defenders and sacking Athens twice. Every lover of history knows about the battle of Thermopylae, where three hundred Spartans—with several thousand allies—held off hundreds of thousands of Persians for three days, while the Persian navy was stymied at Artemisium. The king lay waste to Athens, but the sea battle at Salamis broke the back of his forces, and his navy disintegrated before his eyes. He left his general on Greek soil—to be destroyed the next year at Plataea, along with thousands of subjects—and ran for home. He never returned.

This is where most accounts of the Greco-Persian wars have ended—with the military defeat of the Persians in a few battles on the Greek mainland and their withdrawal.[2] But the meaning of the Great King's disaster reached far deeper than a battlefield defeat and a strategic retreat. There was a change in policy in the Persian court—seventy years of aggressive doctrine was reversed when westward expansion of the empire was permanently abandoned. What are the reasons behind this astonishing reversal of policy? Why was a pattern of attacks four generations old, fueled by a mandate for revenge that Xerxes had inherited from his father, and supported by all the resources of the Great King's court, so quickly broken? The answers are rooted in certain ideas and practices that reached back to the emergence of the Persian Empire from its Near Eastern background, and stretched into military action on the king's own soil that lasted a generation.

In the early 550s BC the Persians under Cyrus the Great—one of the most successful commanders in history—revolted against the Median Empire. He swept out of the Iranian highlands, conquering the Medes and the Babylonians, and in two decades cobbled together an empire that reached from the borders of modern India into Turkey.[3] Cyrus ascended over a mosaic of competing rulers, cities, and empires with roots more than twenty-five-hundred years old, and he lay claim to a complex history in which one basic political-religious leitmotif had been omnipresent: assertions by grandiose rulers of personal magnificence and domination over prostrate subjects. Akkadians, Babylonians, Hittites, Assyrians, Mittani, Medes, Lydians—and myriad others—all took as a given that a ruler close to the gods, a good shepherd and father to his people, would assert his power and demand abject submission from those below. The image of the king was vital here—this was how he appeared to the vast majority of his subjects.

A stele of the neo-Assyrian king Esarhaddon (ca. 680–669) towering over two captives, bound and kneeling before him, illustrates the submission demanded by the awesome image of a great king. The Assyrian *Chronicle* says that Baal, the lord of the land of Tyre, "kissed my feet," "bowed down and implored" Esarhaddon as "lord," and paid tribute.[4] In the next century the cuneiform cylinder seal inscription of Cyrus made the terms of rule explicit: "I am Cyrus, king of the world . . . whose rule Bel and Nebo love . . . all the kings of the entire world . . . those who are seated in throne rooms . . . all the kings of the west living in tents, brought their heavy tributes and kissed my feet in Babylon."[5] Cyrus faced a world that understood what he was after and thought it perfectly natural.

Even Hammurabi of Akkad—a ruler traditionally credited with a comprehensive law code, some twelve hundred years earlier—began his laws with a proclamation of his own elevated status as a man of justice and power, called to rule for the good of his people.[6] Such kings demanded personal acknowledgment of their superior status—along with gifts of riches—from local rulers, who in turn claimed superiority over those below them. Before the Persians, the Assyrians did the most to institutionalize this suzerainty; they created a complex administrative organization with provincial governors, which passed, in varying forms, to the Babylonians and Medes.[7] But Cyrus refashioned this inheritance into an empire of uniquely comprehensive scope and organization. The realm of this so-called Achaemenid Dynasty—reaching ultimately from Egypt to the Indus River—was divided into some twenty satrapies (provinces), each controlled by a satrap (regional governor) who was respectful of local customs but was also bound to the king. This was the Persian Empire, in the eyes of the king the realm of just rule, deservedly magnificent, sanctioned by the gods, and bent on ruthless expansion.

Although the evidence is far from conclusive, Cyrus and the more educated subjects in his court may have dichotomized the entire world between his kingdom—the realm of light and of truth—and the realm of darkness and lies, where he does not yet rule. This worldview may reflect the direct or indirect influence of Persian Zoroastrianism, from the mystic Zoroaster (ca. 628–551 BC), which was in some way associated with the god Ahura Mazda. This dichotomous worldview, syncretized with earlier warrior practices and the desire for submission, would have offered a powerful justification for violent expansion, whether or not Zoroaster himself ever held such intentions.

Texts associated with Zoroastrianism and Persian traditions uphold a certain "right order" of society, founded on a complex hierarchy of castes, to be defended by a good king. A middle-Persian text, the *Book of Arda Viraf,* relates the late fourth-century BC invasions of Iran by Alexander the Great in terms of an evil god; its verses also tell of a Zoroastrian journey through heaven and hell, call for ruthless suppression of domestic insurrection on pain of agony in the next world, and uphold the legitimacy of the Persian ruler against foreign attacks. The *Letter of Tosar*, a third-century AD work that has undergone several translations and editions, demands recognition, by the Turks and Greeks, of the eminence of the Persian "King of Kings."[8] The ideology behind Cyrus had a long history, in many manifestations, both before and after him.

No single set of customs, and no single religion, was made exclusive in the kingdom; Persian kings wanted submission, not homogeneity. They routinely left non-Persians in charge of local affairs, once they had submitted; the attack on Marathon was commanded by a Mede. But Ahura Mazda was the divine symbol of the king's legitimacy, and fire, the use of libations in lieu of animal sacrifices, and depictions of Ahura Mazda to the exclusion of all other gods up to the reign of Xerxes support the conclusion that Darius and Xerxes "may have been good Zoroastrians."[9] The Behistun inscription, a proclamation of power carved into a cliff by King Darius (r. 522–486), lists those kings who hearkened to him, contrasts the Great King's truth with the kings who lied, and connects his rule to Ahura Mazda through his claims to royal ancestry.[10] The chaotic, disunited Greeks—constantly squabbling and quarreling, fighting each other with none becoming dominant—would have been the perfect foil, in the king's eyes, to the order established in his own kingdom.

The conclusion follows that the Persian King of Kings was driven by an ideology—a system of ideas, used for political purposes, and passed on to him by the legacy of his father—that conjoined claims to divine protection with displays of magnificence, demands for submission, and continuous expansion of his rule across the entire world.[11] This was a dynastic ideal with moral force that had guided kings for four generations. Cyrus conquered the Medes in Asia Minor by 550 BC, the Lydians on the coast by 547, and the Babylonians in Mesopotamia by 539. Cambyses (r. 530–522 BC) expanded his rule into Egypt and accepted the submission of the Libyans to the west.[12] Darius claimed the throne in 522 or 521, after ending a revolt by the Medes; he then attacked the Scythians of the Black Sea area in 516, the Thracians in 512, the Greek cities in Asia Minor in 499, including Miletus in 494, and the Greek mainland in 490. When Darius died in 486, he passed the desire for conquest on to his son Xerxes, who suppressed revolts in Egypt and Babylonia before marching against the Greeks in 480. The pattern is clear, even if the precise dates are not.

For these kings, each act of expansion was an expression of legitimacy that strengthened his connections to his ancestors, bolstered the support of the nobility beneath him, affirmed his personal magnificence, and gave his subjects a goal as well as a means to promotion by gaining favor in the king's eyes. All of this suggests that the decision to attack the Greeks was motivated not primarily by strategic concerns—calculations of relative power, for instance, or the need for material resources or taxes—but rather by the ideology of magnificent dominance, and that this ideology, not strategy, would dictate the size, organization, and use of military forces. Such motivations would only be strengthened, should a desire for revenge enter the king's mind. To use purely strategic criteria to understand these events is fundamentally flawed.[13]

Cyrus's conquest of the Lydian kingdom on the coast of Turkey—and the fateful end of the plutomanic Croesus, in 547 BC—brought the Persians into contact with the Ionian Greek city-states on the Aegean islands and the coast of Asia Minor. Cyrus's claim to rule Yaunâ [Ionia], preserved on the cylinder seal, shows that his conquest of the Ionian Greeks was as important to him as his other conquests.[14] Cyrus placed the area under the control of a loyal officer and turned to further campaigns in Mesopotamia. This officer brought the area to heel and allowed the Greek cities a large degree of autonomy as long as they submitted to the Great King. When Cyrus's son Darius invaded Scythia some thirty years later, the Ionian Greeks fought well for the king and protected his line of retreat over the Hellespont. The king must have been pleased to know that the Greeks were loyal warriors; he would need their sailing expertise and their ships in future wars of expansion. But there was dissension in the Greek ranks; some Greek leaders drew the conclusion that the Persians were weak and that a revolt might work.[15] The Persians convinced Greek leaders not to revolt, for the moment, by supporting them as rulers over the Greek cities.[16] Over the next decade, such leaders faced mounting pressures as they tried to maintain their positions.

Whether either side knew it or not, the worldview of the Greeks—and their political systems—differed fundamentally from that of the Persians. The central development was the rise of the independent Greek city-states, or *poleis*, and each *polis* became a self-governing community without allegiance to greater empires.[17] The central idea was not "liberty" in the modern, Lockean sense but rather political autonomy, or self-government, for every Greek in his own *polis*. The Greeks were growing hostile to anyone who tried to rule by nonlawful means—outside of customary and participatory norms—whether an internal tyrant or an outside power. The discourse of the Greeks—itself a product of their autonomy and self-rule—led them to unprecedented examinations of their political life and to direct action against those whose personal power might threaten it. Of course, this discourse became an excuse to attack one's political enemies while lending itself to rhetorical manipulation, but it also limited the power of tyrants and was generally nonviolent within each *polis*. One example may suffice: Miltiades, the later hero of Marathon, had ruled the Hellespont as a tyrant and later faced prosecution in Athens for his actions.[18] Such people were not inclined to become subjects of the Great King.

In short, the deeply rooted Persian ideology of expanding royal supremacy was on a collision course with these budding Greek ideals of self-governance, autonomy, and intellectual inquiry.

The Ionian Revolt and the Onset of War

The so-called Ionian Revolt of the Greek cities on the eastern Aegean and Asia Minor against Persian rule in 499 was born of complex local intrigues, including the collaboration with the Persians by many Greeks. When certain Greek cities in Asia Minor appealed to the Athenians and the Spartans for help in the revolt, the Spartans refused to fight so far from home. But the Athenians sent about twenty ships—not the last time they would get into trouble over a half-hearted commitment. This was enough for Darius to associate them with the revolt but not enough to allow it to succeed; the burning of Sardis—the Persian capital in Asia Minor—gave him a pretext for revenge and a target for his next expedition. The revenge motive was central to the explanation for the war offered by the Greek historian Herodotus; the twenty ships were "the beginning of evils" for both sides.[19] Yet this explanation—a thoroughly Greek view of vengeance as a historical cause—must be held in context with the claim by every Persian king to rule the Ionian Greeks as subjects. Greek revenge did not start the conflict, for Cyrus's expansion into Asia Minor and his demands for submission had occurred far earlier.

Darius's suppression of the Ionian Greek revolt led to the sack of Miletus, a Greek city on the coast of Asia Minor in 494, followed by brutal mopping-up operations across the northern Aegean Sea into the Hellespont, including Lesbos, Byzantium, and Chalcedon. As Herodotus put it, "Once the towns were in their hands, the best looking boys were chosen for castration and made into eunuchs; the most beautiful girls were dragged from their homes and sent into Darius' court, and the towns themselves, temples and all, were burnt to the ground. In this way the Ionians were reduced for the third time to slavery—first by the Lydians, and then, twice, by the Persians."[20]

Peter Green drew out the major consequence of this Persian action: "The Ionian Revolt was over, and the invasion of mainland Greece had, by that fact alone, become inevitable."[21] The success of the king's campaign did not satisfy him—it became the impetus for further action. Success strengthened his support, affirmed his magnificence, and motivated him to further revenge. Yet, as refugees fled, the campaign deeply affected Greek attitudes toward the Persians. Some submitted, either symbolically or in fact (to "Medize," as the Greeks put it) in order to avoid the same treatment, and others hardened into intransigent opposition. Herodotus's account of the drama by the playwright Phrynichus, *The Capture of Miletus*, based on the Persian sack of the city, captures the depth of emotions that the Persian campaigns evoked in the Athenians.[22] If this story is right, the Athenians were moved to tears and banned the play forever.

In 492 BC Darius ordered his first invasion of the Greek mainland, a land and sea expedition that followed the northern shore of the Aegean Sea. Whether he intended to reach the Greeks or merely to secure the northern tribes, his navy was wrecked off the coast of Mount Athos, and his army was forced to retreat by the attacks of native tribes.[23] To this point, the king had engaged in at least seven straight years of war against the Greek cities, following his predecessor's initial contact nearly fifty years earlier, and he now reached into Europe. He had faced setbacks, but no overall failure. His will to continue was unabated. The Greeks, he decided, would submit to the king of truth and light.

The next attack, in 490, moved by sea directly across the southern Aegean, under the command of a Mede, Datis (see Figure 1.1). The fate of the expedition was decided at the battle of Marathon and by the Mede's failure to penetrate the Greek mainland. The Persians went home unrequited, and after honoring their fallen heroes, the Greeks largely turned back to their own affairs. There was no general mobilization by the Greeks to prepare for another Persian invasion.

The decade after Marathon, between 490 and 480 BC, saw the Persians tied down with the death of Darius and the rise of Xerxes,

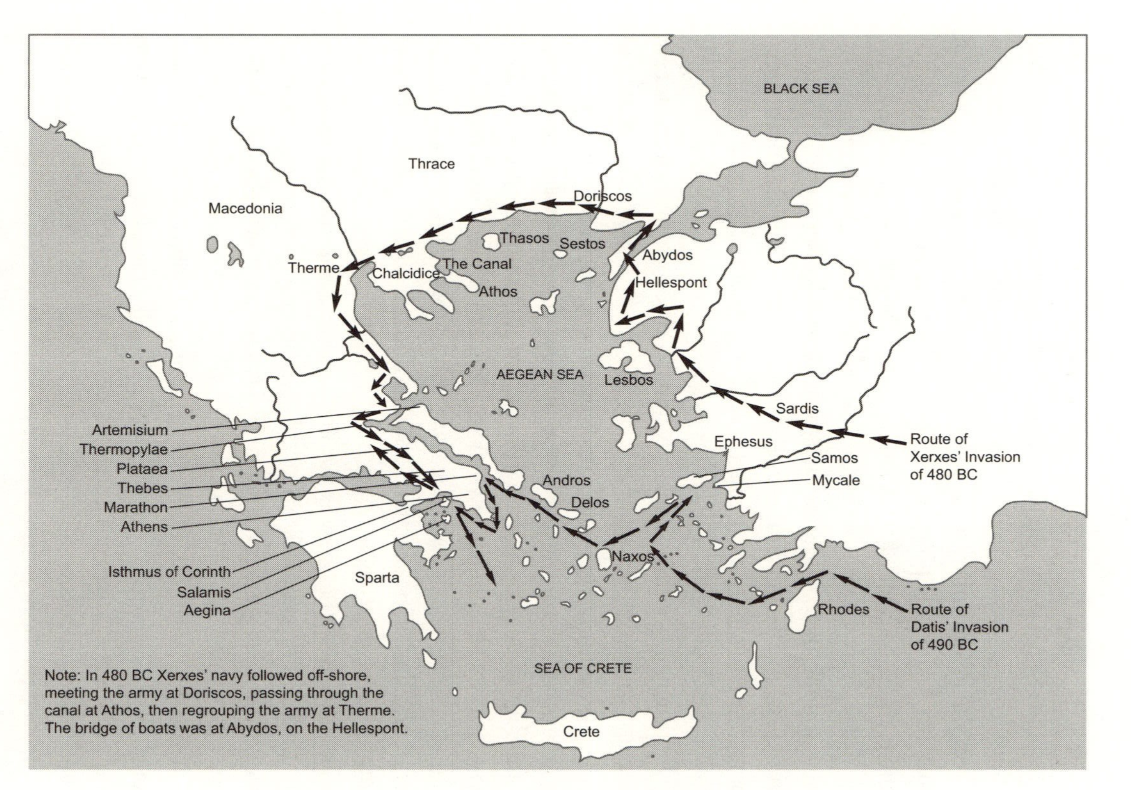

Figure. 1.1. The Greco-Persian Wars: Main Routes of Invasion

who had to suppress revolts in Egypt and Babylonia. Meanwhile, the Athenians focused on ostracizing potential tyrants at home and controlling the sea near Athens by warring against the island of Aegina. When the Athenians discovered new veins of silver in their mines in southern Attica, the politician Themistokles persuaded them to build a great navy—a fleet of triremes, state-of-the-art three-banked rowing ships—instead of distributing the silver by a public dole.[24] The original impetus to build the navy was probably the war with Aegina —not fear of the Persians—although the navy was destined to become the city's bulwark against the Persians and the main military arm of the democracy.

Whether or not any Greeks anticipated the Persian threat, Xerxes took up the legacy of his father's throne, and in 480 BC he marched across southern Europe and into Greece, accompanied offshore by a huge navy. The attentions of the mainland Greeks, under the uneasy leadership of Athens and Sparta, were now firmly fixed on the Persians, and on the reports of their massive army. The Greeks held the king at Thermopylae for three days, forced him to retreat in the sea battle of Salamis in the fall of 480, and finished off the land army at Plataea in the next year.

This much of the story is familiar, and ancient and modern commentators alike generally treat this as the end of the war, but what follows deserves greater emphasis.[25] This time the Greeks—in particular, the Athenians—did not stop at their own shores. Armed with their navy, a powerful offensive weapon, and emboldened by their victory at Salamis, they moved energetically into the Aegean Sea. The Athenians refused a separate peace with the Persians, which would have destroyed the alliance—a strong reason for the Spartans to support the Athenians—and they took an oath with the Spartans against the Persians and their supporters.[26] While the Athenians and Spartans took vengeance against those Greeks on the mainland who had collaborated with the Persians—especially the Thebans—they also confronted Aegean cities that had fallen under the Persians.[27] Herodotus claimed that they first sailed no further east than Delos, but the historian Diodorus of Sicily says that they responded forthwith when the island of Samos off Asia Minor asked for aid.[28] The first

Greek victory on Persian-controlled territory was a rout at Mycale, a promontory on the coast of Asia Minor, opposite the island of Samos, on or about August 27, 479 BC—about the same time as the battle of Plataea.

The Athenians and the Spartans then turned together into the northeastern Aegean, an area that had borne the brunt of Persian attacks in the original revolt, and set out to take control from the Persians. The Spartans soon abandoned their plans to become a sea power, and they retreated to affairs in the Peloponnesus. It was the Athenians who set their sights on total command of the Aegean. They allied with the island polities where they could, coerced them if necessary, and spent a generation driving the Persians off of Andros, Naxos, Samos, Lesbos, Rhodes, and elsewhere; they reached as far east as Cyprus, and to Phaselis on southern Asia Minor. Greek cities revolted. Persian garrisons were attacked, local Persian leaders captured or killed, and a sense of fear rose up in Persia's supporters. Herodotus tells us that the Greek commander Xanthippus, father of the democratic leader Pericles, nailed the Persian governor to a board at Sestos, a city on the Hellespont.[29]

This offensive drive made it impossible for the Persians to return to Greece. The Greeks did not have a singular, thought-out strategy of taking over the Aegean Sea; there was much that was ad hoc in their actions, and much that depended upon the ambitions of particular men, acting from motives that might not align with the decisions of the *poleis*. The Spartan Pausanias and the Athenian Cimon, for instance, fell in and out of favor with their own cities as they vied for control of the sea. The Greeks were deeply enmeshed in political infighting, both within their own cities and between cities. Nor were their actions everywhere a success—the Athenian expedition to Egypt in 455 failed, and it took more than one attempt to succeed on Cyprus. But the end result of this forward projection of power was that the Athenians created a military alliance to defend the Aegean Sea and ended the king's attempts to expand his empire to the west.[30]

Marathon was where Greek freedom was asserted, Thermopylae was where it was defended, and Salamis was where it was saved, but Mycale is where the Greeks were first made truly safe, for the Persians could no longer reach Greece. The final end to the Persian threat can be found not in the Persian defeats in Greece, but at battles in areas controlled by Persia, such as the Eurymedon River on Asia Minor some fifteen years later.[31] Long-term peace between the Greeks and the Persians was likely not achieved until after a second battle at Cyprus, probably after 450 BC, nearly thirty years after the second Persian invasion. If Diodorus is right, the peace followed directly upon Greek successes at Cyprus and Asia Minor: "Artaxerxes the king, when he learned of the reverses suffered by his forces in the area of Cyprus . . . decided that it was to his advantage to make peace with the Greeks."[32] The so-called Peace of Callias is disputed among scholars—the historian Thucydides omits it—but whether or not there was ever a formal treaty, hostilities ended after a sustained projection of power by the Greeks into the Aegean Sea and onto Asia Minor. The zeal for westward expansion never again returned to the dynasty of Cyrus, Darius, and Xerxes.

Why? What ended four generations of motivated attempts by the Persians to expand?

The End of the Ideology of Expansion

The reason why Persian attacks against the Greek mainland ended so abruptly is not to be found in the destruction of the Persian king's wealth or his capacity to wage war. The Persian Empire was astronomically richer than the lands of the Greeks, who held only a fraction of its area and its resources. After the Greeks' victory at Plataea, their commander Pausanias looked over the rich tables of the Persians, in contrast to the simple meals of the Spartans, and spoke of "the folly of the Persians, who, living in this style, came to Greece to rob us of our poverty."[33] The Greeks could never have destroyed the king's resources to the point of physically preventing his return, had he wanted to do so. Nor does the bloodshed on the Greek mainland explain this about-face. Some 200,000 Persians may have died between Thermopylae and the aftermath of Plataea—up to 40,000 drowned at Salamis alone—and few made it home from the army he left behind.[34] But the king's supplies of wealth and men were limitless; he had the resources to raise another army and a tremendous excuse for revenge. But he did not. Why?

A strategic explanation may be found in a severe overstretch by the Persians; they took their massive army further than they could support it, and the loss of their navy left them isolated on hostile soil. The king had to retreat to protect himself and his army's path of escape. But this begs the question as to why the Persians built the army they did, why they overextended it so drastically, why the Greeks held together long enough to confront the Persian threat, and why the Persians never again seriously tried to gain control of the Aegean Sea or the Greek mainland by force.[35]

A more fundamental explanation turns on the nature of the despot's rule, his motivations to form such an expedition, and the contrasting culture of the Greeks. The Persian Empire was the largest, richest, most populous empire ever seen in the West before Rome, but it was not a homogeneous entity. It was a complex arrangement of local and regional leaders, who paid tribute and swore loyalty to the Great King. Its administrative sophistication and relative stability depended on a certain willingness of the people to submit to the king. He ruled a motley combination of Cilicians, Medes, Persians, Lydians, and other tribes, many of whom were there by coercion, others because of ambition, and the rest out of a sense of subordination that began with their local rulers. Slavery was deeply engrained in these subjects; in basic outlook and in practical fact, everyone was a *bandaka* (slave) or *ardu* (chattel or property) of the king.

The idea was translated into the Greek *doulos* (slave) in a letter purportedly from Darius "to Gadaitis my slave," one of his satraps.[36] Herodotus leaves us examples of the king's method of dealing with dissent. One Pythius of Lydia, for instance, who had donated his fortune to the king, asked that one of his five sons be released from service, to remain home with the farm. The king screamed "*you* my slave, whose duty it was to come with me!" and had the son cut in half and nailed to two trees; the army marched between him.[37] These are Greek views of the Persians, of course, and whether they are literally accurate or not, they reflect the essence of the king's rule as it would have motivated the Greeks to oppose him. The basic conclusion is sound: the king's empire was held together by a sense of awe and fear in the king's subjects—awe of his great power, and fear of the consequences of disobedience. The possibility of a usurper was real—Xerxes was assassinated in 465—but the king ruled subjects who expected to look up with awestruck fear at someone above them and would have expected him to expand his rule perforce. For his part, Xerxes had to demonstrate his power and to expand it—the ideology, and the success of his rule, demanded it.

An ideology of violent expansion conditioned Xerxes' entire approach to the campaign.[38] His invasion of 480 BC followed a series of violent events after he took power in 486. Egypt revolted (possibly a few months before the death of Darius), and Xerxes put it down by 484. Then, in 483 or 482, the Babylonians revolted; sources mention two rebels, although it is unclear whether there was one revolt or two. Xerxes was a busy man in his first five years, holding his father's kingdom together; it is quite implausible that he spent those years obsessed with preparing to fight the Greeks.[39] But once he established order and asserted his power, an external war of conquest would serve a wider purpose, beyond revenge: it would defuse revolts, by giving the king's subjects—especially the nobles in his court and the regional leaders—an enemy to focus on and a means to attain prominence in his eyes. Aristotle would point out that "the tyrant is a stirrer-up of war, with the deliberate purpose of keeping the people busy and of making them constantly in need of a leader."[40] The Persian expedition against Greece in 480 was the king's version of an Egyptian Great Pyramid or a Mesopotamian ziggurat: a grandiose project to absorb the energy of the empire, proclaim the king's greatness, and concretize the ideology behind the king's rule. A pyramid is an apt analogy for another reason: power emanated down from the top, bearing down on the masses buried in the foundations, who were too busy marching and working the soil to revolt.

The king's ideology also shaped the views he held of his foes. When he set off in 480 BC, Xerxes treated the Greeks as if they were Near Eastern subjects—how else could he have seen them? He demanded their submission, a policy that had brought order throughout the rest of the kingdom. His expectations of victory and belief in his own power were so strong that his grand advance was noisy and full of braggadocio; he waged psychological warfare through a concentrated propaganda campaign. To prepare the Greeks for their inevitable surrender, he released Greek spies to return home with word of his grandeur: "Go! Tell them what you have seen here!" This was dramatized in the movie *The Three Hundred Spartans*, in which a Greek spy reports that for seven days and seven nights the army moved past him; "I ran out of numbers and still more of them came." The king released Greek grain ships captured en route, gloating that he would eat the grain later. He whipped a river that refused to obey him and built a bridge and a canal that showed off his power; when the Great King commands, water becomes as land, and land as water.[41] His cultivated aura of invincibility was a calculated attempt to demoralize the Greeks with the same kind of propaganda he and his predecessors had always used throughout the empire.

This propaganda campaign is the source of Herodotus's claim that the Athenians at Marathon in 490 BC were the first of the Greeks "to charge at a run, and the first who dared to look without flinching at Persian dress and the men who wore it; for until that day, no Greek could even hear the word Persian without terror."[42] The claim by historians that the Athenians were in fear of the Persians before 490 BC is a product of the propaganda nine years later, which Herodotus wrote into his *Histories* and became accepted as fact. Herodotus's inflated counts of the Persian forces reflect what the Greeks of the next generation thought, but the exaggeration began with what the Persian king wanted their fathers to think before the invasion.

The propaganda was broadcast in two directions, outward and inward—to the Greeks but also into the king's realm. The word went out: the king had commanded an army to gather. It had to be of outrageous size, to demonstrate the magnificence of the king's rule, to focus his subjects onto a great task and away from the temptations of internal revolts, and to foster defeat in his enemy. But the projection of magnificence was a two-edged sword. It led the king to act in ways based not on a good strategic understanding of the situation but rather toward further display of his self-created sense of magnificence.

The massive army was a physical liability, aggrandized past a point that he could control and desperately dependent upon supplies. The very size of his forces suggests that his purposes were not strategic—to end Athenian sea power in the Aegean, for instance—but rather ideological and deeply embedded in his view of his own position as King of Kings.[43] The expedition was also an instrument of internal policy, and, given its scale, the success or failure of Xerxes' rule became dependent upon its success. The stakes were raised; failure of the missions would be his failure and would put his position on the throne in serious jeopardy. Fail it did—and this demanded that he turn his energy toward reestablishing internal order rather than out toward external conquest. When his Phoenician and Egyptian subjects cut for home from Salamis, the possibility of political disintegration became real. The king abandoned conquest of the Greeks, and he strived to reestablish his rule in terms that did not require a new invasion of Greece.

The Deeper Clash of Ideas

The king's purposes, and the motivations of his subjects to follow him, contrasted utterly with the purposes and motivations of the Greeks. Ignorance describes the Greek view of the Persians better than fear, for evidence suggests that neither the Greeks nor the Persians really knew much about the other. Herodotus says that only the Greek geographer Hecataeus knew the size of Persia. The motivations of each were also mysterious to the other. The Greeks interpreted the world from the perspective of an autonomous *polis* contesting with its neighbors; this might explain the curious passage related by both Herodotus and Diodorus, in which the Spartans of a past generation are said to have done a quintessentially Greek thing: they sent a single ship to warn Cyrus not to harm other Greek cities.[44] In contrast, the Great King approached the Greeks from his own ideological framework, as if they were Near Easterners willing to be subject to a greater power; he wanted to know who these Spartans were, so he could force their submission to him.

The differences in motivations reached the average
best of the Persians hoped to achieve favor in the eyes of
King; the majority could hope only to avoid the pain of his
Because all rewards and punishments emanated from his perso
subjects had to base their actions on pleasing him. In theoret
terms, he acted as an *efficient* cause upon them, pushing them to gai
his favor under pain of punishment. But for the Greeks the motivation was less a pain to be avoided than a positive goal to be achieved; as a *final* cause each could claim pride of place in his city. Each had to be persuaded to fight under his leaders, and Greek leaders were continually adjusting their arguments to persuade their fellows. The Persians and their allies had to adjust their thoughts to the whims of the Great King; they were understandably reluctant to speak truthfully in his presence, which left the Great King in woeful ignorance of his true condition. In contrast, Greek political assemblies—whether of citizens inside a city or of leaders among many cities—were raucous, disheveled, and chaotic. They were gatherings of free men, each with an equal right to speak against his fellows, each concerned to gain something for himself and his city. There was nonsense there too, and the basest political motives—deceptions, manipulations, and self-promotion—but such chicanery was open to repudiation by others, and the most powerful among them could one day face his accusers in a jury.[45] The Great King, who ruled by whim, could never understand the strength in such chaos.

As Herodotus put it, the history of the men of Athens was such that "when the Athenians lived under a tyranny [Pisistratus and his sons, who controlled Athens from ca. 560 to 510 BC] they were no better at war than any of their neighbors, but after they got rid of the tyrants they were the first by far. This proves that when they were oppressed they fought badly on purpose as if they were slaving away for a master, but after they were liberated they were each eager to get the job done for his own sake."[46]

The language of final causation is prominent here; a century earlier they fought badly "on purpose" (*ethelokakeon*, "played the coward deliberately") because the "for which" they were fighting was a "for a master" (*despotēi*). Aristotle's later dictum that the free man acts for his own ends, and the slave for the ends of the master, includes his observation that even a slave will work better if given a certain term of service and a goal to be achieved. Herodotus continues: once the Athenians deposed the tyrants, "each man himself [*autos hekastos*] was motivated [*proethumeeto*]" to do the job "for his own sake [*heōutōi*]," as the slave of no one. It was that goal to be attained by his own autonomous action—rather than punishment to be avoided—that distinguished the Greek citizen from the Persian subject. It also distinguished their respective leaders: the Persian king operated by *diktat* and fear; the Greek political actor by public persuasion and claims to personal merit. Every Greek leader had once stood in a battle line or pulled an oar, and each could face a vote of ostracism. Xerxes knew no such things.

Herodotus reveals this crucial difference in two speeches, as the Persians attempted to control the Ionian revolt after 499.[47] The Persian commanders knew they had to regain power over Miletus lest they "be punished by Darius for their failure." So, the Persians summoned certain former Greek leaders (Herodotus calls them "tyrants") who had been pushed out from their cities during the revolt. The Persian commanders said, "Men of Ionia, now it is time to show yourselves true servants of the king. Each of you must do his best to detach his own countrymen from the Ionian alliance." The rewards for loyalty would be entirely negative; the Persians promise that their cities will not be burned and they will not be sold as slaves. In contrast, when the Ionian Greeks assembled to plan their response to the Persian force that had just arrived at Miletus, a Greek commander said, "Fellow Ionians, our fate balances on a razor's edge between being free men or slaves." The Greek alliance did not hold, and the Persians did retake Miletus, along with Caria, Chios, Lesbos, and other city-states.[48] But Herodotus has made the essential terms of the issue explicit.

According to Herodotus, the misunderstanding continued into Xerxes' invasion. As the king prepared to march in 480, two Spartans met up with a Persian commander on the coast of Asia. The commander wondered why the Greeks did not seek the king's friendship and become the masters of all the Greeks under him—a perfect statement of a Persian satrap's position over subjects of the king. The Spartans famously offered an answer that was completely beyond the ability of the Persian to understand: "The advice you give does not spring from a full knowledge of the situation. You know one half of what is involved, but not the other half. You understand well enough what slavery is, but freedom you have never experienced, so you do not know if it tastes bitter or sweet. If you ever did come to experience it, you would advise us to fight for it not with spears only, but with axes too."[49]

They are so few, the king might have thought while approaching Thermopylae in 480; how can they oppose me, with the greatest army in the world? They have no single king, how can they act in common? This Greek view of the king's thoughts is probably right in essence; given his Near Eastern outlook, the king could not fathom free men, living under laws and without fear of a king, willing to defend their homes against all odds, each for his own sake. Given such misunderstandings—which amount to profound a disagreement about the moral nature of man as an autonomous being—he could never develop a strategy to overcome their intransigence.

The contrast between King Xerxes and Themistokles, the Athenian commander at Salamis in 480, exemplifies the differences between the Persian and the Greek views. One might picture Xerxes, sitting on his great throne, looking down upon subjects making entreaties in prostration, versus Themistokles, hands calloused from his past efforts, walking among the Greeks, hearing their abuses and holding them to their agreements. Xerxes had never had to plead for the agreement of his highest general, let alone stand with farmers in an assembly. He was the most powerful by dint of something other than merit; there was no process by which the best could rise among the Persians, and no checks against his malfeasance except the last resort of assassination. In contrast, Themistokles, a Greek politician, orator, commander, and perhaps the first military strategist in a truly political context, had stood in the ranks at Marathon and earned his position through competence. He had to weave the disparate parts of a fractious physical and political landscape—a crowd of bickering voices from autonomous cities with conflicting aims, animosities, and whims—into a whole cloth capable of unity long enough to repel the enemy. And he had to do it under threat of prosecution or a vote of exile, while his city, family, and property were under attack.

The chasm between Themistokles and the king implies deep—even metaphysical—differences in their views of reality. For the king, reality bent to his whim: at his command land becomes as water in the form of a canal, and water becomes as land when a bridge appears. When his bridge was wrecked, the king whipped the river, which must obey like every other subject. The king, sitting on his throne overlooking the naval carnage at Salamis, threw a handful of stones into the water, commanding a land bridge to an island. For the king's subjects, his commands might have come from a more-than-human figure that melded in their minds with the deity, and whose commands can order reality in some incomprehensible way. For the Greeks, any such pretensions by Themistokles would have constituted *hubris*, an impious crime. The Persian king was expected to revel in his wealth; it was a demonstration of his power, which gathered at his whim. But Themistokles had to avoid even the appearance of impropriety; the story of his walking along the shore and disdaining to pick up a piece of gold over concern for a bribery charge captures the essence of his position.[50] The Greek world was not commanded by the voice of any man. Even the Delphic Oracle, an alleged portal to the divine, came to be seen as a mouthpiece for special interests, including Persian propaganda ("fly you fools!" said the Pythian priestess), and it declined in influence after the Persian defeat.

The failure of the Persians to grasp the self-motivated nature of the Greeks, and their equation of Greek disunity with political weakness, left them open to deceptions. According to Herodotus, following the Greek debate about the upcoming sea battle at Salamis, Themistokles deceived the Persians—and put the Greeks into a position of advantage that neither side could escape—by sending a fraudulent spy to play on the king's expectation of a traitor.[51] There was history for this, too; ten years earlier, a former Athenian leader (Hippias) had guided the Persian expedition to Marathon; during the present expedition, a traitor (Ephialtes) had betrayed the Spartans at Thermopylae, and an exiled Spartan king (Demaratus) was with Xerxes' own entourage.[52] The Persians expected a traitor and were disarmed against the deception. They sailed into the trap, and the sea battle became a rout.[53]

The next year, the Persian commander Mardonius sent an offer of a separate peace to the Athenians, to divide them from the Spartans. The Athenians responded: "We know as well as you do that the Persian strength is many times greater than our own; that, at least, is a fact which you need not have troubled to rub in. Nevertheless, such is our love of freedom, that we will defend ourselves in whatever way we can. As for making terms with Persia, it is useless to try and persuade us; for we shall never consent.... Never come to us again with a proposal like this."[54]

After the Greeks had driven across the Aegean Sea, and charged the Persian positions at Mycale, their commander Leotychides yelled out, trying to cut through the dust, smoke, and mayhem to get his words to his men: "Men of Ionia, listen, if you can hear me, to what I say. The Persians in any case won't understand a word of it. When the battle begins, let each man of you first remember *Freedom*, and secondly our password, *Hera*. Anyone who can't hear me should be told what I say by those who can."[55]

More is at stake here than a password and the fog of war; for Persian subjects had no capacity to grasp the freedom of Greek citizens or to match their energy. Decades later, after the Athenians and their allies had brought the war to the mainland of Asia Minor, at the Eurymedon River and elsewhere, the Athenian commander Cimon devised a stratagem to deceive and to panic the Persian land army. After taking the cities of Eion and Scyros, and others at Caria and Lycia, and after a victory in a sea battle off Cyprus, he disguised his own troops in Persian headdress and moved in toward his confused enemy: "In a word, such consternation as well as bewilderment prevailed among the Persians that most of them did not even know who was attacking them. For they had no idea that the Greeks had come against them in force, being persuaded that they had no land army at all ... thinking that the attack of the enemy was coming from the mainland, they fled to their ships, thinking that they were in friendly hands. And since it was a dark night without a moon, their bewilderment was increased all the more and not a man was able to discern the true state of affairs."[56]

The bewilderment of the Persian forces about who was attacking them went far deeper than the mysteries of a moonless night, an outlandish disguise, and a spear thrown out of the dark. The tactics of Leotychides and Cimon were sound; neither the despot nor his officers could fathom the motivations held by the Greeks. Nor could they counter the sophistication and flexibility of the Greeks—in the unorthodox battle line at Marathon, the stand at Thermopylae, the deception at Salamis, and the disguises at the Eurymedon River—which flowed from independent thought and a sense of self-motivated action. As the Greeks projected their energy into the Aegean Sea, the Persians were pushed on the defensive, and the support of their allies in the area collapsed. The Greeks surged outward and filled the vacuum; the Persian king's coastal forces were compelled to defend their positions. The issue was either-or: either the Greek passion for freedom had to be subordinated to the rule of the Great King, or the Great King's desire for dominance had to be put in its proper place.

Xerxes began with the inherited passion for conquest that had motivated three generations of predecessors. But when his army and navy were mutilated by the Greeks and he saw his men sink beneath the waves, he confronted serious personal defeat for the first time. As his Great Pyramid collapsed, the effect on the king was immediate; he set off posthaste to secure his own retreat. His defeat was open and public, and despite his likely attempts to make it appear a victory, he knew that this could be fatal to the dynasty. His position had demanded that he demonstrate his splendor—but at the moment of defeat he reached the point of greatest danger. His task now was to reestablish his position inside his own territory—and this required a permanent change in policy. The legitimacy of his throne had to be disengaged from the conquest of the Greeks.

It may be no coincidence that Xerxes now first mentions the god Arta along with Ahura Mazda, a return to polytheism of a pre-Zoroastrian variety.[57] Perhaps he scrambled to legitimate his position on a new divine basis, with a series of far-reaching decisions about the future of his rule. Such changes would have been consistent with structural changes to the dynasty's foreign policy; we hear no more of military expeditions by Xerxes (perhaps he "retired to his harem" after the defeat), and after his murder in 465 BC, the new king Artaxerxes had no stomach to confront the Greeks. Facing a siege near Cyprus, Artaxerxes sued for peace. Whether this indicates a "decline" in the Persian court may be debated, but the fire behind four generations of Persian expansion against the West was gone, never to return.[58] The Persians became one player among many in the Aegean Sea, an area they could not control with impunity.

It is impossible to be too specific about the nature of the changes at the center of the Persian court, but we may infer a direct relationship between the Great King's motivations, his support among the nobility, his ability to command the population into great ventures, and the fact of his defeat. As long as hope smoldered, the king kept alive his vision of universal rule, and he focused the energy of his subjects onto an outrageously huge expedition. But the totality of his failure—visible for all to see—neutralized the desire for revenge in his mind by making it impossible. He and his officers had to know that another attempt would fail, and that he could not survive it as king. The demonstration of Greek superiority on the battlefield had reached deeply into the center of the king's power. The ideology of expansion collapsed in the wake of concrete failure, and the Achaemenid Dynasty was forced to reestablish itself in terms that recognized the independent existence of the Greeks. With no more promises of great expeditions in the West, there could be no more danger from the defeat those expeditions were certain to bring, to any king foolish enough to try.

Chronology: The Greco-Persian Wars: 547–446 BC

612 BC	Sack of Nineveh by Babylonians and Medes ends the Assyrian Empire
	Babylonians control Mesopotamia (middle and southern Iraq)
	Medes control northern Iraq and Anatolia (Turkey)
	Minor kingdom of Lydia controls Western Anatolia
559–530	Rise of Cyrus I "The Great" and the Persians
550	Cyrus defeats the Medes
547	Cyrus defeats Croesus of Lydia. Persians in contact with Greeks
546–545	Revolt of the Lydians put down by Cyrus's commander
539	Cyrus captures Babylon
530–522	Cambyses rules Persia
525	Cambyses conquers Egypt
522	Revolt of the Medes
521–486	Darius I rules Persia
520	Persian invasion of the Scythians (Black Sea area)
518	Ionian Greeks subjugated
512	Invasion of Thrace (northern Aegean Sea area)
499	Revolt of the Ionian Greeks
494	Miletus subjugated; sea battle of Lade (Ionian Greeks defeated)
492	Land and sea invasion of northern Aegean, wrecked in a storm
491	Darius demands the submission of the Greeks
490	First Persian invasion of Greece: Battle of Marathon
486–465	Xerxes rules Persia
488–486	Ostracisms in Athens
486–484	Egyptian revolt against Persia
483	Babylonian revolt against Persia
483/2	Athens builds navy
481	Xerxes prepares invasion; demands Greek submission; Greek League formed
480	Second Persian Invasion: Battles of Thermopylae, Artemisium
	Battle of Himera (Sicily)
	September: First sack of Athens
	Naval battle of Salamis
479	Athens refuses separate peace
	June: Second sack of Athens
	Battle of Plataea destroys Persian army in Greece
	Battle of Mycale; Persians retreat to Sardis
	Athenian siege of Sestos; Andros, Carystos, Paros taken by Athens

478/7	Athens fortified under Themistokles; builds walls, harbor
	Spartans send Pausanias to liberate Greek cities. First battle at Cyprus
476–463	Cimon commands Athenians in Aegean Sea
474	Themistokles ostracized
470	Athens sends Cimon to liberate Asian coastal cities
	Eion, Scyros; cities in Caria and Lycia taken
465–425	Artaxerxes I rules Persia
462	Failed Athenian expedition to Egypt
450	Cimon battles in Cyprus; takes Citium and Marium by siege
c. 446	Peace of Callias: Artaxerxes sues for peace

Chapter 2

"Only One Omen Is Best"

The Theban Wars, 382–362 BC

The Spartan Military State

Peace between the Greeks and the Persians meant exactly that: an end to war between the Greeks and the Persian king, not universal tranquillity. The Greeks were now free to fight each other with all the energy and passion their vibrant culture could unleash. Hellenic culture was imbued with a warrior premise—a high valuation placed on warrior prowess, proven in highly ritualized infantry combat—that made it very difficult to prevent a new round of internecine fighting. The defensive alliance in the Aegean Sea created by the Athenians to counter the Persians had brought secure trade routes and sound money—but as the Persian threat receded, Athenian attempts to rule the Aegean Greeks as a tribute-paying empire led to mounting bitterness and violent revolts. Antagonisms broke out into violence, and the postwar world coalesced into a polar confrontation that became a war to the death between the democratic, seagoing Athenians and the oligarchic, land-based Spartans. The defeat of Athens in the Peloponnesian War (431–404 BC) left Sparta as the strongest of the Greek powers, but other Greeks formed new coalitions to oppose the Spartan hegemony—as they had earlier opposed Athens. The Spartans feared such alliances and set out to break them. The resulting wars continued for nearly five decades.

Thebes was the most acute danger to the Spartans. As the leading city in Boiotia—the agricultural area north of Attica—Thebes had established a federal league that seethed with democratic ideas, which threatened the very foundations of Sparta's dominance over the Peloponnesus. The Spartan infantry—the most powerful heavy-armed soldiers in the world—marched out to permanently reduce the farmers in "Cow-land"—Boiotia—to subservience. Only one outcome seemed possible. Yet in the winter of 370/69 BC, these farmers dealt a stunning upset to the Spartan warrior elites. They rose up under the greatest of the Greek military leaders, Epaminondas, mobilized a coalition of local supporters, drove the Spartan warriors back into Sparta itself, freed Greeks in the Peloponnesus from generations of virtual slavery, and permanently reduced the power of the Spartan military state.

Few historians today have granted to the wars between Thebes and Sparta the historical importance attributed to the Peloponnesian War.[1] This neglect is largely due to our source for the earlier war, Thucydides, whose penetrating intellect had no counterpart in the next generation. But the Theban Wars were, and remain, highly significant—perhaps more important, in the long-term, than the Peloponnesian War. They ended with the unprecedented, and permanent, defeat of Sparta by an ad hoc coalition of Greek cities, which reclaimed the stolen birthright of independence for thousands of people reduced to subservience by Spartan spears. This magnificent assertion of freedom far exceeded what the Athenians were able to do in their twenty-seven-year war with Sparta. City walls at Messenia still stand in silent testimony to the success of the Thebans and their allies in the Peloponnesus and the determination of the Messenians to resist Spartan domination. In later decades these wars conditioned the Greek response to Macedonian expansion under Philip II and his son Alexander. But for us, the Theban victories have a deeper meaning. They provide an example of how an ideology of slavery was confronted, exposed, and defeated by a forthright projection of military strength. How, and, why, did this succeed?[2]

The meaning of these victories extends beyond the military success into central questions about the achievement of goals by military force. What was achieved by the fighting? Why did these uncouth farmers rise up and succeed where the Athenians could not? What happened in Sparta to make this upset permanent? Sparta was a closed society, and our evidence is incomplete, deeply biased, tempered by anachronistic philosophical assumptions, and often late. Much must be inferred, and as for every case in this book, space does not allow a comprehensive examination. The answers are to be sought, as they were for the Persians, in the ideas by which the participants made decisions and shaped their world, the virtues and political ideals they held paramount, and their views of themselves and their neighbors. But it took an unprecedented catastrophe—a steamroller of a military offense—to make the Spartan failure visible to all. What is the connection between theory and practice here—between the ideals of a tradition-bound military state and its responses to a changing world around it? What was revealed when the Spartans stood face to face, on their own soil, with an unrepentant enemy they had thought to be unworthy of their contempt, whose once unthinkable demands were now inescapable? What happened when they realized they were powerless to reverse the tide that had turned against them?

The story begins in the Spartan heartland, the political, social, and economic center of the Spartan state in the Peloponnesus (see Figure 2.1). This was a fortress impregnable to other Greeks—not a walled city, but a land that stood beyond the power of others, with its own largely self-sufficient economic base, from which an infantry force could march with impunity. An old saying had it that the enemies of the Spartans bury their own dead on their own homeland, while the Spartans also bury their dead on the homeland of their enemies, because no one fights on the Spartan homeland. By dominating the Greeks in Laconia and Messenia and controlling the northern approaches to Sparta, the Spartans at home remained exempt from the consequences of war and in command of vital agricultural resources and labor. They also extinguished all hope of political autonomy for the villagers they had subjugated.

For generations such strategies had been effective. Spartan land armies were shrouded in a myth of invincibility. This "Spartan mirage" —not the image of Spartan supermen that has passed through centuries, but rather the view of the Spartans held by their fellow Greeks[3]—was supported in fact; they had successfully put down all revolts in the Peloponnesus, where they basked undefeated in their political *kosmos*.

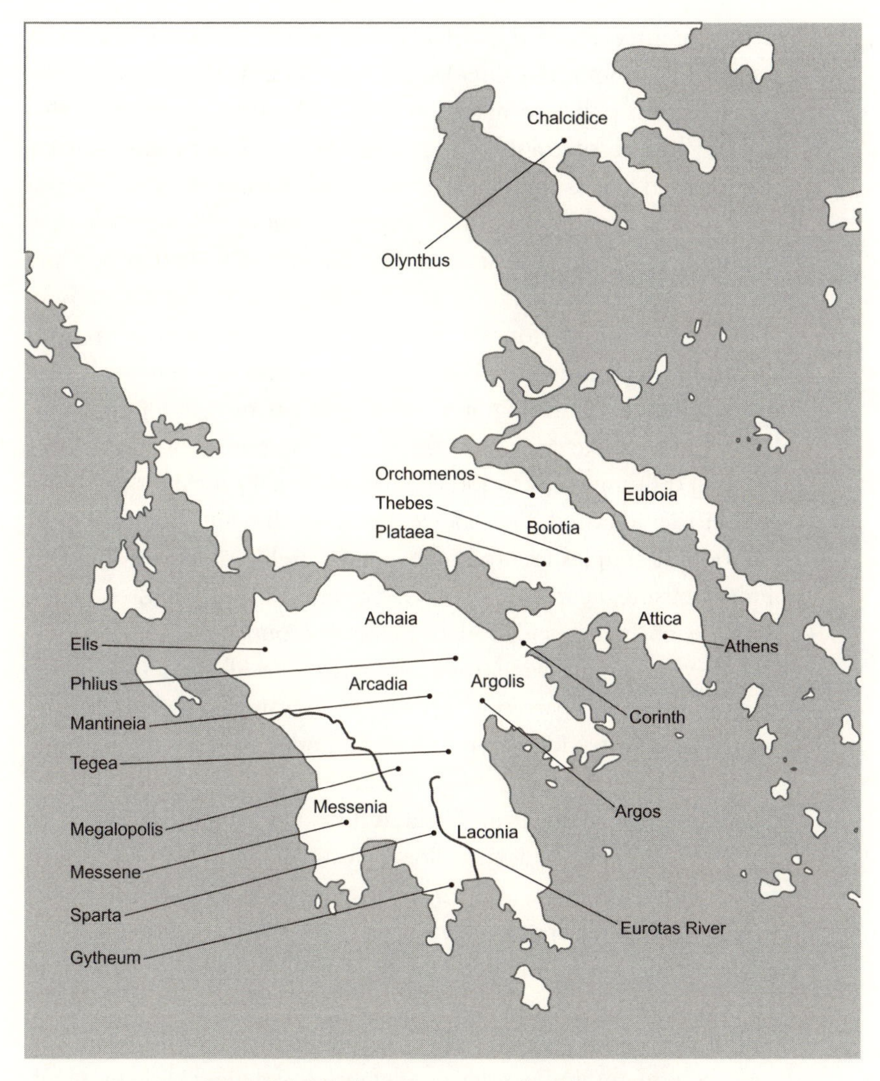

Figure 2.1. Northern Greece and the Peloponnesus

They remained famous for holding off the Persians in a last stand at Thermopylae, and had taken retribution against the Thebans for their support of the Persians after Plataea in 479. An outnumbered Spartan force had smashed through the Athenians at Tanagra (near Thebes) in 457 BC, and twenty-five years later the Athenian leader Pericles convinced the Athenians to abandon their homes to the unstoppable

Spartan infantry. After the defeat of Athens, Sparta was the most powerful city among the Greeks, and the Spartans followed an aggressive foreign policy to maintain that leadership. The myth of invincibility had become a self-sustaining psychological weapon, a force that projected across the Greek world ahead of any soldiers. It was a barrier to action, more solid than any wall, that could make the very thought of an attack on Sparta itself futile. This was Sparta's first line of defense against all challengers.

The deeper foundations of Spartan power were in its system of "education," its militaristic lifestyle, and its ethics of exclusively military excellence. The biographer Plutarch, in his *Life of Lycurgus,* tells us that the Spartan *agogē* was a unique system of rigorous state-mandated training—education is too intellectual a term—that began before birth and lasted into old age. Spartans thought that prolonged separation of couples created more robust children, so men lived in military housing, apart from women. Severity toward children began at birth: each infant was brought before the public authorities, who decided whether he was strong enough to be allowed to live. To toughen him up, a Spartan youth was given a single cloak and made to walk the countryside barefoot, placed under the command of any Spartan citizen who chose to give him orders, taught to be wily by stealing from the villagers and beaten if he was caught, and made to feel proud of his place in a battle line. Spartan children learned enough intellectually "to get by," and Spartan terseness ("laconic" speech) was legendary. The Spartan civic order was based, in essence, on a lifelong dedication to courage and endurance in battle, for the sake of the Spartan state.

All of this had a powerful effect on the Spartans. They were obsessed about maintaining order and defensive of their traditions, which they embodied in the figure of their legendary lawgiver, Lycurgus. His constitution was an unwritten set of rules, rituals, and virtues that anchored the ideals of Spartan civic life in a mythologized history. A Spartan was a member of the military all of his life, eating in common messes, bunking with his fellow citizen-soldiers, and devoid of private property. He owned no slaves but had free rein to boss those subject to the community. A Spartan's view of himself was centered on his military prowess; virtue *was* courage, and the reward of courage was renown and military glory. A Spartan came home with his shield or on it; to live without the shield was outside his frame of reference. The result was a militaristic society, a rigid social hierarchy, and a military *ethos* that survived by the ability to dominate its neighbors.

This Spartan notion of excellence was given voice by one of the few poets to survive from early Sparta. Tyrtaeus earned his glory by motivating the Spartans to achieve their highest virtue by fighting:

> For no man is good in war unless he can endure the
> sight of bloody slaughter and, standing close,
> lunges at the enemy. This is excellence [*aretē*, "virtue"], the best prize
> and the noblest reward to be won by a young man.
> This is a common good for the city and for all the people,
> when a man stands firmly in the front line,
> forgetting shameful flight totally, risking his life
> and showing a steady spirit,
> standing strong by the man next to him,
> and speaking encouragement to him,
> this man is good in war. . . .
> Let a man come now for this, the height of excellence,
> and strive not to hold back from war.[4]

Tyrtaeus had promised that cowardice in battle would be followed by shameful ostracism from the community:

> It is a fine thing for a man to die
> having fallen among those in the front while
> fighting for his city, and it is the most
> painful thing of all to leave one's city and
> rich farm lands to wander as a beggar . . . for he
> will be treated with hostility by whomever
> he meets.[5]

Such poetry—learned as songs by the youth and sung by soldiers to prepare for battle—helps us get a flavor of how the Spartans understood excellence and virtue. Before battles, the Spartans would sing these verses to bolster the mental strength needed for brutal combat.

Such musical poetry would rise in their minds as a set of moral commandments and a powerful injunction to strive for martial success. Passed on orally over generations, such verses took on the status of an internal law governing a Spartan's view of self and others, an *ethos* that defined excellence as courage, steadfast adherence to orders, and the ability to endure pain and slaughter. As a shared ethical standard that reached far deeper than legal institutions, Spartan law commanded them to remain true to the oral code that guided Spartan life; "here we lie, obedient to their words" was the stone left for the three hundred at Thermopylae. The "words" were much more than the particular orders issued before the battle: they were a powerful motivation to fight for Sparta, one's ancestors, and one's honor.

The development of the Spartan military ethos took place on the backs of the Greeks they subjugated. Over the previous centuries, the Spartans had smashed the city of Messene, the political center of Messenia in the western Peloponnesus, moving the people into the country and forcing them to live as peasants. These Messenian Wars ended in a permanent state of agricultural servitude for the Messenians. A hierarchy of subordination radiated from Sparta itself, from the elite Spartan citizens at the center, the "peers" (*homoioi*, lit. "similars," sometimes referred to as the Spartiatae), through the second-class *perioikoi* (the "dwellers around"), to the *heilōtes,* or helots.[6] "Agricultural serfs" is an inadequate term for these Messenian helots, who were not serfs in the medieval sense, but rather subjected peoples who lived in villages and provided food for their Spartan masters. The worst aspect of their condition was that they were stripped of their political identity and denied the fundamental birthright of self-governance due to all Greeks. Conversely, the Spartans were vitally dependent upon the helots—not only for food but as a training ground for Spartan military virtues and a demonstration of Spartan superiority. This Spartan dependence upon the helots—many of whom hated the Spartans passionately—was a potentially fatal vulnerability at the center of Spartan life.

Every year the Spartans formally declared war on the Messenian helots, reciting the formulas needed to invoke the gods and to legitimate their perpetual war against them. The Messenians had spent generations as virtual slaves, scrutinized by a secret police (the *krupteia*) for any shred of self-assertiveness, subject to military decimation should it be found, and without independent political identity or hope. During the Peloponnesian War with Athens, the Spartans decorated two thousand of their Peloponnesian underlings with wreaths in recognition of their valor—then all two thousand disappeared, "tall poppies" eliminated as threats to Sparta.[7] As late as the fourth century BC, the Spartans were still attempting to spread this system of helotage; in 385 BC they razed the recalcitrant city of Mantinea, destroyed it as a political entity, and distributed its inhabitants among five villages. The Spartans had implanted a powerful, militant ideology and a profoundly un-Greek form of institutionalized slavery in southern Greece, and they aimed to protect it with all the vigor and rigidity that their code of military excellence could muster.

Much of the value placed in martial excellence was omnipresent in the Greek world. But important differences between the Spartans and other Greeks—especially the Athenians and Thebans—carried constitutional implications that are important to what follows. The Athenian *politeia,* or constitution, had united the countryside of Attica into a democratic political union. There was no property requirement for political participation; every native-born adult male had full rights of Athenian citizenship. He could appeal the decision of a magistrate in a jury court, speak in the citizen assembly, serve as an official or council member, and defend Athens as a rower in the navy. There were no wars in Attica to enforce the dominance of Athens and no secret police. There was corruption, of course—lies, mistrials, unjust ostracisms, and occasional executions—and Athens' defeat by Sparta in the Peloponnesian War had made clear the failure of the democratic assembly. But the Athenians had reacted by reinscribing their laws and reforming their institutions. On the whole, they took deserved pride in the right of every citizen to speak and to claim a public defense under the laws. Should such ideas spread, they would constitute a threat to the security of the Spartans.

The countryside of Boiotia had rather evolved into a federal union of semi-independent city-states under the leadership of Thebes.[8] Starting from the usual conflicts among rival towns, the Boiotians had developed some degree of harmony with one another, based upon a common defense. Participation in a common council in Thebes was proportional, allocated on the basis of population, and probably designed with a sense of balance in mind. Boiotia was divided into eleven wards; each ward had one leader (a "Boiotarch") and provided sixty members to the council, along with a thousand soldiers and a hundred cavalrymen. Thebes controlled four wards, which gave it a strong, but not total, control over the federation. Internally, the cities had great autonomy. But in external affairs and matters of defense, the union would speak as a unit, under the leadership of Thebes. Facing attack during the fourth century, the federation fragmented but soon rejuvenated under stronger central control. This affirmed the ideals of citizen rule, the leadership of Thebes, and the federation's hostility to the oligarchic Spartans.[9]

Sparta may have had a political constitution of "a normal aristocratic type" for many Greeks.[10] But, in contrast to Athens and Thebes, Sparta enforced a strict military oligarchy over the countryside of Laconia and Messenia. The essence of the Spartan political system was *statism*, a modern concept sometimes expressed as *etatism*. As Austrian economist Ludwig von Mises described etatism, it denotes those political systems that "have in common the goal of subordinating the individual unconditionally to the state, the social apparatus of compulsion and coercion."[11] In her discussion of fascism, a form of statism, philosopher Ayn Rand cited a definition of statism as "the principle or policy of concentrating extensive economic, political, and related controls in the state at the cost of individual liberty."[12] Statism applies to any government with such power, whether a primitive tribal ruler, a theocratic council, or a communist or fascist dictatorship —including a democracy unrestrained by fundamental laws—each of which swallows the lives and fortunes of individuals without regard for their rights. The identification of such governments as statist is relatively new, but the practice is of enormous antiquity.[13]

To some extent, all ancient peoples subordinated the lives of individuals to the community, and none had identified a principle of individual rights. But in relation to other Greeks, the Spartans made this subservience all encompassing, and they enforced it ruthlessly, from cradle to grave. Such a state requires enemies, and for the Spartans these were the helots, along with anyone else who might threaten their dominance over the Peloponnesus. The Spartan system appeared to be a paradigm of order, especially in contrast to the incessant chatter of democratic Athens. But the mirage began with infanticide for the weak and ended on the battlefield for the strong.

The Spartans' mirage of unity masked serious internal disagreements about their approach to the outside world. At least twice—after the Persian wars and after the defeat of Athens—the Spartans had restrained their commanders who tried to create a Spartan empire with an aggressive foreign policy off of the Peloponnesus. The Spartans were split on this issue, and with good reason, because cultural insulation was necessary to maintain the purity of the Spartan system. Yet they returned to such aggressive policies in the three decades after the Peloponnesian War, led in particular by one of their kings, Agesilaos. Paradoxically, and perhaps out of fear, they embraced this policy at the time when Sparta's actual power was waning. Myths and mirages notwithstanding, by the 370s BC the military prestige of the Spartans was far greater than their actual capacity. The disjunction between the appearance and the reality is central to the events that followed.

The social trend that was working against the Spartans was demographic: its shrinking citizen population and the corresponding loss of Spartan infantry. Sparta had always had fewer citizens than the myth had suggested; according to the historian Herodotus, in 479 BC there were as many non-Spartans fighting in their ranks as there were Spartans. A disaster fifteen years later prevented recovery; Plutarch's *Life of Cimon* recounts the effects of an earthquake in 464 BC, in which large numbers of Spartan youth were killed. In the next generation, the twenty-seven years of the Peloponnesian War killed even more. The proportion of Spartans to non-Spartans in the army may have fallen to 1:9 by the late 370s; Xenophon claims that of seven hundred Spartans who fought at the battle of Leuktra in 371, four hundred were killed.[14] For more than a century—and probably much longer—the Spartans had relied on large numbers of allies to fill their military ranks, and their command of those allies depended upon continuous demonstrations of their own superior status.

The Spartans' *ethos* of courage, their obsession with demonstrating it over subjugated neighbors, and their need for material resources provided by others led them to act in ways that made the crisis worse. Spartan civic principles had been institutionalized in the common messes, ideally funded by the city itself so that there would be no undue influence of wealth upon an institution that was born of courage. But the Spartans began to require monetary contributions from the participants, and when some Spartans fell behind in payments, they were downgraded to a secondary status.[15] This may have been a new category, created out of the *homoioi* or "peers," and called the *hupomeiones* or "inferiors." This weakened the infantry by depleting the citizen ranks but also bred resentments among trained Spartans who were demoted into the "dwellers around." According to Xenophon, one downgraded Spartan named Cinadon plotted against the Spartans around 400 BC, and some may have hired themselves as mercenaries for the so-called March of the Ten Thousand into Persian in 401 BC. Xenophon relates that the Spartan peers were vastly outnumbered by their subordinates in the Peloponnesus, especially on the farms, and that when one from among these lower classes was asked about the Spartan peers, he replied that he would gladly "eat them raw."[16]

From this brief overview, tempered by Thucydides' conclusion that fear was a primary cause of the Peloponnesian War—fear that was now directed toward the Thebans—we may infer that the economic, social, and psychological center of Spartan society was in Messenia: economically as a source of food, socially as an underclass foil to Spartan excellence, and psychologically as an affirmation of Spartan dominance. Any loss of control over that area would have devastating economic and political repercussions inside Sparta itself; it would mean the failure of the Spartan system as a whole. The Spartans' fear of revolts was obsessive; they reacted rigidly, in the time-honored way of the spear, which hardened the animosity of the other Greeks and, in turn, intensified the Spartans' own fears and led to more frantic reactions. This vicious circle, intensified by their pursuit of their ethical ideals, set them on a collision course with the ideals of autonomy spreading among their Greek opponents.

In response to their aggressive foreign policy in the three decades after the Peloponnesian War, a groundswell of opposition to the Spartans had grown among other Greeks, all the while their actual power had slipped drastically. The myth of invincibility, though it still restrained those who might otherwise have confronted them, was increasingly hollow at its core. The Spartan ideology of militarism was ripe for a challenge.

An Intransigent Response to an Intransigent Enemy

After the Peloponnesian War with Athens, Sparta set out on an aggressive foreign policy toward the Greeks, promoted first by their commander Lysander and then continued by their King Agesilaos, who ascended in 400 BC with Lysander's support.[17] Sparta engaged in a series of battles with Persian satraps on the coast of Asia Minor, followed by wars in Greece with coalitions of Greek forces, in the so-called Corinthian War (395–387 BC). The negotiated end to this mayhem in 387/6—named the "King's Peace," or the "Peace of Antalcidas" for the Spartan envoy Antalcidas—was intended by Sparta to dismember the Boiotian alliance and to break the leadership of Thebes. Although Thebes had been an ally of Sparta during the Peloponnesian War, the Spartans understood that a resurgent Thebes threatened to bring Sparta's neighbors into alliance, or worse, into the Boiotian federal union itself. The Thebans might even legitimate and defend the political autonomy of the Messenians, which would be fatal to the Spartan state. The Spartans unleashed a series of diplomatic maneuvers intended to weaken Thebes and the other Boiotian cities, by isolating them from their alliances in preparation for military attack. Meanwhile, the Spartans established garrisons in several Boiotian cities.

The setup for the attacks began with a studied ambiguity in the treaty of 386. The peace applied to "autonomous" Greek cities, but what did that mean? The Spartans claimed that the Boiotian cities were not "autonomous," because they were led by Thebes. To prove their independence, the Spartans claimed, the Boiotian cities must sign the treaty separately. This attempt to isolate the Boiotian cities was an interesting position for the Spartans, given their dominance over the Messenians. But because the Spartans had destroyed the city of Messene and moved its inhabitants to villages, Spartan leaders claimed that there was no city of Messene and thus no question of autonomy for them.[18] The Spartan demand for "autonomy" in Boiotia —against the federal leadership offered by Thebes—was a rationalization by Spartan masters who wanted to keep their power over slaves. This was an act of war, intended to allow the Spartan infantry to act against isolated cities. It also masked the fact that a unified Boiotian federal union was growing stronger than Sparta and its allies—a fact that, if discovered, could have deadly consequences for the Spartans.

The most direct of the Spartan attacks against the homeland of the Thebans occurred when a Spartan garrison occupied the Theban acropolis—the so-called Cadmeia, where the federal council met—in 382 BC, under the command of a general who either exceeded his authority or was operating under secret orders. This commander, Phoebidas, may represent conflicting opinions among Spartan officials, once again divided between support for either aggressive action against Thebes or a negotiated settlement and a "stay at home" policy. Phoebidas might have been maneuvered into his position secretly, by an aggressive faction, so that the Spartans could disavow his actions while reaping the benefits of his success. The historian Diodorus offers this account:

> The Spartans took possession of the Cadmeia in Thebes for the following reasons. Seeing that Boiotia had a large number of cities and that her inhabitants were men of outstanding valor, while Thebes, still retaining her renown of ancient times, was, generally speaking, the citadel (acropolis) of Boiotia, they were mindful of the danger that Thebes, if a suitable occasion arose, might claim the leadership of Greece. Accordingly, the Spartans gave secret instructions to their commanders, if they ever found an opportunity, to take possession of the Cadmeia.

> Acting under these instructions, Phoebidas the Spartan, who had been assigned to a command and was leading an expeditionary force against Olynthus [a town on the mainland in the northern Aegean Sea], seized the acropolis. When the Thebans, resenting this act, gathered under arms, he joined battle with them and after defeating them exiled three hundred of the most eminent Thebans. . . . For this act the Spartans, now being discredited in the eyes of the Greeks, punished Phoebidas with a fine but would not remove the garrison from Thebes. So the Thebans in this way lost their independence and were compelled to take orders from the Spartans.[19]

Plutarch opined that the Spartans had condemned the man but approved the deed. Whatever the political intrigues behind the occupation, Phoebidas had placed a Spartan force in an unassailable position over Thebes, which undercut the unity of the federation, ratcheted up the Theban hostility against the Spartans, and further motivated the Boiotian cities to unify against outside threats.

The Spartans were trying to destroy any vestige of democratic expression that might threaten their power, but to do so without provoking a general war that Sparta could not win. They besieged the city of Phlius in the Peloponnesus, for instance, ostensively to punish Phlians for disloyalty during the Corinthian War.[20] But Xenophon claims that certain exiles from Phlius, friendly to Sparta, had complained to Sparta of their treatment; Agesilaos may have used this complaint as a pretext to install his own friends and the friends of his father over Phlius. The result, Xenophon tells us, was the hatred of the Spartans by five thousand Phlians—an attitude that was spreading and deepening.[21]

The Spartan expedition against Olynthus (on a peninsula in the northern Aegean Sea), waged to maintain Spartan leadership, destroyed the Chalcidic federation led by Olynthus—another noble attempt to gain security through political cooperation shattered by the Spartans. Diodorus reflects a historical tradition—probably true—when he wrote that the Spartans reached their greatest power when they subjugated the Olynthians. As a result, "The Spartans, however, had given their constant attention to securing a large population [those subject to the Spartans in the Peloponnesus, both free and helots] and practice in the use of arms, and so were become an object of terror to all because of their strength."[22]

Power leads to fear—the Thucydidean motif appears again. It is impossible to know whether Diodorus was relying on evidence or simply repeating the idea, but there was certainly a reaction by the Greeks against the Spartans. Events here are complex and difficult to unravel, but they can be essentialized for our purposes. The Boiotian War began in earnest in 378, when some Theban democrats killed many Spartan supporters among them, and then rallied to assault the Spartan garrison on their acropolis. The return of the acropolis to Theban control—the "Liberation of Thebes"—was a vital step toward expelling the Spartans from Boiotia, but there were other Spartan garrisons and sympathizers in the area—in Orchomenos, Plataea, and Thespiai, for instance—and the Spartans were determined to support them and their allies.[23]

Another aborted and probably unsanctioned Spartan attack, this time on the Peiraeus, the port city of Athens, in 378, allowed the Athenians to claim that the Spartans had broken the peace treaty and motivated them to assist the Thebans. Diodorus asserts that seventy cities allied with Athens in a common council.[24] The Athenians reformed their Aegean alliances into the Second Athenian Sea League and set out to take control of the Aegean Sea. The Spartans arranged their forces, and those of their allies, for war, and they marched into Boiotia in 377 and 376. The Spartan presence threatened to turn Theban farmers into de facto serfs, growing food for the Spartans to take on their regular raids. The Spartan efforts stripped the Thebans of their harvests for two years, but the Spartans also used their position in Euboia to cut off food imports to Thebes, an attempt to strangle the Thebans.[25] But the tide was turning, and by 376 the Thebans were taking control of Boiotia, and Athens had taken the Chalcidice into its Sea League; Spartan forays in both areas had failed. In 375 a contingent of Spartan infantry was defeated in Boiotia at the battle of Tegea—an ominous harbinger of Sparta's military decline. The Thebans removed a Spartan garrison from the city of Orchomenos, and they continued to drive Spartan supporters out of Boiotia.

At a conference in Sparta in 375, the Greeks tried to renew the King's Peace, but the Spartans and the Thebans again came to loggerheads over the status of the cities in the Boiotian Federation.[26] Sparta continued to try to isolate them from Thebes by insisting on separate signatures of the treaty. Given Athenian opposition to Theban leadership of the Boiotian League—a sign that the growth in Theban power was real—Athens and Sparta came to terms, agreed to divide Greece by land and by sea, and thus isolated Thebes from the peace. The uneasy treaty associated with this agreement—an admission of Spartan weakness—fell apart in further civil unrest. In 373 the Boiotian city of Plataea faced a Theban siege when it was garrisoned by Athenian troops, and the Thebans destroyed the fortifications of Thespiai in Boiotia. Fear of revolts in the Peloponnesus intensified, and the Spartans refused to concede the control of central Greece that would have allowed the Thebans to support Peloponnesian malcontents.

The Greeks again convened, in 371 BC, also under the impetus of the Persian king.[27] Agesilaos again refused to accept the Theban signature on the treaty, demanding the signatures of the individual Boiotian cities. Plutarch, in his *Life of Agesilaos*, related the response of the Theban representative, Epaminondas, to this demand:

> Agesilaos, seeing that the Greeks all listened to Epaminondas with the greatest attention and admiration, asked him whether he considered it justice and equality that the cities of Boiotia should be independent of Thebes. Then when Epaminondas promptly and boldly asked him whether he thought it justice for the cities of Laconia to be independent of Sparta, Agesilaos sprang from his seat and wrathfully bade him to say plainly whether he intended to make the cities of Boiotia independent. And when Epaminondas answered him again in the same way by asking whether he intended to make the cities of Laconia independent, Agesilaos became violent and was glad of the pretext for at once erasing the name of the Thebans from the treaty of peace and declaring war upon them.[28]

By standing up to the Spartan king Agesilaos, Epaminondas was facing down one of the most powerful men in the world and putting the life of every Theban on the line. He was also expressing the reality of the situation, in which the Boiotian cities were largely liberated from Spartan garrisons but still subject to Spartan attacks. Agesilaos was in many ways a fitting protagonist in the fight with Epaminondas. Agesilaos was not originally destined to be a king of Sparta—he was lame in one leg and had not received the exemption, usually granted to future kings, from the rigid childhood training (the *agogē*). His childhood experiences may have enhanced his capacity to inspire and lead Spartan troops; he was one of them, and a personal bond with them might have magnified his moral authority to wage this war. That Agesilaos was using traditional Spartan tactics to expand a dying system of apartheid, while Epaminondas was using innovative methods and principles to lead a groundswell of opposition to that system, brings some clarity to the wider implications of their confrontation and the impending Spartan defeat.[29]

For the moment the Thebans were diplomatically isolated. Diodorus made the point that "there was no city that could legally join them, because all had agreed to a general peace. The Spartans, since the Thebans were isolated [*monōthentōn*, lit. stood alone], determined to fight them and reduce Thebes to complete slavery."[30]

Under Agesilaos, the Spartan army acted as it had always done: to force a traditional confrontation between heavy-armed infantry on the plains of Boiotia and then to punish the Thebans in a set battle on a single afternoon. In 371 BC a full force of Spartan infantry, under the command of their second king Cleombrotos, marched into Boiotia against Epaminondas and an outnumbered army of farmers. In a triumph of intelligent planning and bold execution, the Boiotians solidly crushed the most powerful army in the Greek world, at the battle of Leuktra in 371 BC.[31] The key was the Theban Sacred Band, an innovative group of elite warriors, chosen by Epaminondas not on ethnic grounds, but by merit. This special unit, arranged with extra depth, gave the best soldiers a place to bond with other outstanding fighters on the basis of ability rather than birth. There was nothing comparable on the Spartan side; Spartan peers were not chosen on merit, and they could not brook the elevation of non-Spartan allies into their ranks. The Sacred Band, acting in concert with new cavalry techniques, smashed through the Spartans, put a spear through the Spartan war machine, and demonstrated the power of innovation over the ritualized Spartan organization.[32]

The historian Diodorus is direct about the nature of the Theban victory: the Spartans were "with great difficulty forced back; at first, as they gave ground they would not break their formation, but finally, as many fell and the commander who would have rallied them died, the army turned and fled in utter rout. Epaminondas' corps pursued the fugitives, slew many who opposed them, and won for themselves a most glorious victory. For since they had met the bravest of the Greeks and with a small force had miraculously overcome many times their number they won a great reputation for valor."[33]

The valor now belonged to the Thebans, a point driven home by the death of King Cleombrotos. The Spartans were forced to withdraw, routed by what was in their eyes a swinish, second-rate enemy. There was civil strife in Sparta's main rival in the Peloponnesus, Argos, which faced unrest both from Spartan supporters of oligarchy and from its own demagogues, who wanted to destroy its most distinguished citizens. The strife threatened to allow a clique manifestly unfriendly to Sparta rise to power. Similar strife in Arcadia led to an assault, by the Spartans under Agesilaos, against the territory of Tegea, which had harbored democratic refugees from the strife. This further hardened attitudes against the Spartans and motivated more fighters to challenge the Spartans when the time was right. The Spartans must have felt the noose tightening with every attempt to escape it.

After the battle of Leuktra the Greeks likely gathered for another peace conference, the second of 371 BC and their fourth since 386, and the first not brokered by the Persian king. Although we have few details of the meeting, the Spartans and the Thebans could not have resolved their differences. If the Spartans attended, doubtless Agesilaos did not back down from his demand that the Boiotian Federal League cease to exist. If the Spartans did not attend, the conference might have become a planning session to deal with the Spartans. By this point several cities in the Peloponnesus may have also refused to sign, thus rejecting Spartan hegemony in the wake of its defeat. More disturbances rocked the Peloponnesus, including unrest among the Messenians, the formation of a defensive league among cities such as

Mantineia and Tegea, and an alliance between Thebes, Argos, and Elis (see Figure 2.1). Sparta's hold on the Peloponnesus was slipping fast, Theban power was ascendant, and all parties must have known this.

The defeat at Leuktra shattered the myth of Spartan invincibility and showed the Boiotian allies that it was possible to defeat the Spartans in a pitched battle. But what happened in the next year was more than unheard of; it was previously unimaginable. Epaminondas may have recognized a fundamental imbalance in the Peloponnesus and the need to correct this permanently if peace was to result. Fundamental change would require an armed expedition into the Peloponnesus itself, an action to unite the Peloponnesians into a permanent bulwark against Sparta. Epaminondas may have recognized his own mission at a unique point of transition, in which physical and political conditions were in special alignment. Whatever his thoughts, he seized the reins of history and marched: "The Thebans then led out their army, taking some Locrians and Phocians along as allies. They advanced into the Peloponnesus under the generals Epaminondas and Pelopidas, for the other generals had willingly relinquished the command in recognition of their shrewdness and courage in war. When they reached Arcadia, other allies joined them in full force. And when more than fifty thousand had gathered, they decided to march on Sparta itself and to lay waste all the area surrounding it."[34]

Xenophon, forever hostile to Thebes, writes that, when the Theban army, reinforced by allies from northern Greece, joined up with the army of the Arcadians in the Peloponnesus, the Thebans judged the situation to be under control and prepared to return home. But the troops from Arcadia, Argos, and Elis urged them to strike directly into Spartan territory, citing the lack of men available to the Spartans.[35] Whether Xenophon can be believed here or not, many of those living around Sparta were energized to join the march, and they knew that the Spartans were on the retreat. Laconia and Messenia were approaching full revolt, and backed by a growing cadre of external allies. The military, social, and moral impetus was solidly arrayed against the Spartans.

As Epaminondas plunged his army of farmers boldly into the Peloponnesus, the people in the area—including many former allies of

Sparta—joined him in droves, swelling his forces and projecting a sense of victory in their path. He led a four-pronged military force past Mantineia and down the Eurotas River valley, probably ravaging the port of Gytheum before turning back north, toward Sparta itself.[36] This had not happened in memory, ever, and the experience of attack was new for those Spartans who were not warriors. The city itself had no walls, a testament to the ability of the Spartans to keep invaders off their homeland. For the first time, the Spartans retreated into their city, and braced themselves for the kind of onslaught that was supposed to be impossible against Sparta. Though it had waged many wars, this was a city that had not known the direct effects of war. The circle was now closed, and the war launched by the Spartans was wheeling about and visiting fire upon their land.

Can we picture the Spartans, cramped inside their city, old men, women, and children preparing to defend it with stones and roof tiles, wondering whether they were to undergo what they had inflicted on others—as the Athenians, once blockaded and starving before Spartan forces, had wondered whether *they* would soon suffer what they had done to others?[37] According to Xenophon, the army of the Thebans "went past the city, burning and plundering the houses which they found full of valuables. As for the Spartans, the very sight of smoke seemed unendurable to the women, who had never seen an enemy in their lives; but the men of the officer class, posted in detachments here and there, guarded this city of theirs, which was without fortifications; they looked few and they were few."[38]

"They looked few and they were few"—truth, the great enemy of myth, had laid bare the actual strength of the Spartans. But the mirage had been doubly deceptive, distorting the Spartans' views of themselves as well as the views that others had of them. By forcing the Spartans to face their own vulnerability to military attack, Epaminondas shook the Spartan military state to its roots: the physical defenses, the foundations of its *ethos*, the sense of superiority, and the expectation of victory would not be rebuilt. Those Spartans who had never before seen an enemy must have been particularly shaken; they had known war only in the abstract, as an affirmation of Spartan identity through courage, honor, and glory. Now Epaminondas reduced the idea of war to its essential meaning, to the level of direct perception: smoke, ruination, starvation, defeat, and death. But this is what it had always meant, to the victims of Spartan attacks. It was the Spartans—military practitioners par excellence, but always on someone else's turf—who were now learning the true nature of war.

Then Epaminondas ("whose nature it was to aim at great enterprises and to crave everlasting fame," wrote Diodorus) showed himself to be a commander of the first order.[39] He declined to enter Sparta itself. Despite the personal glory that sacking Sparta would have brought, the carnage would have been horrendous, and he would not have achieved his real goal. Permanent reduction of Spartan power required an end to Spartan mastery over the helots and dominance of the Peloponnesus. It was as if he had read Clausewitz on the "center of gravity" and recognized that the helots of Messene were a "community of interest" hostile to Sparta, and that if they were empowered, Sparta itself would be defeated.[40] He sounded the trumpet and directed the energy of his army away from Sparta.

It is evidence for his leadership skills that the army, motivated for vengeance against Sparta itself, obeyed. He set out for Messenia, the area to the west enslaved under Spartan rule. He set up a military defense and refounded the city of Messene, which the Spartans had so long ago destroyed. Locals rushed in to build walls over the winter; they reestablished a self-governing political center that could act as a bulwark against the Spartans. Slaves whose great-grandfathers had been slaves were now free. Pound for pound, this was one of the most liberating events in human history. These walls—much of which still exists—were built to last, by people who did not intend ever to be slaves again.[41] When Epaminondas left a few months later, the balance of power had shifted fundamentally.

A Greek sense of proportion—perhaps conceived mathematically by a Pythagorean like Epaminondas, who might have come to think that the right should not dominate the left—would favor not the outright destruction of Sparta but rather a balance between Sparta and a new Messene.[42] A year later, in 368, Epaminondas returned to the Peloponnesus, tightened the noose further, and founded Megalopolis—the Great City—as the capital of a new federal league in southern

Arcadia.[43] The Messenians and Megalopolitans each now had a *polis*—a political community with a physical and social center for self-rule—with walls shielding them from their former masters. Sparta regained some measure of prestige by defeating the Arcadians in the "Tearless Battle" of 368; however, the Messenians came to aid the Arcadians, and with a permanent impediment to Spartan ambitions in its north, the material and ideological center of the Spartan state was shattered.[44] Xenophon was so shaken by the reestablishment of Messene and the founding of Megalopolis that he omitted these events from his account entirely. But, from this point forth, he writes of their existence as a fact—an inescapable fact.

As a military event with deep political consequences, the march of Epaminondas through the Peloponnesus is defined by one fact: he had not fought a single pitched battle. He had come close to Sun-tzu's ideal of victory: "Subjugating the enemy's army without fighting is the true pinnacle of excellence."[45] The loss of life was minuscule; the results mammoth. "They were defeated," wrote Diodorus of the Spartans, "at Leuktra first ... and later, when they fought at Mantineia, they were utterly routed and hopelessly lost their supremacy."[46] It was after the battle of Mantineia, in 362, that all Greek cities except Sparta recognized the independence of Messene. In the mid 350s Athens guaranteed Messenian protection in an alliance.[47] The freedom of the Messenians from Spartan helotage was permanent.

An *Ethos* of Autonomy versus an *Ethos* of Dominance

Greek writers could attribute military victory—and defeat—to many things: bravery, chance, divine portents, natural conditions, deception, superiority in numbers. But the character of the commander tops them all—and the virtues of the army are his. The writer Diodorus, probably drawing on the historian Ephorus, lauded the philosophical acumen and moral stature of Epaminondas:

> Epaminondas indeed far excelled not merely those of his own race but even all the Greeks in valour and in shrewdness in the art of war. He

> had a broad general education, being particularly interested in the philosophy of Pythagoras. Besides this, being well endowed with physical advantages [*phusikois*], it is natural that he contributed very distinguished achievements. Hence, even when compelled to fight with a very few citizens against all the armies of the Spartans and their allies, he was so far superior to those heretofore invincible warriors that he slew the Spartan King Cleombrotus, and almost completely annihilated the multitude of his opponents. Such were the remarkable deeds which he unexpectedly performed because of his astuteness and the moral excellence that he had derived from his education.[48]

It is an important claim by Diodorus—in an encomium probably taken from the historian Ephorus—that Epaminondas achieved all he did in a way that was "natural" (*phusikois*, "by natural things"), which means more than physical strength. His elevated qualities were part of his nature—neither contrivances nor attempts to put on a front for the crowd—and Diodorus gushed forth with praise of his excellences: experience in war, a good grasp of its intricacies, and personal bravery. But experience can be gained, military skills were ubiquitous, and many are brave. Epaminondas had more elevated, all-encompassing excellences:

> For it seems to me that he surpassed his contemporaries not only in shrewdness and experience in the art of war, but in reasonableness and greatness of soul ... if you should compare the qualities of these [other generals] with the generalship and reputation of Epaminondas, you should find the qualities possessed by Epaminondas far superior. For in each of the others you would find but one particular superiority as a claim to fame; in him, however, all qualities combined. For in strength of body and eloquence of speech, further more in elevation of mind, and contempt of lucre, fairness, and, most of all, in courage and in shrewdness in the art of war, he far surpassed them all. So it was that in his lifetime his country acquired the primacy of Hellas.[49]

It was the wholeness of the man—the astuteness and moral excellence that Epaminondas gained from his education, integrated with his other qualities—that Diodorus and his sources respected. Exaggeration aside, we are justified to infer that his capacity to innovate was grounded in his abstract understanding of Pythagorean geometry, as well as fourth-century advancements that questioned, for instance, whether the right side is naturally dominant over the left side of any figure. This may be why he changed the battle line at Leuktra and strengthened his left side—which routed the Spartans. It was perhaps to restore a balance between left and right that he turned his army away from Sparta's left toward Messenia and created a political center in Messenia. With such ideas, he may have devised orderly city plans for Messene, designed a proportional federal constitution for Boiotia, and established a peace with the Spartans on the basis of a balance of power.[50]

This rational outlook ran counter to much in Spartan culture and is exemplified in the attitude Epimanondas held toward omens. Greek commanders routinely called upon soothsayers to sanction their actions—but a commander need not accept the omen if it conflicted with his battle plans. Before Leuktra, Diodorus claimed that Epaminondas, when confronted by one soothsayer who promised doom, rejected his prophecy: "Only one omen is best, to fight for the land that is ours."[51] After another omen, many soldiers opposed marching against the will of the gods, but Epaminondas acted as if "considerations of nobility and justice should be preferred as motives to the omen in question. Epaminondas accordingly, who was trained in philosophy and applied sensibly the principles of his training, was at the moment widely criticized, but later in the light of his successes was considered to have excelled in military shrewdness and did contribute the greatest benefits to his country."[52]

This praise of Epaminondas sums up an essential difference between the Ionian intellectual revolution, centered on Athens, and the tradition-bound culture of the Spartans. On the one hand, the Athenians created a magnificent edifice of intellectual innovation; on the other hand, the Spartans were concerned primarily with one virtue, courage, and they cultivated it amid claims to religious piety and obedience to tradition. But this put severe limits on their ability to think creatively. To change the arrangement of their troops or the organization of their political constitution was not in their nature. We hear more than once that the Spartans were constrained by religious festivals from assisting their fellow Greeks.[53] It was the Athenians—by land at Marathon and by sea at Artemesium and Salamis—who changed the order of battle and deceived the Persians. Epaminondas may have been furthering an intellectual revolution that could prove as deadly to Spartan culture as his tactical innovations were to their battle line. By cutting through the Spartan infantry elite and then through its agricultural breadbasket, Epaminondas cut an irreparable slash through the center of Spartan society and made its decline undeniable and thus irreversible. The first lesson of the Theban defeat of Sparta is one of intellectual innovation over inertia.

But beyond the leadership of Epaminondas, the questions remains why local populations—and the helots themselves—joined so readily, marched so energetically, and supported Epaminondas so enthusiastically. Part of the answer is found in how the Spartans treated their fellow Greeks and the moral reaction this caused. In the agonistic culture of the Greeks, one fights those who are worthy of being fought. One does not fight slaves; one shows them the whip and demands that they obey. A passage in Herodotus illustrates this perfectly. The Scythians are facing a slave revolt and fighting a battle with them, until the Scythian commander admonishes his men to put down their spears and pick up their horsewhips; "when they see us coming, they will remember they are slaves."[54] It worked; every one of the slaves was reminded of his position, and the battle ended.

In his discussion of Spartan violence, Simon Hornblower concludes that the Spartans were hated by the Greeks precisely because they treated fellow Greeks as if they were helots—meaning, as if they were slaves.[55] But the helots were Greeks, too—despite "endless symbolic reminders of underclass status"—and their hatred of the Spartans was rooted in their growing recognition that autonomy and freedom were their rights, too. There had been a history of slave revolts against the Spartans in the Peloponnesus, most memorably before the Peloponnesian War when Messenian rebels held a base on Mount Ithome for ten years. The Athenians sent help to the Spartans to put down the revolts, but if our evidence is accurate, the Spartans refused the help when it arrived, perhaps because they knew that the Athenians would sympathize with the Greek rebels seeking freedom. The Spartans were unable to defeat them decisively and were forced to grant them safe-passage off the Peloponnesus—the Athenians helped them to settle at Naupactus. Some forty years later, the Athenians and the Spartans made a temporary peace, in which the Athenians agreed to assist the Spartans in the event of a slave revolt—but no such guarantees were made in the other direction.[56] The Spartans had good reason to fear revolts—for their subjugation of the Peloponnesus was unseemly, and everyone knew it.

It was the Spartans' own actions that raised such enmity among other Greeks. According to Xenophon, in 373 the Spartans sent a commander to Corcyra, off the west coast of Greece, to protect Spartan interests against Athenian incursions. As hunger mounted inside Corcyra, the Spartan commander decreed that deserters from the city would be sold as slaves. As a crowd gathered, the Spartan "actually had them driven back with whips. However, those in the city would not allow them back again in the wall, considering them no better than slaves."[57] The Spartans' own troops fell in morale as the commander struck one of the officers "with a stick and one with the spike of a spear." Such arrogance—whipping fellow Greeks as if they were slaves—bred huge resentments born of moral outrage among surrounding populations. Plutarch relates a case three decades earlier in which a Spartan struck an Athenian wrestler with his stick; the Spartan commander Lysander then berated his fellow Spartan for not knowing how "to govern free men."[58] Epaminondas showed the Messenian helots that they could reassert themselves as free, autonomous men.

The political differences between Boiotia and Messenia are emblematic of the difference between the goals of the Thebans versus those of the Spartans—and suggest that Agesilaos did not understand the reactions of people in either area. Thebes held leadership in an area imbued with ideas of self-government, under a federation that was dissolved under Spartan pressure in 386 and reformed politically in 378. People there largely desired the leadership of Thebes. This leadership had to be asserted perforce, but there were no large-scale revolts against the Theban claims, and most opposition was incited by the Spartans—from garrisons at Plataia and Orchomenos, for instance.

When Agesilaos burned crops in Boiotia, he was attacking farmers directly, not attacking Thebes in order to set the farmers free. His claim that Thebes was enslaving the Boiotians was seen by later writers as hollow—as it was by many of those forced to endure Spartan attacks. But when Epaminondas entered the Peloponnesus, he already had the support of Argos and Arcadia, and the success of his march caused his army to swell while the Spartan forces shrank. Sparta held control not by riding a wave of popular support but by constant pressure from atop, wrapped in an image of power. Like Xerxes, in his own way Agesilaos failed to grasp either the nature of the society he was attacking or the power he held; when the image of defeat replaced the image of power, his support was gone.

The Spartans' obedience to tradition armed them to act with merciless energy—albeit rigidly—when threatened by such revolts. Agesilaos stands as a fitting protagonist to his Theban nemesis. We read of his obsessive attempts to control the Greek world, his obstinate efforts to undercut the peace between the Greeks and to destroy the Theban and Chalcidic federations, his refusal to accept the political independence of Messene, his reliance on traditional infantry tactics, and his disgraceful treatment of fellow Greeks. As his reputation declined after the battle of Mantineia in 362, Agesilaos pressed on with atavistic attempts to restore a system that was at serious odds with the ideals of citizen government that were sweeping the Greek world. Agesilaos, age eighty, was unable to give up the militaristic *ethos* that was his life, even as the Theban forays into the Peloponnesus wrecked the economic center of the Spartan military state and provided a visible demonstration of Spartan weakness. His flight to Egypt was likely an attempt to fill the Spartan treasury, empty since its agricultural base was liberated—but he had to seek salvation off of the Peloponnesus, as any minor power would do.[59]

The Messenian liberation was more than an economic failure; it neutered the *ethos* that had motivated the Spartans to defend their position of mastery. The myth was that when courage was needed the Spartans were always supreme. But the efficacy of their courage as a psychological weapon demanded continuous demonstration of its success. Defeat undercut their position, energized their enemies, and emboldened former allies and slaves. The social strength of the Spartan mirage—that only one in ten of those manning the Spartan army were Spartans—was destroyed when they came to realize that "the Spartans were few, and they looked few." The Spartans held this much in common with the Persian king, whose claims to magnificence rang hollow after his defeats at Salamis, Plataea, and Mycale. The Spartan myth of invincibility evaporated after Leuktra, Messene, and Megalopolis—as Xerxes had seen the myth of his personal superiority shattered after Salamis—and the visible bankruptcy of the Spartan *ethos* revealed their claim to a higher status as a fraud. They lost the support of allies needed to mount an effective campaign.

In the face of these changes, writers such as Alcidamas of Elea upheld the independence of Messene on moral grounds, on the basis of the right to freedom for all men. In contrast, rhetorical defenses of the Spartan side were based on claims to conquest anchored in mythology.[60] The Spartans were on the wrong side of the moral and intellectual debates; all that was left was their image of strength. And what had that been used for? We can imagine the thoughts of Epaminondas's troops when they saw the meaning of the Spartan system of helotage up close. There is shame in a free man who bows to slavery, but the Spartans had enslaved unwilling Greeks by war on their own territory. Where did the shame then lie? Parallels between the thoughts of the Theban troops and those of the Americans who saw the death camps of Germany in 1945 link these armies of liberation, a point made graphically by Victor Davis Hanson.[61] The growth of the forces under Epaminondas as he marched; the influx of exiles into reformed Messenia, and their energetic building of walls; the unwillingness of the Athenians to oppose Epaminondas as he marched home past Corinth; the desires of the Arcadians for protection from the Spartans—all of this suggests that the Greeks knew that the Spartan system was inimical to life as a free man and that the tide had turned against this system.

The effects of the liberation of the helots reached even more deeply into the center of Spartan society, into a direct attack on every Spartan's identity as a warrior. The Spartan's obsession with maintaining his position over others had become a blinder on his own self-understanding. The slaves he dominated became his masters, commandeering his *psyche* and metastasizing into the ever-present fear of an uprising that could threaten the Spartan world. He had to maintain the mirage of superiority for himself as well as others; any chink in the armor of the myth could undercut his control over the helots. But the Spartans of the late fifth and fourth centuries no longer lived up to the myth. In 425 BC, during the Peloponnesian War, the Athenians had established a beachhead on Spartan territory on the western coast of the Peloponnesus and captured some 120 Spartan soldiers on the island of Sphacteria. Thucydides expresses the astonishment of the Greeks that they had surrendered rather than be killed: "This event caused more surprise among the Hellenes than anything that had happened in the war. The general impression had been that Spartans would never surrender their arms whether because of hunger or any other form of compulsion; instead they would keep them to the last and die fighting as best they could. It was hard to believe that those who had surrendered were the same sort of people as those who had fallen."[62]

The myth was that they would never surrender; the reality was that they did—as they were later routed at Leuktra and chose retreat before the Boiotian onslaught. During the last invasion by Epaminondas, in 362, Agesilaos had to post children and old men on roofs as the final defense against a siege. Sparta was saved when the army returned —the Spartans still had allies—but the Thebans had penetrated the city of Sparta itself. The message was clear: Sparta was now one city among many, in a land that was open to attack. Like the Persians, the Spartans now had to turn inward—not to protect a single despot, but to reestablish some footing for their social order, their food supply, and their moral universe. The real victory over Sparta was not in the demonstration of their defeat in the eyes of their enemies but in their own realization that they did not live up to the myth. Knowledge is power, and self-knowledge is the most efficacious and powerful of all.

Epaminondas would invade the Peloponnesus four times in all and would die in battle in 362 BC at Mantinea, trying to keep the leadership won by Thebes. The Mantineans were now charting a course without Theban hegemony—but this was the course of a free people, made possible by Epaminondas's confrontation with Agesilaos and the Spartan elite. The flight and passing of Agesilaos marked the end of an era. He embodied the essence of the Spartan system—and its undeniable failure. In later centuries others would strive to return Sparta to glory, but the dominant method remained not innovation but atavistic attempts to throw back the clock to the traditional, mythologically based constitution of Lycurgus—under Agis IV, for instance, a descendant of Agesilaos.[63]

Our sources are sparse but consistent: the Spartans never again invaded Boiotia. Events of far wider scale—fired by Macedonia and Rome—would soon swamp these puny Greek wars. But even among the Greeks, the Spartans were no longer important. For Alexander to assert his power, it was the Thebans he had to smash—when they tried to evict the Macedonian garrison from their acropolis, waiting for help from other Greeks that never came. To hold his power, Alexander isolated the Spartans from the new League of Corinth—the Peloponnesian cities were of no mind to see the Spartans regain their power—and then largely ignored them. Upon the death of Alexander in 323, it was the Athenians who led a final rebellion. But, a generation earlier, in the moment before the Macedonian onrush, Greek history had taken one last noble turn. Under Theban leadership, Spartan helotage was broken by the actions of other Greeks, not by a foreign dictator. A Whig interpretation of history might suggest that the Greeks wanted to remove this blight from their own history, before their political eclipse. History is not so prescient, but Epaminondas and his troops were just as effective.

What Epaminondas shows us—and what begs comparison to the Persians, and to other cases in this book—is that such wars are powered from an ideological center, for both aggressors and defenders, which relies upon an economic and social base for its material sustenance and its affirmation. This is the intersection of theory and practice. For the Spartans, this economic center was their hold over their Messenian helots, but when the Spartans were defeated and their helots found a political voice, more was lost than someone to do the dirty work. The Spartan *ethos* and its ideological center—the system of ideas that placed them at the top of a social hierarchy and that anchored their excellence in physical dominance—was discredited, its failure in action made undeniable. "They were few and they looked few"—and they knew it. With the fire of their military *ethos* cut off from its fuel, the Spartans now had to face the universe without the shield, as pawns in a world that had left them behind.

Chronology of the Theban Wars with Sparta

All dates BC

ca. 460	Revolt of Spartan helots on Mount Ithome
431–403	The Peloponnesian War: Athens defeated by Sparta
395–387	The Corinthian War: Sparta versus Greek coalitions
386	The First King's Peace (The Peace of Antalcidas)
382–379	Spartan Occupation of the Cadmeia, the acropolis in Thebes
	Sparta tries to destroy the Boiotian Federation
378	The Boiotian War breaks out
	Athens allies with Thebes, versus Sparta
	Sparta, under Agesilaos, invades Boiotia twice
	Spartan attempt on the Athenian port, the Peiraeus
376	Spartans under Cleombrotos march ineffectually against Thebes
	Thebans defeat Spartan garrison at Orchomenos
	Athenians defeat a Spartan fleet at Naxos
375	Second King's Peace; Thebes is diplomatically isolated
	Spartan contingent defeated at the battle of Tegea in Boiotia
	Spartans aid Phokians in Boiotia versus Thebans
	Control of the Chalcidice taken by Athens
374	Thebes razes fortifications at Plataea and Thespiae
372	King's Peace renewed; Spartan expedition against Thebes
371	Battle of Leuktra; Spartans defeated by Thebans in Boiotia
	Peace Conference in Athens; Agesilaos erases Thebes from the treaty
370	Arcadian Federal League is formed; Spartan expedition to Arcadia
	First Boiotian invasion of the Peloponnesus
	Restoration of Messene
369	Second Boiotian invasion of the Peloponnesus
368	Spartans defeat the Arcadians (the "Tearless Battle")
	Founding of Megalopolis (omitted by Xenophon)
367	Third Boiotian invasion of the Peloponnesus
366	Peace of Thebes (brokered by Persian King Artaxerxes)
364	Thebes aids Thessaly; Death of Pelopidas against Alexander of Pherae
363–362	Fourth Boiotian invasion of Peloponnesus under Epaminondas
	Coalitions: Tegea and Boiotia versus Mantineia, Sparta, and Athens
	Battle of Mantineia; Epaminondas killed in a Theban victory

Chapter 3

"I Will Have My Opponent"

The Second Punic War, 218–201 BC

The First Punic War

Ancient Greek historians were masters at reporting grand synchronisms in their affairs: implausible parallels between widely separated events, connected by an idea. Such is the story of the battle of Himera, which goes like this: in September of 480 BC, on the very same day that Xerxes mounted his golden throne over the bay of Salamis, one Hamilcar of Carthage, with 300,000 men and thousands of ships, rowed into battle off Sicily, near the town of Himera. He was as confident of victory over his Greek enemies as was Xerxes—and just as shattered by his defeat. The precise dates of these battles are unverifiable, as are the inflated statistics; Herodotus wrote that "the Sicilians say" the battles were fought on the same day."[1] But the Sicilians had good reason to link their own victory with that of the Greeks to the east, because their victory—against the Carthaginians, colonists of the Phoenician subjects of the Great King—was also a victory for the independence of the Greek city-states. For two hundred years the Greek cities in Sicily and southern Italy remained unbowed before any external authority, until a new power from the north thrust itself into this unruly area and subdued Carthage, Sicily, Greece, and everyone else.

Rome's rise attained global proportions with the three Punic wars against Carthage—the first, over Sicily (264–241 BC); the second, for control of the Mediterranean (219–202 BC); and the third, a straight-up massacre (149–146 BC). These wars were fought with all of the logistical complexity, emotional intensity, and pitiless bloodletting of a modern world war. The antagonists were different in kind. Rome, an inland political force driving toward control of Italy, and Carthage, a maritime power with a preeminent position on the sea, faced in different directions, and it seems incongruous that they should bleed each other white for generations. But they did, and the questions this raises are of timeless import. Why did the Carthaginians start a new slaughter twenty years after their defeat in the first war, yet remain at peace for two generations after the second? Why did the harsh terms imposed by the Romans after the first war lead to new attacks, while the harsher terms of the second did not? Why, in the second war, did Hannibal's mauling of Italy not reverse the Roman will to fight, but a couple of battles in Africa ended the Carthaginian war effort?[2]

It is tempting—if not canonical—to say that early Rome's greatest rival was Carthage, but this is not self-evident. During Rome's early rise serious threats to Italy had always come from the north. It was the Celts and Gauls who threatened Rome itself; the city was sacked in 387 BC, and the Romans never forgot it. The historical tradition strongly suggests that the first relations between Rome and Carthage were peaceful, a consequence of divergent interests and geographic buffers; the Romans had no navy to conflict with the Carthaginians, and Carthaginians did not harass Italy north of Naples. The historian Polybius records treaties with Carthage as old as the founding of the Roman Republic in 509 BC, and this may have renewed an agreement with the Etruscan kings of Rome. Other treaties followed, perhaps in 348, and 306 and 279.[3] The terms of these treaties are obscure and unverifiable, but some of them must have been attempts to come to terms over the fractured areas of Sicily and southern Italy.

In the 330s BC the Roman gaze turned east—beyond the shores of Italy—when Alexander I of Epirus, brother-in-law of Alexander the Great, crossed into southern Italy. The result was *amicitia,* "friendship" —a state of nonwar, at least—between Rome and the Greek cities. In 280 this hands-off relationship failed when Pyrrhus of Epirus came storming into southern Italy from Greece, winning "Pyrrhic" victories that were costly beyond the prize ("one more such victory and all shall be lost!") before pursuing new vistas of mayhem in Sicily.[4] Following his departure, the Romans stepped more actively into southern Italy, a wild and woolly area awash in traders, mercenaries, gangs, and would-be tyrants who could create chaos for the larger powers.[5] Neither Carthage nor Rome could ignore such local affairs, especially if either thought that the other was using a crisis to build its power.

The first truly world-scale war in the West was ignited by two small, nondescript towns, important only for their strategic locations. Rhegium on the toe of the Italian boot and Messana on Sicily had been taken over by separate bands of mercenaries who were brutalizing the local populations. The mercenaries in Rhegium had been sent by the Romans to keep order—the Romans captured, scourged, and executed them; there could be no doubt that the Romans kept their promises to allies. But the mercenaries in Messana posed a more difficult problem. They had captured the town for Agathocles, tyrant of Syracuse, around 312 BC. There was a struggle after his death in 289, and Hiero, the new tyrant of Syracuse, blockaded them into their harbor. They appealed for help to a Carthaginian fleet north of Sicily, who established a garrison in Messana—an ominous development for Rome, should Messana ally with Syracuse, a powerful ally of Carthage. Some of these mercenaries also appealed to the Romans, who saw a chance to limit the power of Carthage by supporting them against Syracuse.[6]

The crisis in Messana became a political and moral dilemma for the Romans. To assist the mercenaries would be inconsistent with Roman actions in Rhegium—but to fail to assist them could create a more dangerous situation on Sicily and in Rome. Polybius explained the issue in terms of a split between established principles of dealing with others and the pragmatic demands of the *plebeians*, the common people of Rome, who were swayed by promises of rewards from their political and military leaders, the two consuls:

> Even after long consideration, the Senate did not approve the proposal to send help to Messana; they took the view that any advantage which could result from relieving the place would be counter-balanced by the inconsistency of such an action. However, the people [plebeians in Rome], who had suffered grievously from the wars that had just ended and were in dire need of rehabilitation of every kind, were in-

> clined to listen to the consuls [the two highest political officials]. These men, besides stressing the natural advantages I have already mentioned which Rome could secure if she intervened, also dwelt on the great gains which would surely accrue to every individual citizen from the spoils of war, and so a resolution in favour of sending help was carried.[7]

This tension between the plebeians and the senatorial aristocracy was of great import to Roman decision making. The *populus* of Rome could influence the elections of the consuls—the two political leaders who were also military commanders—and could thus put enormous pressures on the senators. The Romans compromised. A force under Appius Claudius transported Roman troops overseas for the first time, captured the town, and forced Hiero to retreat into Syracuse. Carthage lost its garrison in Messana, and Hiero became an ally of Rome for the rest of his life. The outcome was a victory for those Romans who wanted to expand into the Mediterranean, as well as for the plebeians who wanted glory and spoils. But this abandonment of consistency had a price, for the Romans now needed a permanent presence on the island, to keep order and oppose the Carthaginians.

These actions put the Romans into direct military contact with the Carthaginians. The First Punic War (262–241), fought over Sicily, brought the Romans into their first overseas conflict, led them to build their first navies, and gave them their first overseas territories. After eighteen years of debilitating effort, both sides were exhausted, but the Roman aristocracy dug deep into its financial coffers, and the Roman people exploded with energy in the expectation of loot. They built a huge new navy, destroyed the Carthaginian fleet off Drepanum, and forced the Carthaginians to sue for peace—albeit without defeating their infantry at the fortress of Eryx, in western Sicily. The long-range tenacity of the Romans won the war of attrition, a tenacity founded not upon economic wealth—Carthage was not poor—but on a drive for dominance, backed by huge manpower reserves in Italy. Under the Treaty of Lutatius, the Carthaginians agreed to return prisoners without ransom, to give up Sicily, and to pay a huge indemnity. This was bad enough, but the Romans then upped the terms, demanding higher payback more quickly.[8] The Carthaginians swallowed these demands, along with the humiliation and anger they engendered.

In 241 BC the First Punic War was over, but three years later Carthaginian mercenaries were rampaging against Carthage in the so-called "Truceless War," a revolt that became a siege of Carthage.[9] The Carthaginians elevated Hamilcar Barca—commander of the undefeated infantry on Sicily—into command, and he slaughtered the rebels and saved Carthage. On Sardinia, some of the mercenaries asked the Romans for aid, which was first refused—but, in a dramatic reversal, the Romans annexed the island along with Corsica. The Roman action was an egregious violation of the surrender agreement, but the area was in chaos with thousands of unemployed soldiers, and Rome needed to step into the power vacuum, to bring order to Sicily and its environs, and to protect the coast of Italy. To have allowed Carthage to rearm and move back into the central Mediterranean was out of the question. With Sicily and Sardinia cut off from them, the Carthaginians now had two, and only two, options: to reduce their ambitions to those of a single city, dependent upon Rome for protection—or to direct their energies west, through Africa and into Spain. There were no other options.

The Decision to Wage a Second War with Rome

In the aftermath of the First Punic War, many Carthaginians—and we need to ask which ones—attached themselves to Hamilcar, and they moved west through Africa, enlisting mercenaries as they outflanked Rome and moved into Spain. This effort to establish a political, economic, and military presence in Spain has been taken as evidence for a groundswell of antagonism against the Romans among the Carthaginians, which became the official policy of Carthage with little or no internal opposition. According to Polybius, there were three causes of the second war: the anger of Hamilcar, the unjust seizure of Sardinia by the Romans, and the growth of Carthaginian power in Spain.[10] The Roman seizure of Sardinia was the event that roused the moral indignation of the Carthaginians and triggered their anger into action.[11] According to Polybius, the success of the Carthaginians in Spain enthused them with the prospect of victory, which strengthened their will and capacity to drive forward into a confrontation with Rome.

Polybius has made the enthusiasm of Hamilcar's supporters clear, and events favor this conclusion—but was there a consensus inside Carthage for Hamilcar's actions? The question is important, because, if there was not, then civilian support for the war could be vulnerable. The city's internal politics is obscure, but it is safe to say two things. First, political decisions were conditioned by religious practices that were as foreign to the Romans as to us. Punic culture had a history of child sacrifices, and a founding myth (perhaps a later invention) in which the son of the first king was sacrificed.[12] The Carthaginians all too often crucified their leaders; one was nailed up after the loss of Messana, another after the Truceless War, and another in 149 BC. The connection between religion and politics was deep. Political officials made regular visits to shrines in the mother city, Tyre, and Punic inscriptions are generally religious and lack the political and legal focus found among the Greeks and Romans. An important inscription relocated to Marseilles from Carthage, the so-called Tariff of Marseilles, tells us about two political institutions—but it was found at a temple to Baal, and its prescriptive content is a list of sacrifices, approved animals, and monetary values more akin to Hebraic rituals than Greek or Roman politics.[13] Carthaginian names are laced with religious symbolism; Hannibal means "he who enjoys Baal's favor."[14] Carthaginian politics was permeated with the religious symbolism and practices of the Near East.

The Barca family—from Hamilcar to Hannibal—elevated a pantheon of deities from Tyre into a point of focus for its soldiers, in order to bolster the strength and status of its dynasty. The family rejuvenated the cult of Baal-Shamim—a deity invoked in a seventh-century BC treaty between Tyre and the Assyrian ruler Esarhaddon, whose image we have seen associated with Zeus and contrasted with Baal-Hammon.[15] Hannibal sacrificed to Baal-Shamim as a child before going to Spain, and he propitiated the god at the temple of Melqart (the Greek god Herakles) in Gades before invading Italy.[16] In 216 he took an oath with Philip V of Macedonia that invoked Baal-Shamim and Melqart.[17] This has important political implications, because if the Barcas relied on the "official pantheon" of Carthage, not a "dynastic pantheon," then there would have been little religious-political break between his administration in Spain and the citizens of Carthage.[18] If, however, Hamilcar rather used the Tyrean deities as an innovative dynastic pantheon and appealed to the soldiers by adapting the Greek and Roman gods of Fortune to the "national religion," then his efforts in Spain would have been more sharply distinguished from the political situation in Carthage.[19] The Barcas may have implanted a religious split between themselves and the aristocracy in Carthage—a matter of potentially great import in Carthaginian political support for the war and to the degree of independence the Barcas enjoyed in Spain.[20] If so, then should Hannibal be defeated, the Carthaginians could accept it without challenging the ancestral gods of the city.

Second, Carthaginian politics was rent by disputes, factions, intrigues, and obstruction, and no political consensus inside Carthage should be assumed for Hamilcar's mission. Carthaginians and mercenaries loyal to Hamilcar—who heard the siren call of robust action and sustained their actions for more than twenty years—filled Hamilcar's ranks with an energy that may appear to us as overwhelming support. But it is far from certain that fanatical anti-Roman hatred dominated Carthaginian politics or that Hamilcar's supporters in Carthage could sustain decades of preparation for an invasion of Rome. A historical tradition, promoted by the historian Fabius Pictor and passed on by Polybius, placed Hamilcar in opposition to the Carthaginian nobility and even made him into a usurper of the constitution.[21] Carthaginians who opposed Hamilcar might have supported the mission to Spain in order to remove many of his supporters from Carthage.[22] The Barcas would have controlled the information sent back to Carthage, in order to maintain support for their efforts and their independence. A serious military defeat, a revolt of the African tribes to the west, or a Roman force converging on Carthage might lead to a swift reversal of the political decision in Carthage to fight and a recall of its armies.

Once in Spain, Hamilcar was out from under the immediate scrutiny of his political enemies. Although we cannot be certain how much he was able to ignore the Carthaginians back home, his geographic position alone suggests that he could move with great independence and largely self-sufficient resources. He acted not with rash anger but with the deliberateness of long-range forethought. He moved up the Quadalquivir River valley, consolidated his position, took hostages from Spanish populations, and aimed for the gold and silver resources—which he used to pay the Roman indemnities from the last war, another reason for the politicians back home to leave him alone. He died in 229 BC, drowned in the Jucar River, and his son-in-law Hasdrubal took over. Polybius writes, following the Fabian tradition, that Hasdrubal "governed Spain as he chose without paying attention to the Carthaginian Senate."[23] Between 229 and his murder in 221 he founded New Carthage, a capital on the coast of Spain, which strengthened the Carthaginians' position but also put them back on the radar screens of the Romans, after years of neglect.[24] To avoid war with Rome, Hasdrubal may have negotiated the Ebro River Treaty with the Romans as a means to define a border in northern Spain. This obscure treaty may have been a delaying action by the Romans, who were fighting the Gauls and who wanted to prevent the Carthaginians from moving north out of Spain.

For nearly twenty years, Rome, tied down in other wars, had done little or nothing to limit Carthaginian ambitions in Spain. Polybius criticizes this failure:

> The Romans suddenly perceived that Hasdrubal had gone far towards creating a larger and more formidable empire than Carthage had possessed before, and they determined to take a hand in the affairs of Spain. They became aware that during these years they had been fast asleep and had allowed Carthage to build up and equip a large body of troops, and so they now tried to make up the ground they had lost. For the present they did not venture to impose conditions or to make war

> upon Carthage, because at this time the threat of a Celtic invasion was hanging over them, and an attack was expected almost from day to day ... so they first sent envoys to Hasdrubal and concluded a treaty. According to its terms nothing was said about the rest of Spain, but the Carthaginians undertook not to cross the Ebro [River] under arms.[25]

The reason for this neglect is partly in Rome's strategic position. The Romans faced threats from the north, including the Celts and the Gauls; from the south, including various recalcitrant states in Italy and Sicily; and from the east, especially Macedonia and the Greek leagues. To consider only the east: in 229 BC Rome sent some twenty thousand infantry and two hundred ships to deal with the rising power of the Illyrians under Queen Teuta in the First Illyrian War. By 219 one Demetrius of Pharos had married into the Illyrian dynasty, and Rome needed to keep him in check: thus, the Second Illyrian War. By 215 Carthage had allied with Philip V of Macedonia, which the Romans had to break: the First Macedonian War. Rome may have been forced to delay action against Spain in 226 BC, but the result nonetheless was Rome's worst crisis in almost two hundred years. Such are the consequences when the strategic context grows beyond a nation's capacity to deal with it.

The young Hannibal took command of the Carthaginians in Spain in 221 BC. The key to controlling Spain as a Mediterranean base was to control the Quadalquivir River valley in the south, the Ebro valley in the north, and the coastal route between them (see Figure 3.1). This strategy would ensure land communication from north to south, as well as access to the sea. The town of Saguntum lay on a vital strategic position on the coast, about halfway between the Ebro River and New Carthage. Hannibal squeezed his military vise against this weakness, and in 220 BC a pro-Roman faction in Saguntum appealed to Rome for assistance against the Carthaginians. Rome supported the pro-Roman side, in another small city that was bringing two great powers into war because it stood at the nexus of their vital interests—and because the Romans saw a moral point at stake in supporting an ally, with potentially disastrous results should that support fail.[26]

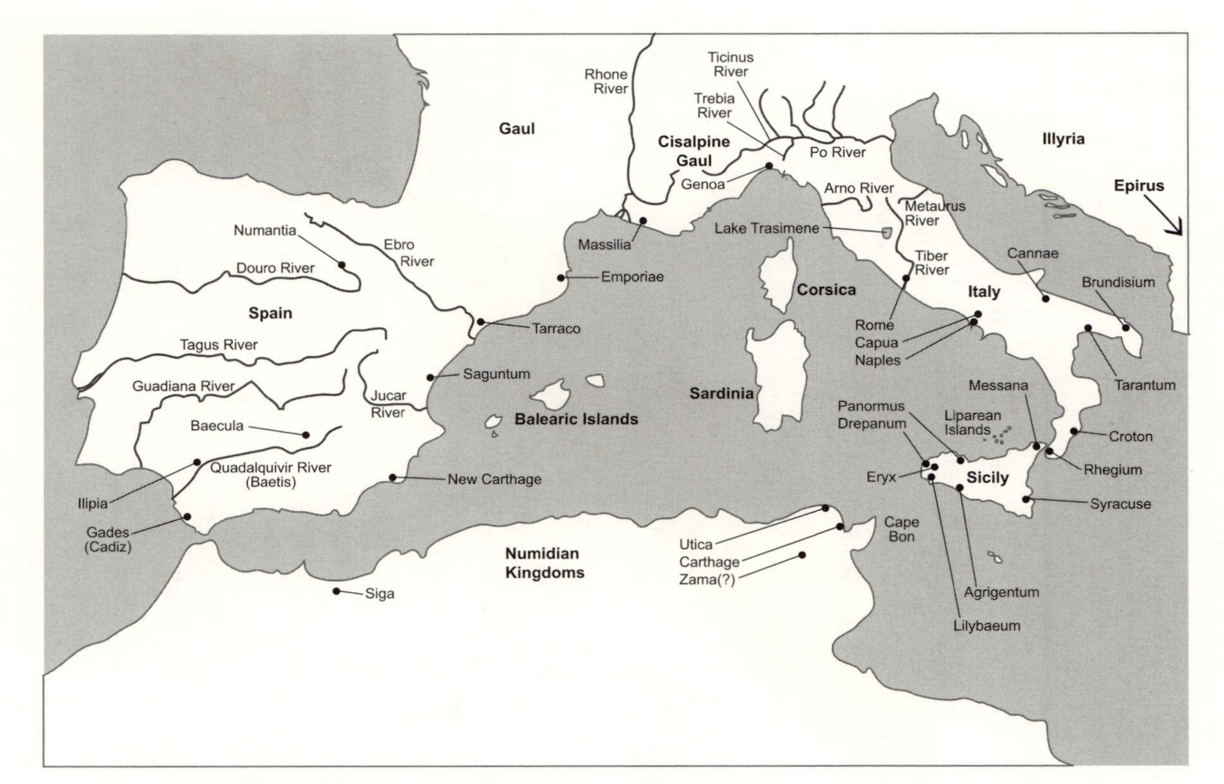

Figure 3.1. The Western Mediterranean: Major Sites Relevant to the Second Punic War

Before or during Hasdrubal's tenure, Rome had entered into a position of trust—in Latin, *fides*—with Saguntum.[27] *Fides*—as had existed between Rome and Rhegium, before the First Punic War—was not formalized as a "treaty" (*foedus*) but was more than a vague "friendship" (*amicitia*). *Fides* was a complex term with a range of meanings that implied a solemn moral commitment, connected to justice, perhaps an oath backed by a certain divine sanctity. As Cicero wrote in the first century BC, "The foundation of justice is good faith [*fides*]—that is, truth and constancy to promises and agreements."[28] As regards to Saguntum, the Romans took a commitment to support such allies seriously and were once again willing to risk war because of that commitment. Although Polybius denies that the Romans debated the issue—he saw the pro-Saguntine position as beyond debate—there must have been disagreement from those who thought that Rome should not be involved in such distant intrigues. The Byzantine compiler Zonaras recorded a senatorial debate over Saguntum between Lucius Cornelius Lentulus, the *pontifex maximus* (chief priest in Rome), and Quintus Fabius Maximus, leader of the Fabius family and its supporters.[29] It is plausible that two members of the Scipio family—who would later lead the Roman expedition to Spain—had received the request from the Saguntines and that they advocated sending help against opposition from Fabius in the senate, who was liable to oppose such far-flung plans. The upshot was that the Romans sent envoys to warn Hannibal not to act against Saguntum—but they did not immediately dispatch a force.

The Carthaginians also supported their position with claims of justice. According to Polybius, Hannibal claimed to be protecting the Saguntines, relying on a pretext that the Romans had caused the deaths of some leading Saguntine citizens. Therefore, "he sent home to Carthage asking for instructions on how he was to act, in view of the fact that the Saguntines were relying on their alliance with Rome to commit wrongs against some of the peoples who were subject to Carthage."[30] Because Hannibal claimed to be protecting Carthaginian subjects from Roman partisans in Saguntum, the Carthaginians were unwilling to make a decision to restrain him. Like any semi-autonomous theater commander, Hannibal made certain that only limited information made it back to Carthage and that the things he needed to do would be done before the politicians back home could say no. This control of information could have added to the misunderstandings in Carthage over the coming war. According to Livy, Hannibal forestalled action in Carthage "by writing to the leaders of his party there and urging them to prepare the minds of his supporters and prevent them from allowing the opposing faction to make any conciliatory gesture towards Rome."[31] After the war, a Carthaginian embassy to Rome claimed that Hannibal had made war "on his own initiative."[32] Nevertheless, the decision for war—if nothing more than a refusal to recall Hannibal—was a political decision made in Carthage. The key to ending the war would be a reversal of that decision and the recall of Hannibal.

For the moment, the decision was in Hannibal's hands: should he avoid an immediate confrontation over Saguntum, or should he move before Rome could react? If the Romans established a garrison at Saguntum, Hannibal would lose control of the Ebro Valley, allies would defect en masse, morale would collapse, and a siege of New Carthage would be a certainty. Tumult in Africa would follow, and perhaps the fall of Carthage itself. Saguntum became an issue of life or death to Hannibal. He captured the city after a bloody siege of some eight months. The resistance of the people of Saguntum may have stiffened, fighting house to house against hardened soldiers, waiting for Roman relief that never arrived.[33] The toothless Roman demand, instead of restraining Hannibal, had motivated him to act quickly—and had stranded the Roman supporters in Saguntum while harming the Roman position.

While the Romans were debating the matter—probably embroiled in political infighting over the decision and waiting for the new consuls to take office in March—Hannibal acted with tremendous energy and totally preempted the Romans. He marched his huge army—some ninety thousand infantry, ten thousand cavalry, and elephants—out of New Carthage in late spring or early summer of 218 BC, over difficult land routes that avoided local resistance in Emporiae, Massilia, and hostile Gallic tribal areas.[34] He was at the Rhône River with his elephants by September and was approaching Italy by the time

Roman forces reached Spain. Hannibal's four-month head start had set the fighting in Italy rather than in Spain. Roman plans for an attack on Carthage were scrapped. By failing to establish a proper peace in 241 BC, ignoring a potential enemy's growing power, and then issuing an ultimatum they could not enforce, the Romans consigned themselves to fifteen years of war in Italy.

Hannibal's plan was to cause massive defections among Rome's Italian allies. It was, in many ways, a superb plan that struck to the center of the Romans' political and economic relationships with their allies. He knew that tensions between Rome and the area of Campania, especially Capua, could lead the Campanians and others to the south to form an Italian league and that this would gut Roman power in Italy. He also knew that the Romans had difficult problems with the Celts and Gauls—although some were potential allies—and that Philip V of Macedonia was no friend of the Romans. If he could gain support from such areas and take control of a good port on the Italian coast, he might be able to permanently reduce Rome's power. Hannibal had accurately identified the center of Rome's strength—but, as we shall see, he did not sufficiently understand it.[35]

Cnaeus Scipio went on to Spain to consolidate the Roman position in the north, while his brother, the consul Publius Scipio, took his force back to Italy to face Hannibal. But Publius soon returned to Spain, recognizing that the key to beating Carthage was not to face Hannibal in Italy but rather to destroy Carthaginian control over Spain and prevent reinforcements. The Scipiones followed the basic policy of projecting power off of Italy against Carthage's western power base, Spain. Meanwhile, the senate in Rome was dominated by Fabius Maximus, arch advocate of a defensive policy on Italy.

Hannibal's blitzkrieg into Italy was brutally successful at killing Romans and their allies; in battles at the Ticinus River (north of the Po River), the River Trebia, and Lake Trasimene, in 218 and 217 BC, he slaughtered tens of thousands of Roman soldiers, and showed the Romans that they could not meet him in open battle. After Trasimene, one Roman consul was dead and the other cut off from Rome. To resolve the constitutional crisis, Quintus Fabius was named dictator for the standard term of six months and given control of the army.

The major question facing the Romans was how to defeat Hannibal, and they saw two options: to entice him to divide his forces and then attack, or to avoid battle, cut off his food, and starve him. Fabius followed Hannibal's army south, watching him ravage the countryside but avoiding a set battle. The frustration of the Romans grew as they watched the smoke from their burning farms and considered the imminent defection of allies, while stymied by their commander—who some say "praised the enemy"—from giving battle.[36] A series of complicated political maneuvers followed, intended to reverse the Fabian strategy of delay and to empower an army to attack Hannibal.

Fabius laid down his office in late 217, and the frustrated Romans elected two new consuls, who took control of the army in March 216. The two consuls were Cornelius Terentius Varro—whom historians have unfairly labeled a demagogue and a "butcher's son"—and Lucius Aemilius Paullus, a member of the Aemilius family who had married into the Scipio family. The Romans raised eight legions—nearly fifty thousand troops—and an equal number of allies, a massive effort.[37] Hannibal was outnumbered two to one, and more than eighty thousand soldiers marched under generals who promised results to the plebeians. They got results indeed; at the battle of Cannae at least fifty thousand Roman soldiers and allies were killed or captured, in Rome's worst military defeat ever. The loss to the Roman senate was catastrophic—some 177 members.[38] History has blamed the commander Varro for his incompetence, but Polybius—a grandson of Aemilius Paullus—wrote the history. The truest cause of the defeat was those Romans who took their plebeian supporters into the battlefield against a foe of far greater ability, rather than devising a strategy to defeat Carthage itself.

Cannae vindicated the Fabian strategy of delay, which Rome largely followed until 210 BC. The Aemilius and Scipio faction was weakened, its leaders dead on the battlefield, its plans to defeat Hannibal in ruins. Fabius's alternative was based on a solid conclusion: that Hannibal could not be beaten in a pitched battle. He would starve, however, if deprived of support from allies. This led Fabius to avoid nearly all confrontations with Hannibal; his strategy was *cunctatio*, "delay," and he became Fabius Cunctator, Fabius the Delayer. This strategy saved Rome from massive slaughter in the short-term; Livy quoted the poet Ennius, who called Fabius the savior of the republic.[39] But this strategy did not win the war; it only avoided defeat for the moment, and at the price of the slow strangulation of the countryside. The war bogged down into the agony of attrition, with interminable marches and countermarches, feints and retreats, and short-range gains and losses. An entire generation of Romans grew used to a foreign army on its home soil.

Following Cannae—with an unmatched enemy force rampaging near Rome, a Roman army obliterated, and the consul Paullus dead—the city of Capua defected to Hannibal, who was courting an alliance with Philip V of Macedonia. Indeed, Hannibal's tactical successes continued for years: in perhaps two battles at Herdonea (in 212 and 211), he caused some thirty thousand Roman casualties, and more successes continued into 208. Hannibal was winning over some of Rome's Italian allies; as time went on, many, including twelve Latin colonies, refused to provide troops, and census figures for 209/8 show an ominous decline.[40] Polybius records 700,000 men at arms available to Rome, but even if inflated, their overwhelming numerical superiority did not allow them to drive Hannibal off Italy.

While the city of Rome contemplated an attack by Hannibal that never came, Carthage itself—the center of Carthaginian political, religious, and social power—was unthreatened by any Roman soldier.[41] The Carthaginians at home heard reports intended to maintain their support for the war. The Romans had done nothing to force the people of Carthage to reconsider their decision to support Hannibal in a war waged on foreign soil. Something other than the false alternative of Fabian delay versus plebeian foolishness would be needed to end the war on Italy.

Publius Cornelius Scipio Secures Spain

By 211 BC opponents of Fabius—led by the Claudius, Scipio, and Aemilius families—had regained some influence in the senate, because of their successes in Spain and plebeian discontent with the failure of

Fabius to end the war. But in 211 the two senior Scipiones were killed in Spain, and the Romans were reduced to holding a defensive line in northern Spain—the very place they had begun seven years earlier. In the same year Hannibal marched to within three miles of Rome—as a feint, to relieve the Roman siege of the town of Capua—but turned back, never intending an attack on Rome itself. In the frantic elections of 210 BC, Rome needed a general who could break the impasse, and there were few able and willing to seize the opportunity.

One of these was a newcomer, Publius Cornelius Scipio, whose father Publius and uncle Cnaeus were the Scipiones killed in Spain, and whose father-in-law Lucius Aemilius Paullus was killed at Cannae. Scipio raised his voice to lead the army in Spain. He was twenty-five years old, too young, too brash, and technically unqualified for command—he had never held an office higher than aedile, a junior position. But he had married the daughter of the powerful Paullus, which had strengthened the ties between the Aemilius and Scipio families; he was their last, best hope to counter attempts by the Fabius clan to control the senate and to regain prestige for the Scipio clan and their supporters.

The senate granted Scipio's request, giving him command over the armies in Spain. As a private citizen (*privatus*) he had no political office; he was given military command over provincial armies (*proconsular imperium*), but he was unable to speak for Rome and could not negotiate a treaty with Carthage. But he had an army and a plan, and once in Spain, he would have the authority to do what his father and his uncle could not.

Scipio was not alone in recognizing that Hannibal's army needed support from Spain, and that to force Carthage to defend its interests in Spain, the Spanish tribes had to be set against the Carthaginians. This was a goal similar to Hannibal's in Italy—but Spain was not Italy. The Carthaginian relationship with the Spaniards—as with their African neighbors—was not political but was one of personal loyalties with chieftain tribes. The Spaniards would far more willingly transfer the yoke to new bosses than would the Italian allies, if their chiefs were to submit to a greater power. Strategically, the Carthaginian generals were geographically divided, which made them vulnerable to a focused attack at a weak point.[42] Scipio must have understood the importance of the coastal route in Spain, as well as the need for speed; a slow consolidation from north to south had been his father's strategy for more than six years, and it had not worked. A fast, direct seizure of this route—and the Carthaginian capital—was the only way that the war in Spain could be brought to a favorable end quickly. He could get the glory for ending it.

Scipio landed in the north, at Emporiae, and he wintered at Tarraco. In the spring he struck, and in a tactical masterpiece of planning and execution he attacked New Carthage directly, with a lightning-fast, coordinated land and sea assault that took control over Carthage's western capital. In one fell swoop he achieved tactical surprise over a difficult coastal route, put three Carthaginian armies on the defensive, gained thousands of prisoners and hundreds of talents of metals, and established his powerful reputation in Spain. Several Spanish princes —such as Edeco of the Edetani, and Indibilis and Mandonius of the Ilergetes—switched sides, taking with them the Spanish soldiers conscripted into the Carthaginian armies. He obtained promises of loyalty from Spanish tribes by releasing hostages held by Hannibal. The Carthaginian generals were now tied down with revolts and forced to reestablish their positions. Although the historical tradition attached to Scipio implies a sudden impulse in his attack—with the bravado and the rashness of youth—the length of the march, the perils of starvation, exhaustion, and three enemy armies, and even the need to time the attack with the movement of the tides, bear all the marks of cool deliberation upon a lesson learned from the death of his father, executed with audacious speed.[43] The capture of New Carthage fundamentally altered the relations of power across the western Mediterranean, establishing a position bordering on the legendary for Scipio—and it did so without the loss of life typical for a major battle.[44]

The fighting continued for several years, but amid the confusion the Carthaginians never bested Scipio in battle. His rigorous training of his army and his tactical innovations came to fruition against Hasdrubal at the battle of Baecula. Scipio overcame a strong defensive Carthaginian hill position with an unexpected pincer attack, and

Hasdrubal was forced to flee with remnants of an army. The three Carthaginian generals—Hasdrubal, Mago, and Hasdrubal son of Gisgo—had an emergency meeting; Mago went to the Balearic Islands to recruit a new force. Hasdrubal fled to the north with what was left of his own army and Mago's. He wintered in Gaul and crossed into Italy in 207. The only effective Carthaginian army left in Spain was that of Hasdrubal son of Gisgo—a political opponent of Hannibal and his brothers.[45]

Scipio did not try to prevent Hasdrubal's march to Italy, a decision that placed Italy in danger. Had Hasdrubal joined with his brother Hannibal, the Romans would have faced another round of bloody attacks. But to stop Hasdrubal, Scipio would have had to sacrifice his offensive capability in the south of Spain for a stronger defensive line at the Ebro River in the north. This had been his father's failed strategy. Scipio's aim was to break his enemy's power for good, not to hold him in check for the moment. The decision was right; Hasdrubal was killed in northern Italy, in part because of the weakened condition of his army. Hannibal was left without support, and Carthaginian generals were in disarray.

There is an important inverse parallel between the situations in Spain and in Italy. In Italy, Roman military commanders faced continuous political pressures and meddling by various groups. While Fabius, for instance, was avoiding unfavorable contact with Hannibal, other leaders were inciting the soldiers to demand immediate attacks. The result was a serious lack of coordination and a climate of indecision, a situation that played into the hands of Hannibal, a lone genius who acted with singular thought and purpose.

Conversely, in Spain, the Roman armies were united under a single brilliant commander, who operated without immediate political interference, while the Carthaginian forces were divided among three generals. Hasdrubal and Mago, the brothers of Hannibal in the Barca family, had been reinforced, probably in 212, by Hasdrubal son of Gisgo, a political enemy of the Barcas who was intended to restrain their power. After Scipio's successes, reinforcements from Spain arrived under the command of one Hanno, from a family long at odds with the Barcas. Hanno the Elder, crucified after the First Punic War, had been replaced by Hamilcar, father of Hannibal and Hasdrubal. The selections of Hanno and Hasdrubal son of Gisgo resulted from political discord in Carthage and were a way to control the Barcas, not to defeat Rome. Yet Carthage's position required constant, focused attention to unreliable Spanish tribes. The divisions between the three Carthaginian commanders left them at perpetual risk. In contrast, Scipio had no rivals for command in Spain, and no divided purposes. He pitted an integrated Roman force against a divided enemy.

Scipio's tactical abilities ascended to an even higher level of sophistication at his last major battle in Spain, at Ilipa (or, Silpia), ten miles north of modern Seville. Scipio also knew that the Spanish soldiers were patently unreliable; his father had been killed when Spanish troops, positioned in the wings against fellow Spaniards, had defected in battle. Scipio instead placed the Spaniards in his center, to face crack Carthaginian troops, and put his best troops on the wings. He then accomplished a commander's dream: he pinned his enemy's best troops by ordering his own center to hold their position without a charge—something not previously done by a Roman commander. Should the irresolute Hasdrubal son of Gisgo order his best troops to charge, he would expose them to attack between Scipio's wings. Scipio's wings then executed a complex tactical movement, which reversed the positions of his cavalry and his infantry and allowed his best forces to envelope Hasdrubal's troops from two sides. By placing the Spaniards in his center and using the Romans and Latins on the wings as the primary attack forces, and arranging his forces in a way that freed up the lines to make such complex maneuvers, he immobilized Hasdrubal's strongest troops while bringing his own best forces in from two directions. The battle became a rout.

Scipio—like Themistokles at Salamis and Epaminondas at Leuktra—was victorious because he could think outside the usual way of doing things. Scipio's generous treatment of the Spaniards led to more Spanish defections. Masinissa, commander of the Numidian troops, returned to Africa to fight for the throne of his dead father against his rival Syphax. In one of the strangest events of the war, Scipio sailed to Africa, to attempt to secure an alliance with Syphax, He dined with Syphax—and with his Carthaginian adversary Hasdrubal. Hasdrubal countered Scipio's diplomacy by giving his daughter in marriage to Syphax. (The Numidian problem would not be solved for the Romans until the pro-Roman king Masinissa was installed.)

There were further operations in Spain, and Mago was still building an army on the Balearic Islands, but the Spanish war was all but over, and Carthage could no longer count on Spain as a resource. Hannibal was still in Italy, of course, but Carthage was otherwise effectively reduced to Africa. The Carthaginian people had to face the fact that they would have few options, should the Romans make it to the shores of Africa.[46] They were probably already considering the withdrawal of forces from the Mediterranean.

Scipio Africanus and the Offensive Alternative

In 206 BC Scipio returned to Italy as a hero.[47] His election as one of the two consuls for 205/4—and thus to the highest military command—was assured, despite his young age and his lack of official experience. But his proposal for an attack on Carthage faced opposition in the senate. Supporters of the Fabian policies still wanted to keep Rome focused on Italy, and did not want Scipio to end up with full credit for ending the war. Examination of these political conflicts would take us far beyond the scope of this book. But a few words will be helpful in grasping why Rome made certain decisions in these wars against Carthage. As the Romans had come to dominate the Italian peninsula and the Mediterranean Sea, they faced a strategic context that was widening at a blinding speed. This strategic complexity, along with the passionate competitions between family interests inside Rome, had made the formulation of a coordinated policy toward Carthage—or anyone else—extremely difficult.

Rome was controlled by the heads of no more than twenty *gens*—a *gens* being very roughly an extended family or clan—whose leaders were known by the names of their families, such as Fabius, Claudius, Scipio, Cornelius, and Aemilius.[48] Along with their supporters, dependents, and others, these leaders formed shifting groups driven more by political opportunism than by platform. But two general policy positions emerged from the political battleground: a defensive, "Italy First!" desire to protect the Italian mainland, and an aggressive desire to project power against Rome's enemies overseas. The Fabius clan—led by the powerful senator Quintus Fabius Maximus—was associated with the first of these. He appealed largely to the senatorial interests in Rome, and throughout the war he opposed aggressive action that might leave Rome at risk. When Scipio returned from Spain, Fabius argued for a defensive posture to oppose Hannibal in Italy and against an expedition to Africa.

In contrast, others—the Scipio, Claudius, and Aemilius clans in particular—realized that Sicily, Greece, and Spain were vital to Rome's interests and generally favored action against Rome's overseas enemies. They often appealed to the plebeians against the senatorial aristocracy. In the First Punic War, Appius Claudius—a consul in 264 BC—commanded the first expedition to Sicily. In 260 and 259 brothers from the Scipio family held successive consulships and pushed to build a navy, again advocating war in Sicily. Cornelius Lutatius Catulus—a member of the Cornelius *gens*—commanded the navy that defeated the Carthaginians off of Drepanum. During the Saguntum crisis Lucius Cornelius Lentulus debated Fabius. During the Second Punic War, members of the Aemilius and Scipio families would fight in Spain, Sicily, and Africa.[49] When Scipio argued for a war against Carthage, he adopted a "Fight in Africa" position as a route to victory and glory for his family name.

Livy reconstructs the debate between the "Fight in Italy" and "Fight in Africa" camps as a confrontation between Fabius Cunctator and Scipio. Livy designs a speech for Fabius that lays out a comprehensive and coherent position against the commitment of military resources off of Italy. Some of this borders on the comic: "With no threats from hostile fleets you sailed along the coasts of Italy and Gaul, putting in at the friendly town of Emporiae; you landed your men and led them to friends and allies at Taracco through a country which held no hint of danger. From Taracco your route led past a chain of Roman strongposts; on the Ebro were the armies of your father and uncle.... You captured New Carthage at your leisure, for not one of the three Carthaginian armies attempted to defend their allies; as for your other achievements—and I do not belittle them—they are in no way to be compared to a campaign in Africa."[50]

Scipio had actually solved a serious Spanish problem that had simmered for nearly ten years. He had thus established the strategic context needed for overall victory. But the military concerns of Fabius were not baseless; he advocated a defense of Italy as a matter of strategic priorities. He regarded such a defense as vital to secure the greatest value of all, Italy and Rome, before attacking the enemy's capital. Scipio's plan would leave the city of Rome defenseless before Hannibal: "Situated as we are now, even apart from the fact that public funds cannot support two separate armies, one in Italy, one in Africa, and no resources are left for maintaining fleets and furnishing supplies, the magnitude of the danger we run is surely patent to everyone. Licinius will be fighting in Italy, Scipio in Africa: now just suppose—which god forbid: I shudder to speak of such a thing, but what has happened once may happen again—just suppose, I say, that Hannibal is victorious and marches on Rome, are we then, and not before, to recall you from Africa, as we did Fulvius from Capua?"[51]

What Scipio should do about this, according to Fabius, follows logically from these premises: "Tell us no more that when you have crossed to Africa Hannibal will surely follow you; cut short these devious ways; direct to where Hannibal at this moment is, and fight him there. You want the victor's palm for ending the war with Carthage? Remember none the less that it is only natural to defend your own before attacking what is another's. Let there be peace in Italy before there is war in Africa; let us feel safe ourselves before we threaten others ... beat Hannibal here first, then cross the sea and capture Carthage."[52]

Scipio answered Fabius point by point, laying out a strategy for defeating Carthage. He cut past the rhetoric of belittling his Spanish successes; "And God knows it would be just as easy, should I return victorious from Africa, to belittle those very things of which the danger is now being grossly exaggerated, in order to keep me at home." Scipio wanted the glory of victory; these are good ambitions, he maintained, because they are good for Rome. But he had no truck with Fabius's plan to chain him to Italy:

> Provided there is no impediment on this end, and you will hear at the same moment that I have crossed the sea, that Africa is blazing with war, and that Carthage is already beset—yes, and the very sound of Hannibal's fleet preparing to sail.... Yes, Fabius, I shall have the opponent you give me, Hannibal himself; but he won't keep me here; I shall draw him after me. I shall force him to fight on his native ground, and the prize of victory will be Carthage, not a handful of dilapidated Bruttian [southern Italian] forts.... Italy has suffered long; let her for a while have rest. It is Africa's turn to be devastated by fire and sword. It is time a Roman army threatened the gates of Carthage, rather than that we should again see from our walls the rampart of an enemy camp. Let Africa be the theatre of war henceforth; for fourteen years all the horrors of war have fallen thick upon *us* ... it is her turn to suffer the same.[53]

Fabius was locked into a short-range perspective, focused on the threats of the immediate moment and misled by his own rhetoric against Scipio. For Fabius, the goal was to defeat Hannibal's army, and victory would mean driving him out of Italy; an attack on Carthage would be a dangerous digression. But Fabius failed to recognize that, even though tying down Hannibal's army in Italy might be necessary to defeat Carthage, it could never be sufficient, for it did not reach to the center of the enemy's war effort, Carthage, and the willingness of the Carthaginian people to allow the war to continue. The Fabian policy was now an impediment to victory and peace. By taking the war to Africa and forcing his recall, Scipio would break the false alternative between avoiding Hannibal and meeting him in battle. Scipio adopted a goal-oriented attitude toward his enemy's central source of strength, directed not toward the removal of a threat from Italy but toward a permanent end to the cause of that threat. He would then, of course, gain the glory, as would befit a victorious general.

After much rancor and a constitutional problem given Scipio's age, control of the military was granted to the two consuls, one to be assigned to Italy, the other to Sicily. Because the other consul was also the chief priest (*pontifex maximus*) and could not leave Rome, command of Sicily could go only to Scipio. The Romans assigned Scipio to the African invasion without assigning him to invade Africa—the responsibility would be his. But the "impediment on this end"—senatorial opposition—had been removed.

One of Fabius's most salient points was that sending resources to Africa would weaken the defense of Italy and of Rome itself. This concern was far from invalid; Hannibal's army was still in southern Italy, still led by Hannibal, and still able to turn Rome's allies. Scipio did his part to maintain the defenses of Italy by neither levying soldiers nor taking public money. His command was technically over two legions on Sicily, soldiers sent there in disgrace after taking the blame for the disaster at Cannae in 216 BC and a defeat at Herdonea, in 212 or 210. He solicited seven thousand volunteers in Italy, and used contributions from allies as well as his own money to create the fleet he needed.

In 205 BC, twelve years after Hannibal's invasion, Scipio sailed to Sicily with a force that had cost Rome virtually nothing. His plan bore fruit almost immediately. The Carthaginian commander Mago sailed from the Balearic Islands toward Italy with twelve thousand infantry and two thousand cavalry reinforcements. He landed at Genoa unopposed (Livy's passage is the first mention of Genoa in history), but reports of Scipio's expedition induced the Carthaginians to recall twenty of Mago's ships to Carthage. The retreat had begun.

Scipio used his time in Sicily to outfit his ships, build up his forces, and reenlist thousands of exiled soldiers who were itching to regain their honor and their homeland. By the time he got done with them, he had one superbly motivated army. In 204 BC, Scipio, then barely thirty—Hannibal's age when he had crossed into Italy—launched a powerful attack on Africa. This was not a raid—the commander Laelius had done that the previous year—but an invasion. Scipio may have again gained tactical surprise by keeping secret his landing point in Africa, or changing it at the last moment.[54] From his first foothold, he staged a simultaneous nighttime attack on two enemy camps, of the Carthaginians and their Numidian allies; the accounts say that he feigned an accidental fire in one to draw the soldiers out unarmed from the other and, as a result, killed enough of them—some say forty thousand—to force the Carthaginians to recruit fresh forces. Following a battle on the Great Plains, some seventy miles west of Carthage, the prince Masinissa was placed on the throne of the Numidians, thus restoring a kingdom friendly to Rome on Carthage's west. Livy writes that the Carthaginians were "profoundly shaken by the news of the capture of Syphax."[55] The Carthaginians, now isolated by land and by sea, were forced to contemplate a two-front attack on their own city.

Now the Carthaginians confronted the reality that Utica was under siege. Like the Spartans who saw Epaminondas approaching, like the Athenians who saw the smoke from the Spartan attacks on their farms, and like aristocrats in the American southern confederacy who would one day see their slave plantations burned, for the first time in their lives the people of Carthage, who had supported war for years on the soil of others, saw their own homeland facing annihilation. Livy recounts the psychological effect of this vision: "Wild reports that the Roman fleet with Scipio in command had arrived ... caused universal panic in Carthage. Nobody had any clear idea of the number of ships the messengers had seen and the size of the raiding force, and sheer fright caused them to exaggerate everything they had heard; the result was a wave of terror, quickly followed by depression at the thought of the grievous change in their fortunes. Only lately had they had an army before the walls of Rome ... yet now the tide of war had turned and they were doomed to watch ... the siege of Carthage."[56]

The Carthaginians did precisely what Scipio had predicted: in a panic, with Africa ablaze, a group of officials went to Scipio's camp, threw themselves prostrate before him (as their eastern customs required), and sued for peace.[57] They recalled Hannibal from Italy, ending sixteen years of war on the Italian mainland. Scipio sent the peace terms to Rome for ratification; Carthaginian envoys begged for peace in Rome. Meanwhile, in one of the most ironic turns of the war, the Roman commanders were ordered to prevent Hannibal from leaving Italy.[58] He was the only person who could have derailed the peace—and his arrival in Africa threatened to do just that. He offered terms to Scipio, which renounced all foreign territory while retaining the right to a fleet. His arrival probably energized a "war party" in the Carthaginian senate—perhaps the last vestige of resistance to Roman forces. But peace would not hold if Hannibal and his forces were not bested in war, and were thus allowed to claim that victory could have been theirs, had they only fought on. Scipio rejected the terms, and after a brief struggle for alliance with the Numidians (Hannibal allied with Syphax, and Scipio with Massinissa), clashed with Hannibal at the decisive battle of Zama in 202 BC. Scipio's tactical innovations, along with Numidian cavalry reinforcements, overcame Hannibal's eighty elephants, and permanently ended Carthage as a great power. "To decide the issue, the two most famous generals and the two mightiest armies in the two wealthiest nations in the world advanced to battle," wrote Livy, who tells us of forty thousand Carthaginians dead or captured next to fifteen hundred Roman dead. Hannibal was unambiguously defeated, in full view of the Carthaginians. In the end, Hannibal's return to Carthage was vital to the peace—as a demonstration of his total defeat, on Carthaginian soil, to the Carthaginians directly.[59]

The peace demanded that Carthage disarm, relinquish all claims outside Africa, agree never to wage war without Roman permission, make restitution over fifty years, and accept a reduced status as a minor power. These terms were in some ways harsher than the terms of the last war; the indemnities, for instance, were for a longer duration, and the Carthaginians would not have the mines in Spain to pay them. But they accepted Scipio's peace, which allowed them to create a prosperous city without an empire. The anger of the Barcas was gone; the advocates of war were silenced. The Romans allowed Hannibal to live in Carthage—given his surrender—and Hannibal may have used his soldiers to develop olive groves, before accepting his last official position in 196 BC and to fleeing in 195 to the court of the Seleucid king Antiochus III in Tyre.

On his return to Rome, Scipio was given the title of *Africanus*, "Beater of Africa," possibly the first Roman named after the place he subdued. The peace he administered lasted fifty years, a tremendous accomplishment. Carthage never again waged war against Rome.

Why Was Carthage Defeated?

The outcomes of the first and second wars between Rome and Carthage are fraught with paradoxes. The first war ended in 241 BC with an agreement by the Carthaginians to surrender, to pay indemnities to Rome, and to renounce their Mediterranean holdings. The Romans did not besiege Carthage—but Carthage's own mercenaries did. The Carthaginians elevated their military commander to high office, who slaughtered the mercenaries, built a new capital on foreign soil, paid off the indemnities in eighteen years from foreign revenues—and then started an even more brutal war on the pretext of anger over their treatment in the last war. The Second Punic War began with Carthage's invasion of Italy, followed by a fifteen-year war of attrition that brought no fighting to Carthage. The Romans landed on the African coast, and the Carthaginians sued for peace without even a siege of Carthage. They elected their commander to high office—who renounced the war and six years later ran into exile. They paid indemnities for fifty years, for which they had few foreign resources, and during which time they rose to prosperity. They waged no war against Rome and seem to have harbored few hard feelings.

What can we make of this? Why did Hannibal's invasion of Italy fail, but Rome's projection of power in Africa succeed almost immediately? Why did the first peace fail, but the second succeed? There are four issues to consider here: the social and political context of Italy versus those of Spain and Africa; the moral issues at stake in the conflict; the strategic necessity for a Roman invasion of Africa; and the admission of the fact of defeat by the Carthaginian leadership.

The Social and Political Context of Italy versus the Context of Spain and Africa

Hannibal's stab into Italy was intended to shatter the social, economic, and military network of support between Rome and its allies and thus to reduce Rome to one state among many. This was a plan that struck to the heart of Rome's power and sought long-lasting political change.

Hannibal did project power directly into Rome's economic, military, and social center, and he did cause defections from Rome's allies, but he was never able to systematically break Rome's political strength, or the Romans' commitment to fight. At no point did the Romans seriously consider giving up.

The social-political context in Italy—in which all of the cities were fundamentally political in their outlooks, with institutionalized forms of government, and influenced by some blend of Etruscan, Hellenic, or Latin culture—meant that Hannibal's goals would require far more sophisticated political action and greater military power than that required in Spain and Africa.

Carthaginian power in Spain was built on the submission of the Spanish tribal chiefs to a stronger power. Neither the Greeks nor the Carthaginians created political systems outside of limited areas in Spain; for the Spaniards beyond the river valleys and the coasts, there was little history of political life, few political institutions beyond the chief's personal influence and power, and little knowledge of what it meant to live politically. The organization was tribal; it was impossible to apply political means—for example, citizenship, *fides*, political offices, and legal institutions—to such chieftain societies. The Carthaginians applied power downward through local chiefs; their loyalty generally meant the loyalty of the tribe. But any relaxation of pressure or disagreement among the Carthaginian commanders could result in a revolt. In the decades after Scipio left, Roman wars against the Spanish continued—but that was irrelevant to the Roman victory over Carthage, when chaos in Spain benefited Rome.

Despite many cultural differences, a similar situation existed in Africa. The chieftain states to the west of Carthage were not political, but tribal, and despite centuries of influence, the Carthaginians had little political depth there. Carthage and Rome each handled the area by supporting friendly chieftains. The replacement of the pro-Carthaginian Numidian king Syphax with the pro-Roman Masinissa led to an immediate reversal of the strategic situation in Africa, which left the Carthaginians isolated. There was no need to reverse the political positions of the people, as required by the political climate in Italy.

Italy, however, was not Spain or Africa. The allies in Italy gained great benefit from the Romans—particularly in the achievement of the rights of citizenship, a distinctly political phenomenon that was growing in Italy—and Rome's own political system remained solid, devoid of civil wars, tyrants, or usurpers. Rome did use horrific force and took hostages if cities threatened to separate, as was done with Capua, Tarentum, and Thurii. But these emergency situations do not explain the overall strength of Roman power on the peninsula. Attempts to replace the leadership in Italian towns could result in factional political conflict or even civil war. Even though there was a long way to go until Augustus, the Romans had far more institutionalized political depth in Italy than Carthage had in Spain or Africa.

Hannibal could despoil the countryside, spewing destruction and terror, but despite successes in a few areas, he could not cause widespread revolts from Rome. Livy was surely motivated to promote the cause of Republican ideals in the age of Augustus, but his reason for Hannibal's failure to win over the Italian allies plausibly represents the situation two centuries before his own time: "Nevertheless, not even the panic caused by these depredations, not even the flames of war on every side of them, could move Rome's allies from their allegiance. And why?—because they were subject to a just and moderate rule, and were willing to obey their betters. That, surely, is one true bond of loyalty."[60]

Fides—loyalty of a kind—was a moral principle that established a political relationship based on submission to a greater power that kept its word to allies. Hannibal was far too smart to simply transpose his own relationship with the Spanish and African tribes to Rome's relationship with its neighbors—but what other direct experience had he seen? As a young boy he had gone to Spain with his father, where he used Spanish and African troops to subdue the Spanish tribes. He would have known that Rome had different kinds of relationship with its allies—but he had little direct experience. In the end, had Hannibal succeeded in creating an Italian federation, located in Italy and allied with the Macedonians, the result would have been more Greco-Roman than Punic in outlook—and would likely have fallen into war with Carthage once its reach extended to Sicily.

Moral Issues and Pretexts

If he was to gain the long-term support of the Italian cities, Hannibal had to offer a solid reason why those cities should enter a long and bloody fight against Rome. They needed a cause to rally around, something more than Carthage's past grievances with Rome. Polybius takes seriously the difference between a cause and a pretext, defining the former as something that motivates one side to take up the fight, and that demonstrates the futility of supporting the other. For an example of Hannibal's failure in this regard, we return to this quotation from Polybius, in which Hannibal asked for instructions from Carthage on the prelude to his siege of Saguntum. Consider here the claims to injustice that Hannibal offers as reasons for the attack:

> Not long before, party strife had broken out in Saguntum and the Romans had been called in to arbitrate; and Hannibal now accused them of having caused some of the leading citizens to be unjustly put to death. The Carthaginians, he warned them, would not overlook this treacherous act of seizure, for it was an ancestral tradition of theirs always to take up the cause of the victims of injustice. At the same time, however, he sent home to Carthage asking for instructions on how he was to act, in view of the fact that the Saguntines were relying on their alliance with Rome to commit wrongs against some of the peoples who were subject to Carthage.[61]

In Polybian terms, had Hannibal named a true cause for his action, he would have rallied people to his side. But there was no such movement against Rome. Astute observers would know that Hannibal had no such "cause" in Saguntum. His claim that the Romans were brutalizing the Saguntines makes sense only insofar as factions inside a divided city may act with mutual brutality. The eight-month siege strongly suggests that the population was largely opposed to the Carthaginians. The atrocities that followed would not have helped Hannibal's reputation. But Hannibal, being "full of a war-like spirit" and "entirely without logic and under a violent anger," relied on pretexts, "as is apt to happen to men who disregard the proper course of action because they are obsessed by passion."[62] Despite the exaggerated moral language, Hannibal's claim was a pretext designed to legitimate a military siege against civilians whose crime was to live in the wrong place at the wrong time. The Romans here took the high road, by claiming that they were protecting Saguntum against an imminent Carthaginian attack—which was obviously true—on the basis of *fides* with the real victims of such attacks.

Neither Hannibal's claims to be avenging the indemnities of 241—which had been paid off in 221 from foreign revenues that had spared the Carthaginians from economic devastation—nor his anger at the seizure of Sardinia and Corsica could hold much force. This was old business. There was, in other words, no legitimacy in Hannibal's attacks that could be found in a redress of grievances—he was simply continuing a war with an old enemy.

If Hannibal wanted the neighbors of Rome to make the decision to oppose Rome, he needed to provide them with a cause to fight for. He needed to motivate people in Italy into outrage against the injustices of the Romans. A few cities accepted this, but no groundswell swept the peninsula, and few foreign leaders linked their own interests with the defeat of Rome enough to justify the scale of invasion that victory would have required. The general reaction of the people in Italy to Hannibal's invasion of Italy was to dig in their heels and to strengthen their resolve. The defeat at Lake Trasimene led to calls for another confrontation with Hannibal—but no Roman embassy asked for terms from the Carthaginians. A Carthaginian fleet at Ostia could not have done what the Roman fleet did off Utica. The Romans argued about *how* to fight back, not *whether* to do so.

Nor were the Carthaginians in Carthage as motivated as they needed to be to win a war with Rome. A reader gets little sense of a fight to the death inside Carthage; they were not fighting for the salvation of their city. They had been led into war by a commander in a far off location, in a country the majority had never seen, through information that was manipulated to increase support and undercut the opposition. There had been no attacks on Carthage, and many of the nobility had grave reservations about the war from the start. When Scipio finally brought Roman power to Africa, there was panic rather than resolve, and a peace embassy rather than a defense.

In the end, the Carthaginians at home never cared enough about Spain itself to apply the force needed to keep it in subjection—and they did not reinforce Hannibal enough in Italy to demonstrate an all-out commitment to *his* war. No Carthaginian fleet sailed into Ostia to support Hannibal. In contrast, Romans of all types cared deeply about Italy; its loss would be the loss of everything, and they were willing to expend all of their resources to defend it. There was literally no place else to go. This fact set the bar very high for Hannibal; Rome itself would have had to be defeated, solidly and unambiguously, its leaders dead and its allies broken. Such a feat was beyond the capacity of Hannibal to achieve without serious support from Carthage, and from foreign allies.

A trenchant analysis of the causes of the Second Punic War by Donald Kagan interprets the end of the first war in terms that presage his analysis of the Treaty of Versailles that followed World War I in 1919: "The peace that concluded the First Punic War ... reflected relationships of power between Rome and Carthage at a moment when Carthage was unnaturally weak. Unless the Romans took steps to destroy its enemy or cripple it permanently, Carthage had the capacity to recover its strength and to become formidable again. The peace imposed in 241 and 238, therefore, failed in this most basic way. It was faulty, also, because the Carthaginians were deeply dissatisfied with it, and many of them were determined to undo it whenever the opportunity arose."[63]

Kagan's view of the peace is based primarily on power relations, and in such terms, the peace after the Second Punic War must have succeeded because Carthage was unable to fight on. Its capacity to fight was gutted: Numidian kings were selected by Rome, Spain was out of reach, and without a navy the Mediterranean was lost. But there are numerous cases in history in which guerrilla wars have continued despite such weakness: tribes in Spain would fight the Romans for years; a Numidian usurper would battle the Romans; slaves, Jews, and Britons would revolt. Where was Carthaginian anger? The surrender was long-term and total. Why? Kagan's "realist" interpretation cannot answer this question, because it fails to properly emphasize the efficacy of ideas that can lead people to fight against all odds—or to accept the kind of peace that was grounds for war twenty years earlier.

The Carthaginians fought the second war for reasons much deeper than being "dissatisfied" with the first peace and able to act on their dissatisfaction. An ideology of power was created by a highly motivated political leader and his group, anchored onto a religious element, grafted onto claims of unjust treatment, infused with feverish hatred, and given a specific direction of action apart from political control. Rome's power over Carthage was a fact after 202; the Carthaginians had no capacity to fight effectively. But had they been motivated to do so, they would have fought nonetheless—and their inferior status would have created pretexts of injustice lasting years. But no such thing followed the Second Punic War—the defeat of those advocating such action was total, and they admitted as much.

The Necessity for a Roman Invasion of Africa

It remains a fact that, as long as the Romans fought a delaying war in Italy, mired in insipid senatorial debates and unwilling to confront Carthage directly, the war dragged on, Rome remained in a state of terror, and Italy was bled dry. But to march out and assault Hannibal's army immediately was not the solution. In the end, victory required strategic intelligence, and awareness of the central locus of the enemy's will to fight. The invading army was a means to this end, not the end.

The words of military historian Captain B. H. Liddell Hart are worthy of recall, as H. H. Scullard saw fit to repeat them in his biography of Scipio Africanus. Liddell Hart cut to the heart of the difference between the approaches of Fabius and Scipio: "Those who exalt the main armed forces of the enemy as the main objective are apt to lose sight of the fact that the destruction of these is only a means to the end, which is the subjugation of the hostile will. In many cases the means is essential—the only safe one, in fact; but in other cases the opportunity for a direct and secure blow at the enemy's base may offer itself, and of its possibility and value this master-stroke of Scipio's is an example, which deserves the reflection of modern students of war."[64]

The object of victory is to reverse the enemy's decision and commitment to wage war. In a major war, this has implications on a social level: will the population accept the decision or not? Such acceptance would require accurate knowledge of Hannibal's failure to subdue Rome and gain the loyalty of its allies. After Hannibal's success at Cannae, Hamilcar's son Mago brought news of the victory to Carthage along with a pile of gold; the Carthaginians reacted with delight, except for a faction opposed to the Barcas, who recognized that the war was not yet over and that revolts against Rome had not broken out in Italy.[65] The war against Carthage was won decisively when Scipio sailed against Africa itself and demonstrated directly to the Carthaginian people the failure of the war and their personal vulnerability to a Roman attack. The collapse of Carthage into fear was the climax of the war; the decision to sue for peace was the admission of helplessness. The fact that the people of Carthage accepted the decision—rather than crucifying their leaders and fighting on—strongly suggests that the decision had broad popular support.

But was the attack on Africa truly necessary? Certainly the war in Italy would have eventually ended without Scipio's attack on Africa. Hannibal's strength had weakened in his last four years; his inability to instigate defections by Rome's allies, and the cutoff of overseas reinforcements, forced him into a guerrilla position that attrition would have beaten. But the Romans had good reason to reject such a policy. It would be a poor plan indeed that counted on the weakness of an adversary as formidable as Hannibal, and that subjected one's homeland to a merciless war of attrition while waiting for him to make a mistake. Every little move against every little town—such as receive a footnote in a history book—is a disaster for the people involved.

But, history suggests that such a plan could not have won a permanent peace. Had Rome entered into negotiations with Carthage, Hannibal's soldiers in Italy would have been in the same position as those in Sicily who had served under Hannibal's father: undefeated, but subjected to a humiliating peace with little legitimacy for them. Hannibal's troops had to be recalled to Africa, in a clear admission that Spain was out of reach, Africa was unfriendly, and their capital was under attack. They had to be defeated on the battlefield in order to make the Roman victory undeniable. Had this not occurred—history suggests a "what-if" based on the previous war—the people in Carthage would not have realized the extent of their defeat, the claims of unjust treatment would have returned to the soldiers, and the next war would have been even bloodier than the last. Undeniably defeated, the Carthaginian people turned back to what they had always done best: trade. They left imperial rule to those who did that the best: the Romans.

The Admission of Defeat by the Carthaginian Leadership

The defeat of Carthage in 202 was a fact—as was their defeat in 241. But an end to the war came when the Carthaginians sued for peace and thus admitted and accepted the fact of defeat. Carthaginian officials had made such a decision in 241—but the military leadership and large numbers of the population then pushed their energies in a direction that led to a greater confrontation decades later. Neither the Carthaginians in Carthage nor the soldiers on Sicily directly witnessed the unambiguous defeat of their best general—on the contrary, he had succeeded in avoiding defeat on Sicily. Two things happened in 202 that had not happened in 241: the clear military defeat of the Carthaginian army under its best commander, and the decision of that commander to accept the defeat. Without such an admission, the pretext that the war might still be won could have continued.

Hannibal himself was crucial here. It is not necessary to see the Barcid efforts in Spain as an independent state in order to acknowledge that Hannibal had driven events forward largely under his own initiative. Such a commander—given wide latitude and huge resources, in places physically distant from political control—must, and did, act energetically and on his own authority. But this also tied the overall outcome of the war to him in a way that had no direct parallel in Rome. We do not speak of "Scipio's War," or "Fabius's War," in the way we speak of "Hannibal's War." This gave him enormous prestige

—and meant that the end of the war could come only with his death or his open admission that Carthage had lost. Hannibal may have blamed his failure in Italy to the lack of reinforcements from the home, stymied, he says, by one Hanno, another political enemy.[66] But the end of the war came when, according to Livy, a Carthaginian citizen spoke against the peace, and Hannibal physically dragged him from the rostrum.[67] Hannibal then apologized—and protested that he knew the ways of the soldier but had no experience in the "rights, laws, and usages of the city and the forum." Only his open recognition that the war was over could elevate the laws and the forum over the military camp.

Epilogue: The Massacre of 146 BC

The peace of 202 had one bad provision, which would one day lead to disaster. The Carthaginians were told that they had to accept the permanent presence of a Numidian kingdom to their west and that they could never wage war on Numidia—or anyone else—without the permission of Rome. They accepted those terms and lived by them. But the boundaries between Carthage and Numidia were not made clear, and border disputes arose almost immediately. In the 150s BC the Carthaginians asked the Romans to mediate claims that the Numidians were encroaching into Carthaginian territory. Rome sided with the Numidians, and to protect their land and their lives, the Carthaginians were forced to rearm—something that a proper verdict by the Romans could have prevented. After an embassy to Carthage, in which the conservative Roman Cato the Censor was shocked at Carthage's prosperity—an affront to his conservative agrarian ethics—Cato began ending his senate speeches with *delenda est Carthago*: "Carthage must be destroyed!" The people of Carthage had chosen prosperity over empire, but Cato could not grasp this.

Scipio Nasica, a descendant of Scipio Africanus, disagreed, but when the Second Celtiberian War in Spain ended in 151 BC—and the Carthaginians paid their last war indemnity—Carthage became the substitute for dealing with Rome's real problems. The Carthaginians posed no threat, and the Romans were getting everything they wanted: Utica had gone over to Rome, the Carthaginians executed their leader Hasdrubal to demonstrate their good intentions, and the new leader Hanno sent a good-faith envoy to Rome. The Romans were not listening. They demanded that the Carthaginians disarm; Polybius describes the people of Carthage carrying thousands of weapons to Utica for disposal.[68] The Romans then sprang an unfathomable, unprecedented ultimatum on the Carthaginians, which had been decided before they disarmed: the Carthaginians were to abandon their city, their ancestral altars, their hearths, and the homes in which their grandfathers had lived and to accept a future of agricultural life in the country. They refused, in shock, and mounted a desperate defense. They and their city were annihilated.

The Third Punic War of 149–146 BC was not a war. It was a massacre. Rome was wrong; the peace of Scipio Africanus was good, and the Romans could have preserved it by just mediation of the Carthaginian complaints. The Romans appointed a successor to Masinissa in 149; they could have ended the Numidian attacks. It is to Romans' eternal shame—there is no credit due here—that they slaughtered a former enemy that had accepted peace and was living by its word.[69]

Readers tempted to interpret the thesis of this book as the need for total destruction of an enemy's population centers should consider the decades that followed the Second Punic War, when former enemies were at peace, with the needless sacrifice of that peace in the destruction of Carthage—and the civil unrest and violence that followed in the next generations for the Romans.

The deepest factors that made war between Rome and Carthage unavoidable are starkly revealed in the massacre of 146 BC. The thesis of W. V. Harris—that Rome was motivated not primarily by defensive considerations, but for power, glory, and material rewards—remains essential to the broad context of Rome's rise. Susan B. Mattern has shown that Rome's desire for dominance explains much in Roman evidence. Studies of Roman art have shown an almost obsessive focus on military dominance and images of power.[70] In some form or other, the mores of a warrior society were omnipresent in the ancient world, but they had special potency in Rome, where the connection between military success and political office could summon a huge network of commitment and resources to any fight. The Romans would never give up the drive for dominance, and they always had the means to fight once more—because the fire in their society burned hotter with every success. Further, Mattern shows why the views of time and space held by the Romans may differ fundamentally from those held today. Attempts to explain the conflicts primarily in strategic terms may err in relying on a worldview that would have been foreign to the ancients.

Was there was anything that the Carthaginians could have done to avoid the final wrath of the Romans? Perhaps Hannibal's invasion of Italy had engraved itself in Roman memory in a way that only total annihilation could erase. But there is no need to stretch toward pretexts. Carthage did accept a reduced status after the Second Punic War, did appeal for Roman mediation against the encroachments of the chieftain peoples from the west, did demonstrate its renunciation of hostility toward Rome, and did send its weapons out of the city. But this did not stop the Romans from turning their full might—and their own fullest measure of perfidy—against the Carthaginians. The Second Punic War remains the example of a successful victory; the Third was a needless and unforgiveable slaughter.

Chronology of the Punic Wars

Background to the First Punic War

480	Battle of Himera, off Sicily
	Xerxes' invasion of Greece
330	Alexander I of Epirus invades Italy
280–275	Pyrrhic Wars
280	Mamertines (Sicilian mercenaries) take Messana
275	Rome takes control of Tarentum
270	Rome restores Rhegium (from Campanian mercenaries)
264	Appius Claudius supports Mamertines against Hiero II of Syracuse
263	Hiero allies with Rome
262–241	First Punic War
263–262	Engagements begin on and around Sicily; Roman siege of Agrigentum
243	Roman senate asks aristocracy to break deadlock (promised indemnity from Carthage)
241	Roman naval victory off Drepanum; Carthage negotiates peace
	Rome annexes Sicily; office of peregrine praetor created to govern Sicily
241–238	Truceless War of Carthaginian mercenaries
	Rome takes Sardinia and Corsica

Background to Second Punic War: Carthaginians under Hamilcar fortify Spain

237	Hamilcar moves into Spain
229	Hasdrubal succeeds in Spain; New Carthage (Kart Hadasht)
226	Ebro River Treaty; Rome to remain north of Ebro River
221	Hasdrubal murdered; Hannibal acclaimed commander
220	Rome grants *fides* to Saguntum
219	Hannibal besieges Saguntum
218–201	Second Punic War
218	Rome issues ultimatum to Carthage; Hannibal marches on Italy
	Romans routed at Ticinus and Trebia rivers
	Roman base established at Emporiae, Spain
	Rome garrisons Tarentum and Thurii in Italy; hostages taken
217	Romans routed at battle of Lake Trasimene; Quintus Fabius, dictator
216	Romans slaughtered at Cannae; Capua defects from Rome
215–205	First Macedonian War; Philip V allies with Hannibal

215	Heironymus succeeds at Syracuse; Hippocrates succeeds in 214
	Spain: Scipiones beat Hasdrubal at Ibera; revolt of the Numidians
214	Rome repels invasion by Philip V of Macedonia
213	Rome besieges Syracuse; falls in 211
212	Tarentum, Herakleia, Metapontum, Thurii defect to Hannibal. Roman siege of Capua
211	Hannibal fails to relieve Capua; feint march on Rome
	Two Scipiones killed in Spain; Rome holds Ebro River line
210	P. Cornelius Scipio given *procunsular imperium* of Spain
209	Rome recovers Tarentum
	Scipio takes New Carthage
207	Battle of Metaurus River; Hasdrubal dies failing to reinforce Hannibal.
	Spain: Battle of Baecula; Hasdrubal loses half his cavalry
206	Scipio victorious at battle of Ilipa
205	Scipio debates Fabius in Rome; Scipio moves to Sicily
	End of Macedonian war; Rome withdraws from Greece
204	Scipio in Africa; Carthage allies with Syphax; Scipio with Masinissa
	Battle of Tower of Agathocles: Carthage ambushed
203	Siege of Utica renewed; battle of the Great Plains; Hannibal recalled
	Syphax captured; Masinissa given the reunited Numidian kingdom
202	Battle of Zama destroys Hannibal
Background to the Third Punic War	
200–197	Second Macedonian War; Romans defeat Greeks at Cynoscephalae
196	Hannibal leaves Carthage for Tyre, and court of Antiochus III
183	Deaths of Scipio Africanus and Hannibal
172–168	Third Macedonian War; Macedonia made a province
172	Carthaginians complain about African intrusions
154–151	Spain: Second Celtiberian War (The "Fiery War")
152	Rome sides with Numidians; Carthage rearms
	Cato versus Scipio Nasica on the war
150	Numidians massacre Hasdrubal's army at Oroscopa
	Carthage crucifies Hasdrubal, sends peace envoys to Rome
149	Rome declares war against Carthage
150–148	Fourth Macedonian War; Annexation of Macedonia
150–120	Polybius, historian, flourished

149–146	Third Punic War
149	Carthage surrounded by Africans, Romans, and Hasdrubal's army
148–146	Achean War: Rome versus Corinth (under Critolaos)
146	Destruction of Carthage and Corinth

Chapter 4

"A Prince Necessary Rather Than Good"

The Campaigns of Aurelian, AD 270–275

The Background to Aurelian's Campaigns

In the centuries after the Second Punic War, the Roman Empire rose to unrivaled dominance over the Western world. After a tumultuous transformation from a republic into a monarchy in the first century BC, Rome brought tens of millions of people into the longest period of peace, law, and stable government yet experienced. But the imperial *Pax Romana* was based upon the undivided power of one man, and only an unchallenged transfer of command to a new emperor could preserve the peace beyond his lifetime. In AD 192 the succession failed—the first failure since the death of Nero in AD 68—and the world was shattered by civil war. An uneasy political order was restored, until the chaos returned in 235. Over the next forty years the subordination of political authority to military force reached its logical conclusion as the empire devolved into military rule, usurpations, and foreign attacks. Emperors were made by the acclamations of troops, who asserted their decisions against rival claimants on the battlefield. The goal of every emperor in this frenetic time was to maintain the loyalty of his troops and defeat challenges to his position—but few could do so beyond the range of a few weeks, months, or years.

This was the situation in AD 270, when a cavalry commander known as Aurelian rose to the imperial purple. He spent his first year in frantic defense of Italy and the city of Rome. But in 271 he launched a whirlwind of offensive military campaigns, boldly asserting Roman power in ways not seen in decades. He aimed first at barbarian tribes in the Black Sea area that had attacked Italy, but then directed his army eastward, into Asia Minor, Syria, and Egypt, before marching more than three thousand miles west, into Gaul (see Figure 4.1). Within three years he had pushed back the immediate threats to Italy, reestablished imperial control over the eastern and western provinces, ended usurpations among troops, and built defenses to protect the city of Rome. The progression of Aurelian's victories—from northern Italy into the Danube River basin, into the east and then west—suggests a well-thought-out, integrated plan of action, with each step directed toward a single goal.[1] Whether he conceived his campaigns in this way—and it is not obvious that he did—when he was murdered in 275 he had done more in five years to restore the integrity of the empire than his predecessors had done in thirty-five. How did he accomplish this important feat? And why did it not last? Such questions lead us beyond the strategic aspects of these victories into their broader social and intellectual contexts.

The problem of interpretation begins with a deep mismatch between Aurelian's reputation and his positive results. The literary sources for Aurelian's reign are woefully incomplete, often late, and at times contradictory if not fictitious—but they can be forthright in their opinions of his wanton cruelty. He was, according to the writer Vopiscus, "a prince who was necessary rather than good." Eutropius called him a man of "ungovernable temper, and too much inclined to cruelty."[2] But these writers also agree that Aurelian's march into the Near East resulted in few pitched battles and many celebrations, as city after city threw open its gates without a fight, and locals welcomed the return of Roman order and protection.

These narratives suggest that, far from acting with studied rapacity, Aurelian strove for recognition of Roman authority in terms that hearkened back to traditional ideas of trust and submission to authority, and that he enticed many areas to return to Roman control without bloodshed. He restrained his army from gratuitous rampages and resorted to destruction only when usurpers failed to submit. Aurelian's achievements compel us to reevaluate his military campaigns but also to consider the social, religious, and intellectual backgrounds

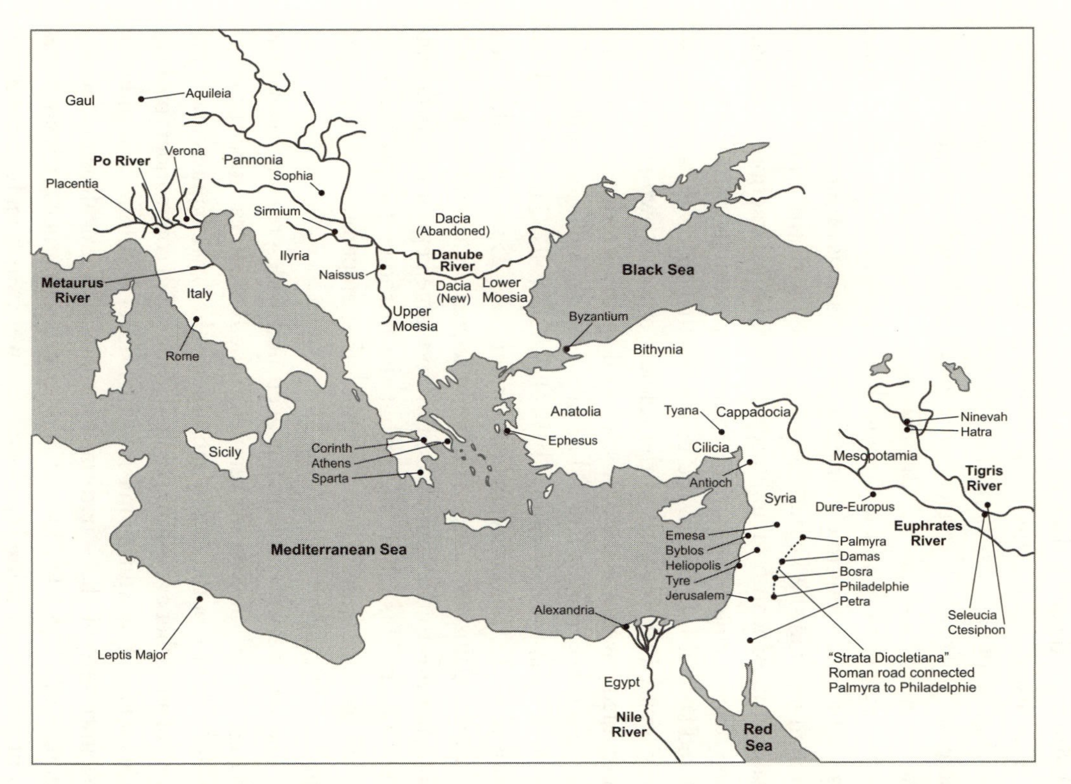

Figure 4.1. The Roman Empire, ca. AD 275: Major Sites Associated with the Emperor Aurelian

within which he acted. Why did he achieve the successes that he did? Did his reputation contribute to his success? Did he take steps to make the reunification of the empire permanent? What was the basis of his power, and did the nature of that power set inherent limits to what he could accomplish with it?

The history of the third century before Aurelian's rise can receive only a few paragraphs here. In AD 192 the *Pax Romana* ended in violence when the deranged emperor Commodus was murdered by the Praetorian Guard, and this elite army unit auctioned the throne to the highest bidder. Constitutionally powerless, the senate in Rome had little choice but to confirm the acclamations of soldiers. But the Roman army in the Danube area disagreed with the choice and acclaimed an African-born military man of Syrian extraction, Septimius Severus, *imperator*. Septimius marched peacefully into Rome, defeated several usurpers in the civil wars that followed, and successfully focused the loyalties of the soldiers onto his own person. Such undivided power was the only safeguard against a return to war—and its source was not in Rome but in provincial military camps. The Severan dynasty—Septimius (193–211), Caracalla (211–217), Elegabalus (218–222), and Severus Alexander (222–235)—defended their positions against challengers; brought legal, administrative, and religious reforms to the empire; and by and large kept order at the empire's heart. This all affirmed what the historian Tacitus had once called the dirty secret in the empire—that emperors could be made outside of Rome.[3]

The empire was undergoing deep and widespread demographic changes, which altered the temper of the Roman leadership at every level.[4] The rise of the equestrian order—the "noncommissioned officers" of an earlier age—into political leaders and commanders brought provincials of middling rank into high-level positions and undercut the authority of the senatorial aristocracy.[5] Caracalla's citizenship decree of 212 greatly widened opportunities for equestrian advancement and further shifted the locus of power away from Rome and into the provinces. It is an indication of such changes that every Roman emperor from Claudius II in 268 until Theodosius in 378 was from Illyria (excepting Carus, 282–283, from Gaul).[6] Some foreign tribes enjoyed client-status with Rome and were trained in Roman armies; conversely, Roman armies could be manned with troops from the periphery of the empire who might otherwise be enemies.[7] Newly enfranchised officers could interpret their positions in ways very different from officials trained in republican politics.

As a result, over three centuries the fundamental bases of Roman political power had changed. Force had replaced republican ideals of lawful succession, imperial decrees were law; and whatever was left of Roman constitutional principles was drowned in the acclamations of the soldiers. In such a climate, a successful emperor would need administrative, legal, and religious solutions of continental scale, legitimated by a claim to divine sanction, and backed by a fearsome imperial presence with unchallenged control of the army.

The Severan dynasty ended with the murder of Severus Alexander in 235. The decades to follow brought a series of crises, usurpations, and war, in which emperors were made in military camps, and every imperial claimant was impelled to defend his position against rivals. More than fifty usurpers can be discerned in the fifty years after 235. Pretenses notwithstanding, Rome was no longer a political unity, but a shifting collage of armies and cities. The best that most emperors could do was move from threat to threat, each crisis creating the conditions for the next. The very idea that military force should be subordinated to the control of civilian political institutions, which conducted themselves by means other than force—an idea vital to the Roman peace—was lost.

As a result of the political fragmentation and the need to guard against internal usurpation, the Romans were locked into a reactionary military position, without a coherent plan beyond the crisis of the moment. Motivated threats appeared from several directions. In the East, the Sassanian Empire of the Persians had replaced the Arsacids and was flexing its muscles in Armenia and Mesopotamia. The Persians invaded Mesopotamia between 239 and 243, in campaigns that ended in the death of the emperor Gordion III, who was probably killed by the challenger Philip in 244.[8] Another Persian invasion in 253—which followed the deaths of the emperors Philip in 249 and Decius in 251—pushed as far west as Antioch, putting the Persians in contact with the Aegean Sea. The confrontation with Rome reached its apex when the emperor Valerian died in the hands of the Persian warlord Shapur in 260, and the Persians again drove west. With Roman military power in the East seriously degraded after 260, the Romans empowered a local kingdom, the desert city of Palmyra, to stand against the Persians—a policy with important consequences to which we shall return.

The most immediate danger to Italy came from tribal societies west of the Black Sea, which threatened Italy through the Balkans.[9] In 249 the commander Decius, sent by the emperor Philip to suppress a rebellion on the Danube frontier, marched back, killed Philip at Verona, and claimed the throne. This left the frontier undefended and allowed a Gothic force to drive in and reach the Aegean Sea. Decius responded but was in turn killed in 251 at the battle of Abrittus—the first confirmed death of an emperor fighting barbarians. His successor Trebonianus Gallus paid tribute to the Goths but was killed by his own troops. The attacks continued and culminated in the horrific invasion and looting of the classical city-states in 267 by an organized coalition of tribes from the Danube valley and Black Sea areas. Byzantium and Ephesus were pillaged, as were Athens, Corinth, and Sparta. The wondrous Temple of Artemis at Ephesus and the agora in Athens were both torched. The Mediterranean world was in shock. The killings of Decius by the Goths in 251 and Valerian by the Persians in 260 mark a divide in Roman history, after which the emperor himself may be defeated by foreigners.[10]

To the west, an independent Gallic empire had arisen, ruled by the usurper Postumus in the late 250s, and leading to Tetricus (271–274). This institutionalized political entity bore little hostility to Rome—at least in contrast to the Black Sea tribes—but it had cut Rome off from Britain and from a land route to Spain and was eating into the tax revenues needed by the empire. The size of its forces, and the weakness of the Romans' defenses in Italy, made it a potential threat to Rome.

If the reader is overpowered by the complexity of these events, be advised that this is only touching the surface of a mercilessly chaotic environment. By 269 the Roman Empire was effectively divided into three independent kingdoms—Gaul, the central Mediterranean, and the East—that were each potentially at war with the others. The lower Danube River area was a base for attacks that threatened Greece and Italy as well as shipping routes to the Black Sea. The perennial problems of defense, food shortages, and plague became all the worse when the desert kingdom of Palmyra threatened Rome's grain supply in Egypt and trade routes to the east. The emperor Gallienus was killed by his own troops in 268, and even though Claudius II Gothicus (268–270) won at least two victories against the Goths, the crisis in the Balkans and the Aegean was unresolved. Meanwhile, another group, the Iuthungi, a tribe from the Black Sea area, allied with the Alamanni, and a loose coalition of other tribes invaded Italy itself. This was the first military invasion of the mainland of Italy since Hannibal in 218 BC, nearly five hundred years earlier. Roman roads, once arteries of prosperity, became avenues for looting and destruction. The situation was beyond critical.

Aurelian's Offensive Victories

In AD 270 a man of Illyrian origins, Lucius Domitius Aurelianus, emerged from these chaotic struggles to take control of the army and the empire.[11] The writer Vopiscus labeled him *manu ad ferrum*, "hand on the iron," which distinguished him from another, less forceful Aurelian and made an accurate commentary on his skills and priorities. Although it remains impossible to reconstruct a clear chronology of his life and reign, we can draw some inferences about his background. He was in many ways typical of those who now dominated the Roman military: the son of a family from the Balkans that may have been enfranchised by Caracalla's grant of citizenship in 212, he had struggled his way into the highest ranks of the army and may have held positions of command over the cavalry. Although writers have often granted him a humble background, he was probably born of the equestrian class and thus represents the new breed of emperors, born of some means but with a less than senatorial status. His exposure to Roman ideals was probably limited to his army experiences; the order he knew was the discipline of a military camp. But his reign bears the mark of an intelligent capacity to apply this obsession with discipline to the empire at large.

He was probably promoted by the emperor Claudius from a position as commander of the Danubian cavalry to overall cavalry commander, perhaps to reward his assistance in hastening the death of the previous emperor Gallienus. Aurelian shined in Claudius's victories over the Goths. The emperor died of the plague in 270, and his successor and brother Quintillus ruled for a matter of weeks. Although Quintillus had the support of the senate in Rome, Aurelian had the troops of the Danubian army. This was no contest; Quintillus chose suicide, if he was not murdered.

After accepting the acclamation of the troops, and probably receiving a deputation of senators, Aurelian marched against the Iuthungi, an eastern European tribe (first mentioned at this time) that had moved rapaciously into Italy through unguarded passes in the Alps. Loaded with loot, they turned around when they heard that Aurelian was coming, and he cut off their retreat. They sued for peace, and in a calculated display of Roman pomp and power, Aurelian, arrayed in his purple robes, greeted their ambassadors with his army arranged in a giant crescent—the power of the Romans encircling their inferior opponents. In a scene designed to demonstrate Rome's invincibility, the ambassadors stood in awe before the emperor as if before a god and asked him to renew an alliance. Their terms included a payoff by the Romans to guarantee the peaceful intentions of the barbarians—something they had come to expect. But Aurelian promised them war, not money. They provided hostages and some two thousand troops for Aurelian's army, and he allowed them to retreat.

The sense of grandeur in the scene reflects Aurelian's attempts to reestablish the power of Rome and the emperor, even if the scene is embellished by our source, Dexippus of Athens.[12] Dexippus, author of the *Skythika* (a history of the Scythians, from the Black Sea area) may have fought against the Heruli, and perhaps even organized the defense of Athens against the onslaught. It was in the interest of Dexippus to amplify Roman power, for this was all that stood between the Greeks and more attacks. Aurelian stood as a god over his enemies

—an action with implications into the future—and demanded that they acknowledge his superiority. The days of Trebonianus Gallus were over; Aurelian's solution was to return Rome to its position as the power that makes peace—not that begs for it—by restoring a "submissive" and "awestruck" attitude in the defeated.[13] Dexippus participates in building a reputation for Aurelian that casts the Goths into contemptuous inferiority and projects a sense of unopposable Roman power into the minds of friends and foes alike.

Aurelian went on to Rome, where he accepted the obligatory thanks of the senate and probably exterminated a few enemies. But he quickly returned to the battlefield—probably in the spring—against the Vandals and their allies in Pannonia (roughly present-day Hungary). With his person off of the Italian peninsula, the Iuthungi invaded Italy again, sweeping in through the mountains, looting near Milan, and facing an open road to Rome itself. Word of their impending second invasion reached Aurelian on the Danube, probably after March, when the Alpine passes were open. He came back just in time to be soundly defeated in an ambush at Placentia.[14] The barbarians may have spread out into the countryside, to better loot it, and the Roman soldiers probably did likewise, both as a consequence of the ambush but also to stop what could become a guerrilla war. Roman roads were once again pipelines of invasion and terror.

The inhabitants of Rome fell into a panic; religious books were consulted and purification rituals followed; only those believing in the partisan interventions of the gods held any hope of escape from the coming slaughter.[15] Aurelian faced opposition and perhaps usurpation in the senate, which he later put down ruthlessly. But for the moment he rose to the occasion, re-formed his forces, mobilized northern Italy, and defeated the barbarians at the mouth of the Metaurus River, probably at Fanum and Ticinum. What was left of them again retreated from Italy. He had saved Rome from the immediate threat—which must have boosted his prestige in the city—but the future looked bleak. Aurelian's defense of Rome illustrates the ad hoc nature of the Romans' response to such threats; their leaders were forced to run from attack to attack, unable to bring any kind of general resolution and thus unable to bring any security beyond the present moment. The situation brings to mind Hannibal's invasion of Italy, although the Romans of that bygone era did not live under a foreign-born emperor and were not dependent upon provincial armies for their lives.

But this time Aurelian did not settle for driving the barbarians out of Italy. He must have thought beyond the present moment, looking for a long-range solution to the problem of the Goths, and indeed to the whole set of problems facing the empire—of which the Goths were not the worst. He may have recognized that he could not use his resources properly as long as he was bogged down by reacting to the endless attacks of his enemies. He may have realized that the empire's economic problems could not be resolved until military command was reestablished across the empire—especially against Palmyra, which was stripping Roman control from Egypt and the East, and the Gallic Empire in the west. This would require a ruthless attack on the social and military center of the Gothic tribes, the removal of the usurping leadership in Palmyra, and then a march into Gaul. Such a plan would allow him to unite the power of the empire into a common cause and to reduce the risk of revolts by directing its energy in a way he could control. Besides, he might have thought, this was how a powerful Roman emperor should act—this was a plan worthy of his prestige and one that could bring the world back under Roman rule.

After executing enemies in the senate and dealing with serious issues internal to Rome, he set his sights on the heart of the crisis in the East, Palmyra. In the autumn of 271 he marched, first over the Danube River against the Goths. He invaded their land, killed their king, sacked their towns, and defeated them on a social as well as military level. He caused deep disruption of the Gothic settlements in a way that made his own victory unimpeachable. These were the most decisive Roman victories of the third century AD. No Gothic force ever again broke through the Dardanelles, and it was 125 years before a Gothic army again made it to Italy.[16] Despite its brutality—or perhaps because of it—this campaign was the one with the greatest positive effects into the future. He had demonstrated the overwhelming force that would fall on those who refused to accept Roman rule—and would soon demonstrate that he was also willing to accept into trust those who did.

But Aurelian still faced a daunting set of problems in the Danube River area. Under the emperor Trajan, some 160 years earlier, Rome had established Dacia, a province north of the river. In Aurelian's day, the province had become inordinately expensive to defend and impossible to control. In a move that is unclear to us, Aurelian withdrew his military and administrative forces from the area—including perhaps those civilians who wished to relocate—and established a border at the river itself. This shortened his defensive line by hundreds of miles, which allowed him to withdraw the two legions north of the Danube—needed for the battles to come—and to leave eight in defense of the line at the river. The old province of Dacia became a pacified buffer zone outside the new border. He then established one or two new provinces in modern Bulgaria south of the Danube, perhaps Dacia Ripensis and Dacia Mediterranea.[17] He also founded a provincial capital with a mint in what is now Sophia, thus asserting his political presence through imagery, while bringing Roman money into the area.

These actions ran contrary to the traditional ideals of a Roman emperor, who was expected to defend and expand Roman territory. Aurelian was not the first to give up land; Hadrian had relinquished territory taken by Trajan in the East, perhaps because Rome was unable to protect it.[18] But such an action could have undercut Aurelian's prestige, in an age in which any semblance of weakness could be deadly. By naming two provinces south of the river "Dacia," Aurelian created a propaganda fiction that he had added to Roman territory; Festus and Eutropius express this view.[19] Aurelian did in fact give up the Dacia that Trajan had conquered, which must have been seen as a retreat by those who understood the issue. It must also have been personally awkward, especially if he had claimed imperial legitimacy back to Trajan. But his resolution suggests that he understood how a strong border, firmly under control, was a greater asset than an overextended territory that was not under control. In a certain sense, he had extended Roman rule—over formerly contested areas south of the river.

Aurelian was now known as Gothicus Maximus, "Greatest Beater of the Goths," and he could celebrate his Victoria Gothica. The historian Ammianus Marcellinus—writing more than a century later, in the 380s, and for whom the barbarians were the "ruin of Rome"—notes that the Goths were stopped for decades.[20] In sharp contrast, the Gothic historian Jordanes omits these victories—good apologist that he was for the Goths. Three generations would pass before these tribes would again invade Italy.

The emperor now moved east. "Aurelian, made more impetuous by his great success, immediately marched against the Persians as if the war were in its final stages," write Aurelius Victor, interpreting Aurelian's motives in terms of an overweening confidence that follows victory and that anticipates an ultimate goal, the Persians.[21] If this is right, Aurelian had achieved an important offensive momentum, both tactically and psychologically, and had placed his enemies—especially the leadership in the East—onto the defensive. Zosimus wrote of his crossing into Bithynia by way of Byzantium and of the return of every city up to Antioch into Rome's orbit without a fight. This is the general tenor of the sources, who minimize any struggle that Aurelian had in returning the eastern cities to Roman rule. If true, then Roman rule had not completely collapsed in the East—although local forces may have taken over many of the police duties once done by Roman soldiers—and populations in Antioch and nearby cities had not wanted to reject Roman authority.[22] This supports the conclusion that Aurelian had moved into a power vacuum and reasserted control over an area that generally welcomed it.

The story in the East centers on Palmyra, a trading city and oasis on the caravan route between the Mediterranean Sea and the Euphrates River (in modern Syria). Palmyra—ancient Tadmor—had a place in early history.[23] It stands on the crossroads of Roman, Greek, Arabic, and Persian cultures and was bilingual into Roman times. Pliny the Elder, writing in the first century AD, acknowledged Palmyra's position between the Romans and the Parthians, noting that "as soon as there is discord, [the city] is always the focus of attention of both sides."[24] The emperor Hadrian renamed the city for himself, and Septimius Severus gave it the status of a *colonia*—a recognized position in the empire. Some of its leading citizens rose to senatorial rank, and those of equestrian rank were anxious to proclaim it. A Palmyrene inscription by one Aurelius Vorodes describes him as a *hippokos* (an equestrian) and a *bouleutēs* (a councilman).[25] At least one family in Palmyra could boast of a number of senators, including Septimius Odenathus, cited in an inscription, who lived under the emperor Philip (244–249).[26] In other words, the people of Palmyra—like many eastern cities—were members of the empire and were proud of their Roman citizenship and their status. The Palmyrenes had not opposed the Roman army when it passed through Palmyra several times en route to Mesopotamia.[27]

During the 250s and 260s the Romans actively promoted the rise of Palmyra as a counter to the threat posed by the Persians.[28] But the Palmyrenes treated with both Persians and Romans—and at some point, probably in the early 250s, the Palmyrene leader Odenathus probably approached the Persians with an offer of alliance.[29] Odenathus, spurned by the Persians, rose to Rome's side when the Persians attacked in 253, and it was in large part because of Palmyrene support for Rome that the peace held in the late 250s. The Persians killed the emperor Valerian in 260 and drove west into Antioch, but Odenathus savaged them as they retreated and, in doing so, filled the vacuum caused by the destruction of the Roman army. The Romans showed no sign of being unhappy with this; no Roman emperor visited the area between 260 and 272, which suggests not necessarily a collapse of imperial administration, but a lack of commitment to the area. The Palmyrenes were bringing order, and this was valued by Rome, at least for the moment.

In recognition of his support, the Romans had given the Palmyrene prince Odenathus several titles, and perhaps a formal position, as a governor or a commander. When he died, probably murdered in 267, Queen Zenobia—one of the more effervescent women of this or any other age—attributed high office to her son Vaballathus, who was probably ten years old. By the time the emperor Claudius II died of the plague in 270, Zenobia had built a base of support among intellectuals and officials in Antioch and parts of Asia Minor and Arabia, installed a friendly prefect in Egypt, and created a cultural center in the East. Events to this point can be explained by an extension of Palmyrene interests in the face of a Roman power vacuum—a development that should have been beneficial to the Romans. But Palmyrene control of Egypt was a direct and immediate threat to Rome's food supply, and increasing use of imperial imagery on the coinage became a challenge that Rome could not ignore. From the Roman perspective, Palmyra was building an independent kingdom with a knife poised at Rome's jugular vein.

The nature of the status granted to Odenathus by the Romans is important to our evaluation of the situation. Unfortunately, that status is not clear. Many scholars accept that Odenathus had received the title Dux Orientis (Leader of the East) or Corrector Totius Orientis (Commander of the Entire East) from the emperor Gallienus, perhaps in 261, and that these titles conferred a specific office.[30] A statue in Palmyra refers to him as "Septimius Odainet, King of Kings and Commander of All the East," and there are several commentators and inscriptions confirming these and similar titles. But historian Fergus Millar notes that there is no contemporary evidence confirming these titles for Odenathus; the historical commentators are later, and of the two inscriptions, one was set up by Zenobia, and the second by two generals in AD 271. All the evidence is posthumous to Odenathus.[31] It is also possible that the titles granted by Rome were a local expression, in formal terms, of what the Romans intended to be honorific and not titular—and that Zenobia and her followers elevated them after his death. Perhaps the Palmyrenes did not understand such subtleties of difference.

A better explanation for Odenathus's position is that he died friendly to Rome (murdered by an ambitious woman, on behalf of her son?) and that his widow Zenobia then claimed the titles, backdated them, and transferred them to her son. This could have made her son, Vaballathus, senior to Aurelian when the latter was acclaimed.[32] Zenobia at first placed her son's image on the same coin with Aurelian, but she was soon minting coins with her son's image labeled "Augustus," and herself as "Augusta."[33] Despite many uncertainties, this much is certain about Odenathus: he was never called Augustus. The power of images, inscriptions, and titles was the first line of Palmyra's encroachment on Rome, an advance guard that proclaimed Zenobia's growing influence in Anatolia and Egypt but that also made a confrontation with the energetic, ambitious Aurelian inevitable.

A different interpretation of the relationship between Zenobia's Palmyra and Aurelian's Rome follows. Palmyra may not have been the primary reason for the emperor's advance. The area was long overdue for a demonstration of Roman power, and the Palmyrene actions would have served as a credible pretense for such an expedition. The Persians may have been the Romans' primary target, as Aurelius Victor has already told us. They had killed an emperor just over a decade earlier, and they had raided into Mesopotamia and further west multiple times in the past three decades. The Romans may not have given up their long-standing desire to control affairs in Mesopotamia and Armenia—and whether the Persians ever intended to take permanent control of Mesopotamia and points west would have been irrelevant to Roman ambitions.[34] The Palmyrenes stood between the Romans and their eastern goals, and no Roman advance could be successful if the Romans did not control the areas to their rear. All of this would have left Palmyra subject to Roman military actions to reestablish authority in the area.

If this is so, then Palmyra had not been following a policy of military challenge to Rome. It is likely that trade and not military expansion was the basis for Palmyra's growing influence, along with Zenobia's relationships with intellectuals and officials.[35] The decisive break with Rome did not occur until late, perhaps April 272—right about the time that Aurelian was approaching Tyana.[36] The whole crisis sounds as if it began as a misunderstanding between Rome and a true ally, exacerbated by cultural and language differences and made acute by an overreaching queen. The Palmyrene court—accustomed to the ways of the East and not the Roman West—might not have understood the challenge it was posing to Rome by disseminating the titles and images.[37] But it had transgressed a line with the Romans, by claiming both Roman offices by inheritance—which the Romans would not accept—and a position equal to or above that of Aurelian.[38]

There is also a matter of *fides*—the relationship of trust that was so important to the Romans in their dealings with foreign cities.

Zenobia's court in Palmyra had grown in stature; she was creating a military, cultural, and political capital in the East and was not demonstrating a proper attitude toward Rome. Rome, then, would demonstrate the proper Roman attitude toward imperial cities: good dealings with friends, and merciless destruction of enemies.

Rumors of Aurelian's cruel success against the Goths were his first line of offense against the Palmyrenes. Such victories, combined with awareness by the locals that Aurelian would embrace them if they rejected Palmyra, would have sharpened his own forces and given them confidence in their commanders, undercut the support enjoyed by his enemies, and brought many contested cities over to his side without bloodshed. Many easterners may have once supported Odenathus, but this support did not necessarily transfer to his heirs, especially if Odenathus had been murdered and his widow had created a deadly challenge to Rome.[39] By claiming the title "kings," the rulers of Palmyra may have offended those Palmyrenes who were accustomed to life in Roman Syria.[40] The picture that emerges is one in which much of the population was set to welcome the Romans and to reject Zenobia, should the Romans move with sufficient force and confidence.[41]

Aurelian—with his "hand on the iron"—did just that, acting with overwhelming force against those who resisted but with leniency toward those who submitted. Literary sources generally agree about both sides of this policy: Bithynia and Galatea either remained loyal or gave over immediately, and Egypt may have come back while he was en route.[42] City after city opened its gates to him, renounced its loyalty to Zenobia, and hailed his arrival with celebrations. If the sources are accurate, the deeper he went into eastern territory, the more his army swelled—an indication of the legitimacy that local populations recognized in Roman rule, as well as the unpopularity of the Palmyrenes.

An oft-told story has it that when Aurelian approached Tyana—a city that controlled a pass into Cilicia—a Palmyrene faction closed the city's gates, and Aurelian famously swore that not a dog would be left alive inside. After a traitor opened the city, he executed the traitor ("he who spared not his native city would not have been able to keep trust [*fidem*] with me," wrote Vopiscus) but then commanded his army not to loot the place.[43] The soldiers obeyed but reminded him of his oath. He agreed, pious man that he was, and turned their minds to laughter by directing them to kill all the dogs. Even if born of literary invention, this suggests that Aurelian had the competence of command to control his troops at the moment of greatest release, a virtue earned through harsh discipline and training. The sources tell us of no sackings for the purpose of allowing troops to blow off emotional energy through carnage and revenge. He was much more nuanced in his approach than one should expect of a rapacious commander at the head of a bloodthirsty Danubian army.

The reference to trust (*fides*) suggests that Aurelian had drawn a firm distinction between cities that were properly part of the empire and barbarians who deserved no quarter. Another account has it that he told his soldiers: "We are fighting to free cities and if we prefer to pillage them, they will have no more trust [*fides*] in us. Let us rather seek plunder from the barbarians and we shall spare those whom we regard as our own."[44]

Fides, if not Aurelian's own word, was at least an idea that was still in use. *Fides* as an understanding of trust with a friendly people—as Rome had upheld with respect to Saguntum—was the kind of term that an emperor such as Aurelian, who was reestablishing Rome's dominant position in the East, could invoke. But *fides* has a related meaning, which applies to enemies who had surrendered to Rome: the defeated had to trust the victors not to abuse the limitless power they possessed. It was not a treaty or agreement but a plea by the defeated, based on complete submission to the victors.[45] Vopiscus writes that Syrians keep faith (*fides*) only with difficulty[46]—a view that may reflect an anti-Syrian sentiment from the Roman aristocracy and that bears comparison to claims of "Carthaginian perfidy" in past centuries. But perhaps the Syrians refused to remain in submission. To set up Palmyra as a threat to the trust between the empire and its cities would have given Aurelian the target and the pretense he needed to unleash his forces against those undermining Rome's position. Trust with Palmyra would then require the complete surrender of Zenobia and her court.

Aurelian may have divided the world, in his mind, into those who are "our own," and those who are not. He reserves the title "barbarians" for certain areas of the empire that deserved no mercy—

especially the Goths of the Black Sea area—whereas the "free cities" were those amenable to trust. Greek and Latin writers were active participants in this dichotomy. But many Goths of the Danube area were victims of barbarian attacks and had joined the Roman army to bring the recalcitrant East back into line. Zosimus tells us that Aurelian's army was manned by Moesians, Dalmatians, Pannonians, and Gauls from Noricum and Raetia, as well as Mauretanian horse, and easterners from Tyana, Mesopotamia, Syria, Phoenicia, and Palestine.[47] The categories of "our own" versus "barbarians" may have been accurate given the attacks into the Dacian areas, and those categories certainly served Aurelian's purposes, but the composition of the army and its position in the East were less straightforward.

If Palmyra had been deliberately empowered by Rome to stand against the Persians, then it is not so clear who broke trust with whom when Aurelian marched against the city. It must be stressed that Palmyra had not militarily attacked the Romans or Roman interests—the forays into Asia Minor and Egypt were based on trade or administration—but such encroachments were threats to the power that brought stability to the area. Aurelian may have reinvigorated the idea of Rome as the proper ruler of the East—thus anchoring *fides* onto a foundation of Roman power in the West, not onto an agreement between equals.[48] This would have placed the educated Zenobia and her court into the position of enemies—a worthy target for an army consisting, to some degree, of soldiers recruited from the barbarians. The pretext was that she had claimed a position above her proper station. But the real issue may have been the need to end neglect of the area and to reassert Roman control.

The details of the action in the East are beyond confirmation—but Palmyra made no serious attempt to oppose Aurelian's early advance. Zenobia probably established defensive positions on the Orontes plain, near the city of Antioch—this was the limit of her effective military power. Aurelian defeated the Palmyrene cavalry with a ruse, in which he enticed his enemy's heavy-armed cavalry to chase his light horsemen into exhaustion before he turned and slaughtered them. Zenobia's army retreated toward Palmyra itself; Aurelian's successes had made the Syrian religious center Emesa to the south unsafe for the

Palmyrenes. He entered Antioch as a liberator. Once the pro-Palmyrene supporters had fled, he accepted the city back into the empire and again restrained his army from a customary rampage. The city was, after all, one we should "regard as our own."

Aurelian's arrival altered all factional relationships inside Antioch. At some point, for instance, he mediated a dispute involving pro-Palmyrene Christian supporters of Paul of Samosata, who had been condemned and deposed as a heretic by two pro-Roman Christian synods (in 264 and in 268 or 269). Such antagonism between the aristocracies in Rome and Palmyra—feuding eastern and western Christian factions—is suggestive of the many disputes that were tearing at imperial authority. Aurelian affirmed the authority of the bishop of Rome to decide the issue—probably by letter rather than personal adjudication—and thus legitimated the position of the bishop of Rome over provincial bishops and strengthened the Roman faction.[49] Aurelian doubtless had no idea of the historical precedent he was setting.

The desert route to Palmyra was caravan country, controlled by nomadic tribes who assisted Aurelian's crossing of some eighty miles of desert. This crossing makes it all but certain that he did not have his full army with him when he reached Palmyra's gates. He took control of the lush, rich city without a sustained siege—archaeology does not reveal massive siege walls, but rather less sophisticated bulwarks designed to fend off a smaller Bedouin attack.[50] Cutting off routes to the west was enough to destroy Zenobia's support inside the city. Zenobia made a desperate dash east on a camel, perhaps hoping for support from the Persians. Zenobia—in power for the three years since her husband's death—may have tried to form a new alliance with the Persians, to protect her eastern borders while spreading her influence to the west. The Persians certainly declined. Her capture by Aurelian ended such aspirations and gave the Persians a chance to escape Roman attack. Both sides would have been happy with this resolution. Aurelian was at the geographic limits of his power; to move into Persian was likely not possible, although it may have been tempting. He again restrained his troops and embraced Palmyra as a child that had strayed.

Aurelian executed leaders of the opposition, including the philosopher Cassius Longinus, whom Zenobia blamed for the war.[51] This incident is shrouded in mystery—and the words of the unreliable writer Vopiscus—but it suggests more than a mere attempt by Zenobia to save her own skin. She blamed the war on certain factions in her own court—who, she claimed, had caused the loss of trust with Rome. This too served the interests of the Romans, provided a scapegoat had been found. Aurelian took her captive, probably intending to parade her during his future triumphal entry into Rome; whether or not this happened, the public subjugation of Zenobia in Palmyra left no doubt that she had been beaten.[52] Roman administration in Egypt was renewed through a diplomat, perhaps the future emperor Probus.[53] Aurelian was now Palmyrenicus Maximus with a Victoria Parthica—the latter perhaps a propaganda claim that he had beaten the Parthians. The senate in Rome voted him Restitutor Orientis, "Restorer of the East."

One conclusion defines this eastern campaign: Aurelian had marched to the heart of his enemy's territory, recovered his allies, and captured the enemy leader, while fighting only one pitched battle and mounting no sustained sieges. The number of lives lost was very low, especially by ancient standards, and in a place that was notoriously treacherous and logistically formidable. "With the advance of the Emperor's army, Ancyra, Tyana, and all the cities between it and Antioch submitted to the Romans."[54] Only at Antioch was there serious resistance—due in part perhaps to the faction of Paul of Samasota—and that ended when Zenobia's army withdrew. The presence of overwhelming force, for the legitimate purpose of restoring Roman order, had empowered his allies, demonstrated the deep divide between the Palmyrene leadership and the population, and allowed the eastern cities to give up without a fight. Rather than gratuitous cruelty, Aurelian in the East came close to Sun-tzu's ideal: to win without fighting.

He marched his army back overland, through the Balkans toward Gaul. This was an important decision, for the temptation to move against the Persians must have been great. They had killed a Roman emperor and were the original reason why Aurelian's predecessors had empowered the Palmyrenes and created this problem. Although the Persian kings were weak at this point, with Palmyra reduced they could be expected to take advantage of the vacuum.[55] A fight could not be prevented, but it could be postponed, until Aurelian brought down a more immediate threat to Rome in Gaul.

Heading west, he again dealt ruthlessly with revolts by the tribe of the Carpi, perhaps inspecting the Danube fortifications en route to the western provinces. Meanwhile, a usurper in Palmyra forgot his place; he may have hailed a certain Septimius Antiochus as king, and perhaps massacred the Roman garrison.[56] Syrian perfidy! Aurelian charged back to Palmyra too quickly for opposition to organize. He tortured the usurper to death and reduced Palmyra to a subordinate position from which it never recovered. He might have yet concluded that the pointed destruction of the recalcitrant leadership would take care of the problem, although it is plausible that he dealt harshly with the population, perhaps allowing the Temple of the Sun to be ransacked, and then restoring it. The ruins of Palmyra today attest to the reduced status of a once powerful city but do not suggest that the town was systematically destroyed. Aurelian may also have dealt with a revolt by a Greek paper merchant in Egypt.[57]

Aurelian then turned upon the Gallic West, where a series of usurpers had established an independent, politically organized empire that included Britain and Spain. By the time Aurelian gave his attention to this Gallic Empire, Postumus and two successors were dead, and Tetricus—acclaimed in 271, the last Gallic leader of note to face Rome—was trying to prevent the disintegration of the empire. Spain had not recognized Tetricus, and control over Britain may have been lost. Tetricus—whose literary reputation is largely of a hopeless incompetent—may have sent a secret offer of surrender to Aurelian. These sources further claim that after Aurelian captured Lugdunum (modern Lyons), and during the battle at Durocatalaunum (at Chalons-sur-Marne), Tetricus betrayed his troops by setting them up for a rout by the Romans. Aurelian would not have wanted such slaughter—he could have used the soldiers—but the victory pacified the area and allowed the Romans to reassert control.[58] It is also possible that Tetricus did not surrender; his weak image may have been drawn from imperial propaganda, and he was strong enough to mint coinage into 274.[59] In any case, Aurelian was now Restitutor Galliarum.

Aurelian, fortifying Gallic cities and restoring their defenses, may have built walls around Dijon—another decision that reveals a policy of leniency and trust toward those loyal to Rome. After assuring the loyalty of the senate in 271, he had also begun walls around Rome before leaving for the East. Once Palmyra was under his control, he had left it and Egypt to commanders, returning in anger only after they repeated their disloyalty. Once the Gallic leadership was removed, Aurelian allowed fortifications to be built. All of this suggests Aurelian's belief that the people of Rome, the East, and Gaul could be trusted—that they were among "our own"—if their insubordinate leadership was removed and *fides* was restored.[60]

True to Roman form, Aurelian returned to Rome in triumph, parading thousands of gladiators, wild animals, and other prisoners. He was acclaimed Restitutor Orbis, "Restorer of the World." Surviving inscriptions in Rome, Gaul, and Africa proclaim his glory.[61] When he had taken power in 270 the empire was effectively divided into three separate and antagonistic kingdoms; five years later, it was more united than it had been in fifty years. Aurelian's offensive strategies had laid the foundations that future emperors would need to restructure the political administration of the empire.

The Religious Aspects of Aurelian's Success

Our sources are unanimous about Aurelian's strategic success at unifying imperial authority. But whether he had a "grand strategy" or thought with a strategic frame of reference is a more difficult question. His actions imply that he did. Aurelian's policy was to reassert Roman power across the empire. He dealt with Rome's enemies in the order of the threat that each posed to Rome: the first objective was the Vandals and Goths (who had attacked Italy); then Palmyra (which threatened Rome through usurpation); then the Gallic Empire (able to strike Italy, though not immediately threatening), and then, perhaps, the Persians (given Rome's long-standing ambitions to control Armenia and Mesopotamia). As long as Rome's problems were considered to be an ad hoc series of disconnected events, each put down for the present moment only to return in the next, the best that Rome could accomplish was an uneasy balance between threats. Aurelian may have done the best possible at the time to break out of this way of thinking.

But such events had their genesis in the broader political developments of the day, which were themselves dependent upon a certain climate of ideas and the practices they engendered. An observer at the time would have been hard-pressed to find more than remnants of the republican ideas that had empowered the Romans of yesteryear to maintain a lawful form of government. The loss of these ideas —and the changing relationship between the army and the political leadership that began in the decades after the massacre of Carthage—left only military victory and the acclamations of the soldiers to decisively demonstrate an emperor's legitimacy. These acclamations repudiated the senatorial order that had, in a bygone era, largely subordinated force to political deliberation and decision. This left the Romans with only two options: single rule, or violent anarchy should imperial rule fragment.

To achieve this Aurelian would need more than a solid military organization; he would need to legitimize a single point of veneration toward which all could be loyal. He began in the military camp, by thanking the god who had assisted him in his victory, an action that could provide soldiers of various ethnic backgrounds with just such a point of focus. But the empire as a whole would need more. Aurelian could assert imperial authority over millions of people only by melding emperor and god in the minds of imperial subjects.

The philosophical climate at the time was conducive to this effort. In an age of uncertainty and danger, people will look for solutions to the problems of life. Aurelian lived in a century when a unified conception of the cosmos and human affairs was manifested in several syncretic developments, which could provide comprehensive solutions by melding divergent phenomena into a new whole. Attempts to create a unified political authority under a single Roman law were one such development. First-rank legal figures under the Severans include Paullus, Papinian, and Ulpian of Tyre, compilers who worked to create a single corpus of laws out of thousands of particular cases. The *ius gentium*—the law of all peoples—was a set of unifying principles, common to everyone despite local practices, with deep roots, especially in Stoicism.[62] All such movements looked for a "One in the Many" that could prevent the fragmentation of the empire into war.

A more fundamental form of divine syncretism arose in various intellectual movements that postulated a single transcendent point of veneration for human beings. Neoplatonic philosophy—a reinterpretation of Plato that reached its apex in the third century—had myriad complex intellectual permutations and rejoinders. But common to all was a central thrust toward the "One," a singular principle, deity, or force, elevated over all of reality, to be reached by various means, from pure mental activity to rituals or magic. A few of the figures participating in such movements can illustrate their wide influence in the East. Apollonius of Tyana (ca. AD 40–120), was a philosopher whose Pythagorean mysticism—based on abstract numbers—served eastern occultist and magical views. Ammonius Saccas (died around 240) was a teacher of both the Christian writer Origen and the philosopher Plotinus (205–270), whose Neoplatonic views were influential among the imperial aristocracy. Plotinus's student Porphyry (ca. 232–305) was from Tyre; Porphyry's later student Iamblichus (ca. 245–325) advocated theurgy, a merger of philosophy with magic, including pagan sacrifices. The Christian leader Paul of Samosata (ca. 200–275) and the philosopher Cassius Longinus (ca. 213–273) were active in Zenobia's circles. These and many other figures had one foot in the philosophical schools of Athens and another in eastern religious thought. Each was teaching that reunion with a transcendental divine One was the goal of human life—they disagreed only about its nature and the means to get there.

Such philosophical views dovetailed with politically influential religious movements. The emperor Septimius Severus came from North Africa, from a family of Syrian extraction, but he studied law in Rome, commanded a legion in Syria, lived in Athens, and developed ties with eastern intellectuals. He maintained an interest in sun worship as well as astrology—a way of connecting the movements in the heavens with life on earth. He married Julia Domna, daughter of a priest of the sun at Emesa, which unified the politics of the imperial throne with the eastern sun cult. Julia commissioned a biography of Apollonius of Tyana. Septimius's son and successor Caracalla established a sanctuary to the eastern deity Serapis in Rome.[63] Elagabalus named himself after the sun god Elegabal—a syncretism of the Greek *Hēlios* with the Syrian *el-Gabal*, which was related to the Phoenician *Baal Shamim* as well as to *Mithra* and earlier deities.[64] When Elagabalus performed a divine marriage between the Roman sun god Sol Invictus Elagabal and the goddess Dea Caelestis, he was forming a syncretic union between East and West and placing both under his control as chief priest. Elegabalus and his strange rituals were rejected by Romans, and his memory was damned. But the Invincible Sun, Sol Invictus, once rid of its overt eastern debauchery, continued to enjoy broad appeal in Rome—and did not contradict the basic premises of the Neoplatonists, the sacrifices of the Danubian troops, or the rituals of the priests at Emesa.[65]

Thirty-five years later, Aurelian rose, and he worked within this climate to establish a celestial-imperial presence that mirrored his own reunification of the empire on earth. The area from which he came—somewhere in Illyria—and beyond it, over the Danube River, was inundated with sun worship, especially among the troops. One inscribed dedication to Lunae Solique deis—the gods Moon and Sun—references the Thirtieth Legion and the emperor Severus Alexander. Another, from an imperial legatus in Apulum (in Dacia), suggests a position of divine guardian for Sol over the Thirteenth Legion. Such dedications were consistent with the decree of Elegabalus that Sol be the guardian of the legions.[66] To the west, even troops from Belgica acknowledged Sol Invictus, among other deities, especially Jupiter. Another dedication publicizes the dedication of the first and tenth cohorts from the province Belgica.[67] Such practices, if universalized and identified with the emperor's presence, could make his rule seem like a structural part of the universe. The military hierarchy of the army would mirror the divine hierarchy to the gods and the One.

Aurelian's encounter with sun worship in Emesa would have been different from what he had seen all his life, and it would be an error to identify Sol with Elagabalus without qualification. But different people could see the Sun in different ways, yet still see the same sun. In Aurelian's mind, his decision to bring sun worship to Rome would have given proper credit to the divine presence that had assisted him to victory. If handled properly, sun worship could secure the loyalty of the Syrians—many of whom wanted to return Elagabalus to Rome—while offering a common ritual of sacrifice for his Danubian troops and the Roman *populus*. His version of this ritual could have allowed the inhabitants of Rome to acknowledge the indigenous Sol as a Roman deity, perhaps even with a nationalistic component set in opposition to the eastern sun god. Despite the political gulf between the East and the West, sun worship could bring some measure of harmony to the Greeks, Danubians, Anatolians, Egyptians, Syrians, and others serving in military units.[68]

The key to promulgating sun worship would be to merge the Greek Elegabalus with the Latin Sol in a way that could be seen as legitimate from various perspectives. The Emesans could continue as they were accustomed with Elegabalus, and the Romans could venerate Sol. Such a syncretism maintains the separate identities of the original phenomena. This would be different from a fanatical imposition of a Syrian cult *over* Roman deities, as Elegabalus tried to foist on the Romans. In Rome, Aurelian established a cult of Deus Sol Invictus, the "Divine Unconquered Sun," as Theos Hypsistos, "God the Highest," based on a notion of celestial hierarchy not radically different from that of many Neoplatonists.[69] Aurelian's coinage emphasized Sol after 272.[70] Aurelian established a new college of priests—the *pontifices dei Solis*—a temple, and quadrennial games, probably at the Circus Maximus. Aurelian was not being radical here—Sol Invictus had appeared on coinage going back to Antoninus Pius (138–161 AD), and the Circus Maximus was one of two places to which the first emperor Augustus had transferred two obelisks from Egypt and dedicated them to Sol.[71] Aurelian's creation of public ceremonies, holidays, and games presaged the devotion shown to Sol by Diocletian and later Constantine, who put Sol on coins prior to his conversion to Christianity—at which, we are told, he saw the sun in the sky in the form of a Greek cross.

The final step in Aurelian's new theocracy was the deification of the emperor himself into the direct political counterpart to the "One" of the Neoplatonists, the singular god of the religious, and the hierarchical command structure of the military. The view of an average person toward the emperor was distant and detached, akin to his view of a god who sits on a seat high above mortal men. Aurelian was perhaps the first Roman emperor to make this position explicit, by claiming a divine status while still alive. His heavenly image was directly connected to Deus Sol Invictus—who was made Dominus Imperii Romani, "Master of the Empire of the Romans"—and given a positive relationship with Jupiter. In coins issued in 274, he claimed the titles Deus and Dominus (God and Master), was attributed with descent from the sun by birth (*natus*), and was possibly the first emperor to wear the diadem and jeweled robes in public.[72]

Students of military history today are not used to discussing religion when considering a military victory. But this is part of the gulf between an earlier, more superstitious age—when Fortune was a necessary ally in any successful battle—and our own day. Aurelian's victory would not be complete until the soldiers of the Balkans, the intellectuals in Antioch, and the *populus* in Rome were venerating the same god, in the sky and on earth. In this sense, Aurelian—and not Diocletian—is the originator of the *dominate*, a term sometimes used by historians to differentiate the period of the later empire from the "first-citizen" rule, or *principate*, of the first emperor Octavian Augustus. The difference between Octavian and Aurelian was the former's claims to authority as first citizen versus claims to rule by a living divinity akin to a master's absolute power over a household. At its root, Aurelian's innovation was "fertilized perhaps by philosophical discourse about the supreme god."[73]

For Aurelian, political authority, divine sanction, personal deification, and military victory were all perspectives on the same imperial project: the victory of a unified imperial power over the empire. As chief priest in Rome, the emperor was expected to preserve the common rituals of the city. Such a project, extended across the empire and enforced by superiority of might, could provide the order needed to preserve imperial rule. When this order failed, by usurpation or doctrinal disputation, the division of the empire, a new doctrine, and new round of warfare became inescapable. The collapse of the ideas needed to maintain a rational system of government was complete—all that was left was the force of an emperor-god.

Aurelian's Wall and the Fall of Rome

The remainder of Aurelian's brief reign reads as a coda that ends in failure. Rome itself, where Aurelian probably spent some of AD 270, 271, and 274, was still defenseless. He faced an array of crises, including Rome's worst riots since the first century BC, and a revolt of the mint workers that led to street fighting.[74] The currency problems were caused by decades of coinage debasements, which had reduced the value of metals in the coins to below 2 percent of their stamped value.[75] The agony of the Romans was an accurate response to barbarian invasions, ruinous monetary inflation, threats to water and grain supplies, and dependence upon provincial armies. For relief they looked to a greater power—they were long past the point of defending themselves. Aurelian probably distributed new silver, as well as pork, bread, and wine to the Romans.[76] He had only months to address these monumental problems.

But the deepest source of concern in Rome was doubtless its vulnerability to barbarian invasion; the Italian peninsula could not be permanently defended. With the support of the senate and people of Rome, Aurelian organized the Romans to build defensive walls around the city, probably in 271.[77] They extended west over the Tiber River, to encircle the mills that produced bread, and east to bring the Praetorian Guard into the defended areas of the city—Rome was now garrisoned by an army unit. The walls were finished by the emperor Probus and heightened by Maxentius in the next century. When Constantine marched on Rome in 312, his opponent Maxentius did not rely on the walls but rather marched out to meet his foe. The purpose of the walls was to protect the city until an army could come to its rescue, not to serve as a sufficient defense in their own right.

But there is a deeper meaning to these walls. For centuries the majestic grandeur of Rome's strength had been flaunted in the open roads into the city. It had been six hundred years since the city was last besieged, and for centuries an invasion of Italy had been inconceivable. Every Roman could see this power and this confidence in the openness of Rome's roads, which connected her without fear to the world. These roads were like the shipping routes that had crossed the Aegean Sea when Athens was at her apex, which had brought the goods of the world to the foremost Greek city. The openness of Roman roads was true power, far stronger than mere walls. These roads were lines in the face of a confident city, the sinews of an invincible civilization with a people who admitted to no threats capable of striking their capital.

Now the threats were real, and a new invasion was only a matter of time. Aurelian's walls might have reassured the Romans that another invasion could be held off before penetrating the city itself, although Rome remained dependent on outside sources for its food and water. But the walls may also have had a deeper effect: they could not save Rome, but they could remind every Roman, every day, that he was perpetually at risk. The assault of the Goths onto Italy had been shocking but temporary evidence of Rome's precarious position—as the sight of the Thebans had revealed the paucity of the Spartans, and Scipio's landing swept fear through Carthage—but Aurelian's walls were an open admission of permanent weakness and vulnerability. If any doubt remained that Rome was no longer the source of energy at the center of empire, that doubt was now removed.

If one wished to name a particular event as marking the true Fall of Rome, one might look not to the elevation of King Odoacer to emperor in 476, or to the sack of Rome in 410, or even to the division of the empire in 395. The moment at which the city of Rome became just another town among many in an anarchic landscape, and not a source of law and power, was AD 271, with the decision to build walls around Rome. The political and military center of the empire had been gutted; the rest was a matter of time.

Aurelian had left Rome in 275, whether moving east against the Persians or north against the Goths is unclear, when he was killed, probably in September. We do not know the manner of death; it was not a revolt of the soldiers, but probably an attempt by an aide to escape his financial discipline.[78] The aftermath strongly suggests that the soldiers regretted his killing; perhaps they were coming to realize that the incessant lynching of their commanders was destroying any possibility of lawful order. The soldiers appealed to the senate in Rome to choose a successor—and the senate, terrified of making a wrong decision, sent the matter back to the soldiers. In AD 96, after the murder of Domitian, the senate had found its will and appointed the aging Nerva, an act that avoided a repeat of the civil wars that had followed the death of Nero in 68. But in the two centuries since, the idea that the army and not the senate selected emperors had become commonplace; no one in the senate had the backbone or the skill to reassert senatorial authority over the army. After a few weeks, the elderly Tacitus was chosen—doubtless by the troops.[79]

Aurelian was the epitome of the purely military emperor—the strong "hand on the iron" needed to forestall anarchy and relieve the ominous threats hanging over Rome for the moment. The reestablishment of order under a single commander was a fundamental prerequisite to the broader reforms made by Diocletian in the next decade. In this regard, Aurelian was victorious. But the lack of any alternative to force as the standard of political success, and claims to divine sanction as its symbol of legitimacy, left the fundamental nature of imperial power dependent on the imperial personality. On one level Aurelian motivated his forces to regain the military unity that had once been Rome's greatest strength, and on another level he did not challenge the violent foundations of that unity. The walls around Rome were the concrete consequences of this failure. A military emperor first and last, Aurelian could not use his power to redefine the source of that power in any fundamental way. Nothing Aurelian did—or could do—would stem the tide of changes sweeping over Rome, or bring forth the ideas needed to establish lawful, stable, and peaceful government. Every Roman remained a subject, and all the more so as the divine pretensions and doctrinal positions of the emperors hardened in the next century into a rigid, and ruthlessly enforced, orthodoxy—the new standard of theocratic legitimacy.

Chronology of Aurelian's Campaigns

The *Pax Romana*

31 BC–AD 14 Augustus establishes the imperial system, the Principate

AD 68–69 Death of Nero; Civil Wars over the succession

The Severan Dynasty

AD 192–195 Death of Commodus; Civil Wars of succession

193–211 Septimius Severus

211–218 Caracalla

212 Caracalla's citizenship law

218–222 Elegabalus, imports Syrian sun worship to Rome

220s Rise of the Sassanid Dynasty in Persia

222–235 Alexander Severus

Military Rule

235–284 The Age of the "Soldier-Emperors"

243 Persian invasions of Mesopotamia and Syria

244 Death of Gordion III; Philip acclaimed emperor

249 Murder of Philip; ascension of Decius

251 Death of Decius in battle

253 Persian invasions west toward Antioch
Valerian acclaimed emperor

250s Gothic attacks into the Balkans
Rise of Palmyra as an ally of Rome

260 Persian invasions; capture and execution of Valerian
Gallienus, son of Valerian, acclaimed emperor

267 Death of Odenathus of Palmyra

268 Death of Gallienus; Claudius II acclaimed emperor

270 Death of Claudius II; Aurelian acclaimed emperor

Outline of Aurelian's Campaigns (all dates uncertain)

270 August: Iuthungi begin the first invasion
October: Aurelian acclaimed in Sirmium
November: Defeat of the Iuthungi in Italy

271 Late 270 or early 271: Aurelian's first visit to Rome
February–March: War with the Vandals
March–May: Second invasion of the Iuthungi into Italy; battle at Placentia
Spring: Revolt of the mint workers in Rome
June–September: Aurelian's second visit to Rome; defensive walls begun
Autumn: Defeat of the Goths; Dacia abandoned

272 January–April: Aurelian advances from Byzantium through Asia Minor
Tyana taken
April–May: Antioch and Emesa taken
Summer: Palmyra taken; Egypt under Roman control
Aurelian moves west
273 January: War with the Carpi
Spring: Revolt of Palmyra suppressed
July: Turmoil in Alexandria suppressed
274 April: Aurelian into Gaul; battle of Durocatalaunum
Aurelian's third visit to Rome
Cult of Sol Invictus, priesthood, games, and holiday initiated in Rome
275 Aurelian prepares for war with Persians—or the Goths?
Murder of Aurelian

Chapter 5

"The Hard Hand of War"

Sherman's March through the American South, AD 1864–1865

The Nature of the Conflict

The leap from the ancient world to the modern is a big one. The rise and overthrow of the medieval religious order, the Greco-Roman rebirth, the Newtonian scientific revolution, and the discovery of political liberty resulted in a new and special focus on the sovereign individual. The Rights of Man—wholly foreign to the ancient world—undermined the power of kings, challenged historical standards of status and nobility, and required a new moral conception of the nature and purpose of government. These developments, along with the concomitant rise in weapons technology, were bound to have important effects on the aims and practices of warfare. The Constitution of the United States made the new conception of government real, and the Americans set out to apply it across a continent that was set apart from the baggage of European conflicts. But the Constitution was a product of negotiation between states jealous of their powers, and had failed to apply the American founding principles of liberty to everyone. The elevation of so-called states' rights—in practice, the right to keep slaves—over individual rights sowed the seeds of a titanic struggle between irreconcilable moral purposes in the United States. The Civil War was the price paid for this contradiction. From a dispute that many thought would be quickly resolved, the war turned into a four-year nightmare that butchered more than 600,000.

For three years Union armies rallied under a series of commanders, who wavered between excessive caution and suicidal charges. The failure of Union generals to capture Richmond left the capital of the Confederacy intact, while Union armies were unable to reach deeply into the heartland of the South. Although the southern states began with fewer resources and lost them more drastically as the war continued, misleading reports of successes in distant battlefields, true at first but never indicative of the relative strengths of the two sides, kept unrealistic expectations of a Confederate military victory alive. But the danger to the Union was less a decisive military defeat than the possibility of an irresistible civilian reaction against the war in the north, a moral crisis that could have forced the administration to negotiate a settlement with a politically victorious Confederacy. Had the Union felt a major defeat in the last half of 1864, and had George McClellan been elected president, such a sanctioned division of the United States—and a new war in the next generation—might have occurred.[1]

Not until the Union armies under General William Tecumseh Sherman marched through the center of the South, as part of a comprehensive strategy broadly outlined by General in Chief Ulysses S. Grant, was the Union victory assured. Sherman moved from Tennessee into Atlanta (September 1), across Georgia to Savannah (November 12 through December 22), and then northward through the Carolinas (beginning February 1, 1865). He tore up rail lines, burned plantations, lay waste to the material and psychological foundations of the southern war effort, and demanded that southerners end the war. By April—five months after leaving Atlanta—the war was over. Like the eastern march of Aurelian, Sherman's campaign suggests a close relationship between the demonstration of overwhelming force by a legitimate political power and a loss of will to resist in a civilian population, which delegitimized the local leadership and lessened the need for direct use of force. In each case the commander's reputation for cruelty—which spread in advance of his army—became a historical reputation that has obscured the central meaning of the campaign and its effects on the peace that followed.

Among the intricate social and political factors that made the war what it was, there was one ideological issue at its heart—statism versus individualism—and it took the form of one concrete alternative, freedom versus chattel slavery. Southern state governments claimed the authority to legalize, institutionalize, and defend the ownership of human beings. "States' rights" is a statist doctrine with roots in the Spartan conception of a superior citizen class, applied to the American federal system, modified in the antebellum American South with a racist criterion for selecting the slaves, and strengthened by a deep emotional attachment to the states. Southern leaders saw infringements upon the "rights" of their states—including limits to the expansion of slavery into new territories and the possibility of a constitutional amendment against slavery—as federal aggression. The overall trend of the South from the time of the American Revolution had been to extend the institutionalized enslavement of an entire class of people into new territories. The overall trend in the North, albeit unfolding slowly, was toward a widening scope of liberty. The war would determine which of these two trends would guide America's future.

Although the war was not seen in terms of slavery versus individual rights by everyone, including racists in the North, this was in fact the case, and many southerners recognized it and said so. Alexander H. Stephens, vice-president of the Confederacy, knew it, and expounded upon this at the Georgia convention on March 21, 1861: "Our new government is founded upon exactly the opposite idea [from abolition]; its foundations are laid, its cornerstone rests upon the great truth that the negro is not equal to the white man, that slavery—subordination to the superior race—is his natural and normal condition."[2]

For many in the North, abolition had become a moral issue, which was worth the threat of war.[3] Thomas Wentworth Higginson wrote: "Either slavery is essential to the community, or it must be fatal to it,—there is no middle ground.... War has flung the door wide open, and four million slaves stand ready to file through.... What the peace which the south has broken was not doing, the war which she has instituted must secure."[4]

But most southerners did not own slaves and might not think that chattel slavery was worth defending if named as such; their moral task was to defend their states and their "way of life" against "northern aggression."[5] Many southerner leaders shared these motives. Alexander Stephens argued against secession but went to the Confederacy because his home state of Georgia was seceding. Robert E. Lee himself, who declined an offer of command over the Union armies, said that he could never fight against his beloved Virginia—a focus on a single state that may have prevented him from recognizing the importance of the war in the West. Sam Houston, governor of Texas, saw his state's secession as "usurpation" with "revolutionary schemes"—yet proclaimed, "I went back into the Union with the people of Texas. I go out from the Union with them."[6] Under the tutelage of their leaders and a barrage of newspaper columns, many poor whites came to see northerners as aggressors against their states, and the Union as a tyranny.

The tensions and moral fervor rising on both sides were made all the more ferocious by the deep threats against their values that each side saw in the other.[7] Northern politicians had enacted economic tariffs and protectionist policies that were unconscionable restraints upon southern farmers; southerners properly reviled these measures and claimed the right to nullify federal laws on a state level, an issue that was not directly addressed in the Constitution. This Doctrine of Nullification placed the powers of any state over the authority of the federal government and allowed any state to annul a federal law that was not specifically authorized in the Constitution as state leaders read it. On November 19, 1832, the South Carolina Ordinance of Nullification, enacted by convention, rejected the federal tariff and upheld this power. Although South Carolina repealed the ordinance when Congress rescinded the tariff, Andrew Jackson had asked for a "Force" bill empowering him to use federal troops to enforce the law. This law was passed. These events became pretexts for the South to annul northern laws, but they also put in place the prerequisites for the North to use force against states that attempted to do so.[8]

Despite such tensions, the most compelling issue in this conflict would be not trade but slavery. South Carolina did not ask free states to join the secession—"states' rights" and slavery were inextricably bound, and they would rise and fall together. "The South alone should govern the South, and African slavery should be controlled by those who are friendly to it," said one voice, identifying southern politics with slavery and racism.[9] The crisis came to a head not over demands that slavery be ended in the South—although many northerners wanted nothing less, and many southerners feared that complete abolition was the direction that the country was moving—but rather over its extension into new territories, which could leave the southern states vulnerable to a constitutional amendment. The election of Lincoln, who was opposed to slavery in new territories all the while accepting the constitutional right of the southern states to keep what they had, set the secession process into motion. On December 24, 1860, South Carolina seceded. By the time Lincoln was inaugurated, in March 1861, six more states had followed. On April 12, 1861, the state militia of South Carolina fired on the federal garrison at Fort Sumter and forced its evacuation two days later.

Strategic Integration and Goal Orientation

Like so many wars, this one began with expectations of a quick end. Northern civilians went to the first battle of Bull Run as sightseers, expecting an enjoyable day in the sun watching the Rebs learn their lesson. Both sides suffered a rude awakening and found themselves thrust into a malevolent nightmare that grew voraciously for four years. As the war progressed and became even bloodier, many people came to realize that a permanent division between North and South would lead to permanent hostilities.

At the start the Union had its Cunctator, the delayer General George McClellan. In his Virginia campaign, in the second year of the war, he stood a few miles away from Richmond with some 110,000 men, facing southern forces of some 62,000. Yet he consistently overestimated the strength of his enemy, held back when he could have attacked, and soon withdrew, chased off the peninsula. Certainly a full assault on Richmond would have been a bloody affair, and an

American Cannae was not out of the question. But the results of his strategy were neither a sparing of lives nor a retreat by the Confederate forces, but rather the bloodiest days in American history, the battles of Second Bull Run and Antietam. McClellan stopped Lee's invasion of the North, but the creek at Antietam ran red with blood, and the promise of a swift win collapsed into a gruesome stalemate. McClellan exhibited a desire to avoid a confrontation with Lee that bears comparison to the Fabian strategy toward Hannibal. President Lincoln despaired at getting the man to fight and relieved him of command.

As an organizer McClellan was superb, and many of his troops loved him. The logistics involved in keeping the number of men he had on the field would rival those of a major corporation today, and the commander who could handle the task was an individual of special skills and stamina. But there is more to commanding an army than keeping it organized and supplied. Union navy secretary Gideon Welles records a diary entry for September 3, 1862, following McClellan's retreat after the second battle of Bull Run, as Lee stood poised to drive into Maryland:

> Washington is full of exciting, vague and absurd rumors. There is some cause for it. Our great army comes retreating to the banks of the Potomac, driven back to the entrenchments by the Rebs.
>
> The army has no head. Halleck is here in the Department, a military director, not a general, a man of some scholastic attainments, but without soldierly capacity. McClellan is an intelligent engineer and officer, but not a commander to head a great army in the field. To attack or advance with energy and power is not in him; to fight is not his forte.... Wishes to outgeneral the Rebs, but not to kill and destroy them.... He detested, he said, both South Carolina and Massachusetts, and should rejoice to see both states extinguished. Both were and always had been ultra and mischievous, and could not tell which he hated most.... He was leading the men of Massachusetts against the men of South Carolina, yet he detests them both equally.[10]

This passage illustrates a variety of deadly cracks in the Union position. A general who thinks that his own people are as bad as the enemy is in no position to put forth the demanding effort, in the face of indescribable slaughter, needed to force a victory. He can have no moral certainty in his cause if he is convinced that the heart of his side is as bad as the other. He may, for instance, improperly evaluate the strength of the forces involved—as McClellan drastically overestimated the size of the opposing army; in his mind, 45,000 Confederate soldiers became 170,000.[11] Falling into disputes with Washington politicians, he did not use his superiority decisively against the Confederate capital.

McClellan was not alone. In May 1862, following a narrow Union rescue at Shiloh, General Henry W. Halleck, taking command in the West, moved more than 100,000 men slowly toward Corinth, Mississippi, against an inferior Confederate force. By the first week of June the Confederate army had withdrawn from Corinth, and Halleck set off in pursuit. Halleck—a theoretician more than a field commander—followed the conventional wisdom of taking the fight to the enemy's army rather than driving toward an objective. New Orleans and Memphis had been captured, but Halleck divided his forces into eastern and western armies and then into units needed to guard rail lines and other points of logistical importance. By October 1862, the Union and Confederate armies were back to fighting over Corinth, the place that Halleck had failed to control five months earlier. Massive logistical problems would have stymied any Union force, and success might not have been possible for anyone at that moment, but the fact remains that Halleck's cautious actions did not bring the war to an end.

The overall picture of the Union armies at this point is one of indecision and disorganization among its commanders, with no integrating principle, no clear lines of command directing it to a single purpose, and no unified strategy to focus hundreds of officers on a well-defined goal. Following McClellan's retreat from the Richmond area, General Pope, for instance, took command of a new Union Army of Virginia in June—some forty-seven thousand men—charged with assisting McClellan, protecting Washington, and maintaining the Union position in the Shenandoah Valley. Three months later the army was disbanded, following the second battle of Bull Run, and

Pope lost his command. Lee was set to invade the North through Maryland. After the battle of Antietam on September 17, McClellan was replaced over the Army of the Potomac with Ambrose Burnside, who lasted until January 1863; he was replaced by Joseph Hooker, who lasted until June, when he was replaced by George Meade, three days before the battle of Gettysburg. Burnside confused sacrifice with offense and ordered his men into suicidal charges at Fredericksburg. Hooker was replaced because of fears by the administration that he also, like McClellan, constantly overestimated the forces of the enemy and underestimated his own. The contrast between Hooker and Burnside bears comparison to that between Fabius and Paullus.

A unified, focused northern strategy first emerged in the West, where Union armies achieved their first solid victory with the capture of the Mississippi River. In June 1862 Ulysses S. Grant took charge of the Army of the West, in command over Major General William Tecumseh Sherman and Acting Rear Admiral David Dixon Porter. Halleck became general in chief. Grant organized a coordinated land and sea assault against Vicksburg, Mississippi, a city on the river about halfway between New Orleans and Memphis. By March 1863 Grant had isolated Vicksburg and begun a siege. He faced mammoth problems of logistics and coordination, including hundreds of miles of marches through swamps, but after several attempts, on July 4, 1863, Confederate general John Pemberton surrendered, along with twenty-nine thousand prisoners. With New Orleans, Baton Rouge, Vicksburg, and Memphis in Union hands, the South was solidly divided from the West, and unable to use the river.

In this campaign Grant and Sherman not only succeeded in taking control of the Mississippi but also took steps to solve two major problems that had hamstrung the North. First, the problem of divided strategy was under control, in the territory that Grant commanded. The focused actions of Grant, enacted boldly by coordinated generals, prepared his army to move forward as a single whole. Commanders in the field still had broad discretion, especially in comparison to armies today, but they would no longer act independently of an overall strategy. Second, to implement such a strategy the commanders were coming to terms with the huge problems of logistics that kept armies chained to supply lines. The Vicksburg campaign had placed tens of thousands of troops into enormous bogs, and starvation lurked as the consequence of failure. These problems were solved in an atmosphere of offensive action, in which a different understanding of the goal of the war, fostered by Lincoln, was adopted by the Union commanders.

By early 1864 Lincoln realized that he had found his new general in chief. Grant understood not only the nature of his task, to break the southern will to fight, but also the means to attain it—an integrated, goal-directed strategy. Grant's *Memoirs* demonstrate his thoughts about the improper fragmentation of the Union forces when he took supreme command, and how these had to be united into a single strategy directed offensively at the South:

> The Union forces were now divided into nineteen departments, though four of them in the west had been concentrated into a single military division. The army of the Potomac was a separate command and had no territorial limits. There were thus seventeen distinct commanders. Before this time these various armies had acted separately and independently of each other, giving the enemy an opportunity often of depleting one command, not pressed, to reinforce another more actively engaged. I determined to stop this.... Accordingly I arranged for a simultaneous movement all along the line. Sherman was to move from Chattanooga, Johnston's army and Atlanta being his objective points. Crook, commanding in West Virginia, was to move from the mouth of the Gauley River with a cavalry force and some artillery, the Virginia and Tennessee Railroad to be his objective.... Sigel was in command in the Valley of Virginia. He was to advance up the valley, covering the north from an invasion through that channel as well while advancing as by remaining near Harper's Ferry.... Butler was to advance by the James River, having Richmond and Petersburg as his objective.[12]

Grant unified these forces into a single army, composed of four main military elements, each moving toward a specific objective and acting in unison to achieve a common goal. The two broad strokes of this strategy would be Sherman's army in the west and the south, and the army of the Potomac under Grant in the east. In essence, Grant was to defeat, or at least pin down, Lee in the north, while Sherman swept into the south. On both sides of this vice would be an army, preventing an invasion of the north from the west and the east. The offense would commence as a singular motion. This required a complex series of preparations involving hundreds of rail cars and thousands of tons of supplies. Grant would have to combine two elements: logistics sufficient to keep his army supplied, and the ability to move quickly when the time was right. Events played out in a series of battles between Lee and Grant; the devastation of the Shenandoah Valley by Sheridan's cavalry; the destruction of the southern armies by George H. Thomas at Franklin, Tennessee; and Sherman's advance to Atlanta and Savannah and through the Carolinas.

Grant's strategy was an integrated plan that served a broader goal. The goal was the destruction of neither the enemy's armies nor its material resources—each of which is a means to an end—but the utter defeat of the enemy's will to fight by taking the war into the South itself. The strategy serves this end. Grant, in Sherman's later encomium of his commander, "penned Lee and his Army of northern Virginia . . . for ten long months on the pure defensive, to remain -almost passive observers of local events, while Grant's other armies were absolutely annihilating the southern Confederacy."[13]

General W. T. Sherman's Educational Program

The decisive defeat of the south was brought to bear by Sherman's destruction of the center of the Confederate war effort, southern plantation culture itself. Sherman developed his views of his enemy, and of war itself, in context with the complex strategic and logistical innovations needed to achieve Grant's broader goals.

For the first three years of the war, the dominant strategic approaches of both the Union and the Confederacy were closer to Fabius Cunctator than to Scipio Africanus; armies searched out and attacked enemy armies, and many battles involved head-on military assaults against entrenched positions. Although it is important not to overstate the number of such assaults—statistics show very few deaths from bayonets and sabers—an obscene number of young men met their ends in brutal face-to-face assaults against lines of other men.[14] These confrontations were consistent with the prevailing wisdom that the army of the enemy was always a general's primary goal—a principle accepted by Fabius in his debate with Scipio and appearing in many other contexts—but this was conducive neither to a general's need to maintain the strength of his army nor to a young man's desire to stay alive. Nor could it extinguish the motivations to continue the war that radiated from the southern homeland.

Sherman himself, moving south through Tennessee, was not successful when he followed the wisdom of attacking entrenched positions. Facing the usual grim casualties from assaults on defended positions—as at Chickamauga in 1863—he recognized the bloody ineffectiveness of these attacks. After outflanking the Confederate general Joseph E. Johnston—Sherman referred to Johnston's actions as a "Fabian Policy," and said that the South did him a "valuable service" by replacing him with the less-capable John Bell Hood—Sherman took control of Atlanta after a siege, on September 1, 1864.[15] Hood then turned his army north, into Tennessee, and Sherman had a decision to make: should he follow the enemy army—or should he strike toward the southern homeland?

In his preparations for movement into Georgia in early 1864, he had read southern newspapers, which had denounced Johnston for retreating without giving serious battle. "But his friends proclaimed it was all *strategic*," Sherman wrote, "that he was deliberately drawing us farther and farther into the meshes, farther and farther away from our base of supplies."[16] One southern newspaper editorial referred to Johnston's retreat as a "retrograde movement" that "surpasses the world in strategy."[17] But Sherman knew that Johnston was gaining strength as he moved and that he (Sherman) was losing strength as he followed him. To solve this problem, he refused to follow Hood's attempts to draw him north, and he set his sights back on his true objective, the South itself—another parallel to Scipio, who in 207 BC refused to follow a Carthaginian army out of Spain and toward Italy.

A singular focus on a goal allowed these generals to avoid digressions from their main objectives and to direct their forces toward their main goals.

As Hood took the Confederate army north into Tennessee, Sherman sent a force of more than forty thousand men north, to reinforce General Thomas, and then left Thomas to deal with Hood. He later wrote that he "felt more anxious for General Thomas's forces" than his own, but the defeat of Hood by Thomas at Nashville vindicated Sherman's decision to divide his forces.[18] While Hood was heading north, expecting Sherman to follow him in support of Thomas, and while Lee's army was pinned in Virginia, Sherman steeled himself and his forces for a strike against the center of southern culture.

Sherman's main problem was now logistical: he had only one rail line connecting Atlanta to Chattanooga, a jugular vein upon which his survival depended. Much of the seeming delay in Sherman's actions—the two and a half months in Atlanta from September into November—were spent on problems of logistics. Even Grant, as late as September 1864, did not see how Sherman could solve the problem of supplying his army; he continued to think in terms of taking and holding strategic cities, connected by rails, as the primary answer. In a letter to Sherman, he wrote:

> What you are to do with the forces at your command, I do not exactly see. The difficulties of supplying your army, except when they are constantly moving beyond where you are, I plainly see.... [given new forces sent from the north, to fortify Mobile and Savannah] You could then move as proposed in your telegram, so as to threaten Macon and Augusta equally. Whichever one should be abandoned by the enemy, you could take and open up a new base of supplies.[19]

Sherman responded:

> I would not hesitate to cross the state of Georgia with sixty-thousand men, hauling some stores, and depending upon the country for the balance. Where a million people find subsistence my army won't starve.... But the more I study the game, the more I am convinced that it would be wrong for us to penetrate further into Georgia with-

> out an objective beyond. I would not be productive of much good.... I should keep Hood employed and put my army in fine shape for a march on Augusta, Columbia and Charleston; and start as soon as Wilmington [SC] is sealed to commerce, and the city of Savannah is in our possession.[20]

At this point, Grant and Sherman still assumed that Grant's forces needed to take Wilmington and Savannah from the north in order for Sherman to march up from the south. Sherman overcame this problem of securing his northern flank by relying on Thomas to tie Hood down in the north, by leaving the enemy in the dark about his true objectives, and by foraging rather than maintaining supply lines. But to march through the South created a whole new set of military problems. The overarching problem was that of maintaining the strength of his force as he advanced through enemy territory. Clausewitz had observed seven ways that the strength of an attacking army may be depleted:[21]

1. If the object of the attack is to occupy the enemy's territory (Occupation normally begins only after the first decisive action, but the attack does not cease with this action.)
2. By the invading army's need to occupy the area in their rear so as to secure their lines of communication and exploit its resources
3. By losses incurred in action and through sickness
4. By the distance from the source of replacements
5. By sieges and the investment of fortresses
6. By a relaxation of effort
7. By the defection of allies

Sherman developed innovative answers to these problems. First off, he recognized that the traditional way of controlling enemy territory—taking a position by force, establishing supply lines, and then leaving a portion of one's forces to control the position—progressively weakened an army and led to destruction down the road. At each stage, forces are lost in assaults, in protecting supply lines, and in fortifications needed to guard the rear of a fighting force that is continually falling in strength. By the time the force reaches the goal, it may be too weak to attain it—and too far separated from supplies to regain strength. To get from Tennessee to the sea and then up through the Carolinas, Sherman needed to be as strong when he reached the Carolinas as he was in Atlanta.

To avoid combat losses, he abandoned plans to mount sieges on the way. He would avoid fighting entrenched enemy armies and thereby relieve himself of the need for many heavy guns; he left nearly two hundred artillery pieces behind and dragged sixty-five cannon. As well as lessening casualties, this would reduce the supply needs of animals—one heavy cannon requires two teams of six or eight horses, with all the hay and feed they need. Further, an army of sixty-two thousand men, traveling as a single unit, would take days to pass a single point, spreading its forces out and weakening any particular point. He divided his forces into four corps, which would speed him up, maximize his impact on the South, allow him to make feints to either side, and best enable him to gather resources as he went.

This still left monumental logistical problems. How could he protect his army's rear, and keep it supplied, without leaving such garrisons? He still needed thousands of tons of supplies for his men and animals. In his letter to Grant of April 24, 1864, written while Sherman was in Tennessee, he reported that, when he had arrived, 60 to 85 railroad cars per day were provisioning his army. This was enough to ensure starvation; his army needed 145 cars per day. In the two days before the letter, 327 railroad cars had arrived.[22] When he reached Atlanta, he created a howl by ordering that all railroad cars be used for military supplies; no civilian loads would be allowed to interfere with his goals.

But by this point Sherman recognized the possibility of dropping all reliance on supplies from the rear and living off the land. He wrote that he had obtained, and studied, the U.S. census tables of 1860 and a report of the controller of the state of Georgia, with the population and statistics of every county in Georgia.[23] Sherman knew from the outset that the land he was passing through was rich in foodstuffs; in his letter to Grant of April 10, 1864, acknowledging Grant's letter of

April 4, Sherman wrote: "If Banks can at the same time carry Mobile [Grant's plan called for Banks to take Mobile] and open up the Alabama River, he will in a measure solve the most difficult part of my problem, viz. 'provisions.' But in that I must venture. Georgia has a million of inhabitants. If they can live, we should not starve. If the enemy interrupt our communications, I will be absolved from all obligations to subsist on our own resources, and will feel perfectly justified taking whatever and wherever we can find."[24]

The diaries of Sergeant Rice Bull indicate orders to carry three day's rations at the outset.[25] Foraging squads—not raiding parties, intending to loot, but organized squads—would bring food into the army and could be used for friendly noncombatants. Thucydides had written that reliance on foraging as a means of supplying armies had made it impossible for the Greeks to defeat the Trojans quickly, but Sherman was marching through rich territory, not settling in for a siege of the enemy capital, and the constant motion would work to his benefit physically and psychologically.[26] To feed his army and to deprive the enemy of food was the same action, an integration of means and ends. He reasoned that if the enemy cut him off—by a scorched-earth policy—the reality of war would give him the sanction to take whatever he needed.

Most of all, he had to keep a firm eye on his strategic goal, to get from Atlanta to the sea, and then to move north to meet up with Grant. This goal would take precedence over secondary targets, such as taking a city like Augusta; it was the end point that must not be forgotten. All else was a means to an end.

These observations allowed him to adopt a strategy of "multiple objectives." His goal was the sea and then Lee's army in the north; these were certain. But he left open where he would arrive at the sea; this could depend on circumstances. Major Thomas Osborn wrote that no one—from the generals to line officers—knew Sherman's destination, but "no one had any misgivings about the success."[27] Scipio may have used the same approach when planning the assault on New Carthage, and later when moving by sea to the coast of Africa; the particular terminus could be left open. For both commanders, a firm goal and fluid actions to attain that goal were not incompatible; flexible *tactics* within an integrated *strategy* could achieve an inflexible *purpose*. In this way he could keep the enemy off guard and unsure of where he was going. "I can take so eccentric a course" Sherman wrote "that no general can guess my path."[28]

The seeming simplicity of this idea masks the myriad details, and years of experience, it subsumes; Sherman writes, for instance, how, as a lieutenant in 1844, he had been sent into Georgia "to assist Inspector-General Churchill to take testimony concerning certain losses of horses and accoutrements by the Georgia Volunteers during the Florida War." He had ridden in on horseback and taken note of the country, especially around Kenesaw, Allatoona, and the Etowah River. We have little reason to doubt that these details in his memory served him well twenty years later, when he altered his route to avoid the strongly defensible Allatoona pass en route to Atlanta.[29]

Sherman's plan was consistent with his orders from Grant: "You I propose to move against Johnston's army, to break it up, and to get into the interior of the enemy's country as far as you can, inflicting all the damage you can against their war resources."[30]

Instead of protecting rail lines as means of supply, and guarding them to prevent their use by the Confederates, he would tear them up as he went, thus depriving the South of a means to reinforce its own army, and cutting Richmond off from Louisiana, Alabama, Mississippi, and Florida.[31] He tried to keep this under control and to prevent rapacious looting and attacks on noncombatants. Although some of this occurred as Sherman's orders passed down through the ranks, evidence indicates that it occurred far less than in the past. Sherman's army was not a marauding gang of drunken barbarians. Their politeness toward southern women—Sherman's orders required written record of what they foraged—became legendary, and many a local revealed the location of hidden foodstuffs to Sherman's men.

Sherman had anchored his decision to move forcefully through Georgia on an alternative for southern leaders: to end the war by returning to the Union, or to face the consequences of their decision to use force against the Union. While in Atlanta, he had made quiet attempts to arrange a visit by Governor Joseph Brown and Vice-President Alexander Stephens, to discuss Georgia's withdrawal from the war.

Such a decision would have required Georgians to prevent southern armies from moving into Georgia. Sherman wrote to Lincoln that he would take his army across Georgia by roads, and pay for the supplies he needed, if the visits were fruitful. Lincoln was attentive and supportive to the proposal—but Stephens and Brown did not come. Their failure to respond to Sherman is one of the lost opportunities that southern leaders had to spare their own people from the consequences of secession. It was after this attempt failed that he wrote to Grant, on October 9, of his intention to "make Georgia howl!"[32]

Sherman's commitment to follow this through was made possible by his identification of the nature of his enemy, and his understanding that the operations of the various armies, integrated with the political aims of the North, constituted a unified whole—a seamless integration of means and ends. In a letter to Grant, he showed his own awareness of the relationship between his actions, the southern motivations for war, and the political situation in the North:

> I propose to act in such a manner against the material resources of the South as utterly to Negate Davis' boasted threat and promises of protection. If we can march a well-appointed army right through his territory, it will be a demonstration to the World, foreign and domestic, that we have a power which Davis cannot resist. This may not be war, but rather statesmanship, nevertheless, it is overwhelming to my mind that there are thousands of people abroad and in the south who will reason thus—"If the north can march an army right through the south, it is proof positive that the north can prevail in this contest," leaving open only its willingness to use that power. Now Mr. Lincoln's election which is assured, coupled with the conclusion just reached makes a complete logical whole.[33]

Such a demonstration was the culmination of a distinct change in the character of the war. As John Esten Cooke, a Confederate officer, observed, "Once, under McClellan, they seemed only bent on fighting big battles, and making a treaty of peace. Now they seem determined to drive us to the last ditch, and *into* it, the mother earth to be shoveled over us. Virginia is no longer a battlefield, but a living, shuddering body, upon which is to be inflicted the *immedicabile vulnus* of all-destroying war. So be it; she counted the cost, and is not yet at the last ditch."[34]

Sherman's tactics—like those of the cavalry commander Philip Sheridan, who was set to operate in the Shenandoah Valley—would shock southern society to its roots by the sheer force of his demonstration. This was not an unattended consequence; it was central to Sherman's plan, and it centered on destroying property while avoiding the loss of life. An army burning its way through Georgia plantations is not a compassionate thought, but the creation of peace out of war was not a compassionate process. Sherman knew that the war could not be won as long as southern civilians thought that they were winning the war and were able to send men, arms, supplies, and psychological comfort to their army in the north.

From the outset both sides—but especially southern aristocrats—had suffered from serious misunderstandings about the nature and course of the conflict. The diary of the southern lady Floride Clemson shows how southerners back home could evade the nature of the Union threat. While in Beltsville, Maryland, before moving south, Miss Clemson wrote that "Grant is not even in the papers with his grand 'Onto Richmond.['] I suppose he is stuck in a swamp down there. Gold is about 275 per cent & everything in proportion. We have constant company here, & are having a very pleasant time. Indeed I sometimes think I am too happy, I have so much to be thankful for."[35]

Such thoughts were not atypical. The diary of Lucy Rebecca Buck, a lady of the Shenandoah Valley, a daughter of a slave owner and a "stereotypical southern belle" in the words of her editor, reflects the same lack of awareness of the progress of the war. In an early entry, April 28, 1862, she tempered her misery with hope: "Was miserable—whiled the time away reading northern newspapers. It is exasperating to read their 'canards' and their poor attempts at wit at the expense of the Confederacy and the Secessionists. I wonder if they suppose that we are crushed and discouraged because of a few reverses in the tide of fortune. If they *do* I hope they'll ere long have optical demonstration of the fallacy of their opinions. Father heard that there had been a complete and decisive victory by our army at Yorktown. Oh, for truth!"[36]

Oh for truth indeed! There was no battle at Yorktown, and the southern armies retreated from the area in the next weeks. A year later—May 30, 1863—she was writing of a Confederate victory at Vicksburg. In fact, the town fell to the North on July 4, and in her entry of July 10, she wrote "which, by the way, we didn't believe."[37] Miss Buck refused to accept that her cause was lost. She was sublimely correct, however, in one evaluation: a clear, optical, demonstration of the truth was in order, for both sides. Only after the success of the North in the Shenandoah Valley does despair lead her to cease writing during the darkest hours—something manifested in many southern diaries—and to end her writing for eight years after the South surrendered. Her hopes for a southern victory were silenced.

These diaries reveal a spirit in the South that was a foundation of the rebellion. As a student of history who had lived in the South—the founding superintendent of the school that became Louisiana State University—Sherman recognized that many southern leaders were inciting war in the North without facing the consequences on their own territory. They were mouthing the abstraction "war" and exhorting the honor attached to it, but not experiencing its concrete reality. They were like the citizens of ancient Carthage, who supported the war in Italy until they saw Scipio's ships. It is instructive to look at a map of the South and to observe how little fighting occurred in South Carolina and Georgia. Aristocratic southerners in those areas were detached from reality, and Sherman aimed to bring it to them.

The anti-Union secessionists, Sherman also realized, were in the minority; in many areas, the majority of southerners had not favored secession. The issue in the South was not merely one of whites versus blacks, or even chattel-holding whites versus their slaves. The issue was a southern ideology, dominated by elite leaders who claimed to stand for the values of the entirety of the South—something of which Sherman had experienced and written about.[38] The Civil War has been called a "rich man's war, poor man's fight," in which large southern slaveholders, many of whom were serious instigators of the war, were given exemptions from the draft (the "twenty-slave law"), and plantations grew cotton while their boys went hungry marching through enemy territory with feet bleeding from the lack of shoes.[39]

One southern soldier, High Private Sam Watkins, wrote that, from the day that Tennessee seceded, "almost every person was eager for the war, and we were all afraid it would be over and we not in the fight. . . . But we soon found out that the glory of war was at home among the ladies and not upon the field of blood and carnage and death, where our comrades were mutilated and torn by shot and shell. And to see the cheek blanched to hear the fervent prayer, aye, I might say the agony of mind were very different indeed from the patriotic times at home."[40]

Some Union troops rebuked southern women for pressuring young men off to distant killing fields, "where men are being killed by the thousands, while you stay at home and sing 'The Bonnie Blue Flag'; but you set up a howl when you see Yankees down here getting your chickens. Many of your young men have told us they are tired of war and would quit, but you women would shame them and drive them back."[41]

Others expressed clearly their disdain for the aristocracy that had brought the nation into war. In February 1865 Major Thomas Ward Osborn wrote of the "fine library" in the home of the writer William Gilmore Simms—which Osborn would not object to burning, because "his influence has been very great in carrying on the war."[42]

Major George Nichols wrote: "Tonight we are encamped upon the place of one of South Carolina's most high-blooded chivalry—one of those persons who believed himself to have been brought into the world to rule over his fellow creatures, a sort of Grand Pasha and all that sort of thing . . . [but now the negroes] are making brooms of his pet shrubs, with which they clear the grounds on front of the tents."[43]

Sherman saw vividly the divisions between those pushing the rebellion and the majority who were not so motivated: "They allege they cannot abide us. I know that is the feeling of some, but the masses can. I have associated with rebels & have seen our troops do it under flags of truce, and during lulls in war, but I do admit that Some of them are so embittered that all would be benefited by an eternal separation. They cannot kill us all, but we may them. They must be killed or sent away ... I would like to see the abandoned plantations pass into new hands, even that of negros, rather than to speculators with Contract negros whom they treat as slaves."[44]

Sherman was following a policy, and pursuing a goal, developed after Grant witnessed the determination of the southern armies at Shiloh. Grant, in his memoirs, records his conclusion that the rebellion would not soon collapse. "I gave up," he wrote, "all idea of saving the Union except by complete conquest."[45]

To rip out the source of the rebellion, Sherman set out on what was, in effect, an educational mission. His actions served to connect the abstraction "war" to its concrete referent in reality: immediate, personal destruction. No longer would "war" float in the minds of southerners as an elixir, calling up notions of social superiority, bereft of its real meaning. The smell of smoke would haunt southern civilians as it had haunted the people of Sparta and Carthage—the smell of failure caused by their own willingness to wage war on others. War now meant loss, poverty, shame, and death. Now, knowing its nature, they could reject it.

The meaning of the burning of Atlanta was a demonstration, through the destruction of property, of the very meaning of the war, and the consequences of continuing the rebellion. Sherman's optical demonstration united force and resolve in a way that left no doubt of the outcome, should southerners not accept the Union. Sherman burned property in order to collapse the will to fight and save lives. He wondered, later in life, why he was so harshly blamed for destroying property while other generals killed their men by the thousands. This is a matter of values: who should be blamed, the general who establishes an effective peace at the price of material destruction, or the one who drags a war out for years under piles of corpses? As military historian Victor Davis Hanson has observed, if we count the bodies, we may change our conclusions about who in history has been the true protector of life and of peace.[46]

In a letter to Union General Thomas, Sherman had demonstrated his own awareness of this issue: "I propose to demonstrate the vulnerability of the south, and make its inhabitants feel that war and individual ruin are synonymous terms."[47] Sherman intended to make the vulnerability of the South explicit, to northerners, southerners, and foreigners alike, in the kind of "optical demonstration" that Lucy Buck desired even if not to her liking: the personal destruction of those who had brought war to America. To say that "war and individual ruin are synonymous terms" is to establish a principle of causality in human action, and to refuse to exempt those who made the decision to fight from the consequences of that decision. This demands more than a detached intellectual understanding; Sherman wanted them to *feel* this awful truth. As Sherman put it himself, in a letter to General Halleck, "This movement is not purely military or strategic, but it will illustrate the vulnerability of the South. They don't know what war means, but when the rich planters of the Oconee and Savannah see their fences and corn and hogs and sheep vanish before their eyes they will have something more than a mean opinion of the 'Yanks.'"[48]

In some cases, Sherman realized, southerners would be unwilling to give up their hatred and their desire for war with the North under any conditions:

> The young Bloods of the South, sons of Planters, Lawyers about town, good billiard-players and sportsmen, men who never did work, or never will. War suits them, and the rascals are brave, fine riders, bold to brashness, and dangerous subjects in every sense.... They must all be killed, or employed by us before we can hope for Peace.... In accepting war it should be pure & simple as applied to the Belligerents. I would Keep it so, till all traces of the war are effaced; till those who appealed to it are sick and tired of it, and come to the emblem of our Nation and Sue for Peace. I would not coax them, or even meet them half-way, but make them so sick of war that Generations would pass away before they would again appeal to it.[49]

Sherman did not hide these thoughts behind facades of fancy language, diplomatic doublespeak, or appeals to "prudence" designed to convey something other than what he meant. His words suggest that he thought the reasons for the war, and the terms whereby southerners could avoid destruction, needed to be made explicit. He urged that the following letter (excerpted here) be read to the people of Huntsville, to explain his views and to allow them to prepare for his arrival. He again offered an alternative to the people of the South—to return to the Union, or face the consequences—this time directed squarely at southern civilians:

> I know the slave-owners, finding themselves in possession of a species of property in opposition to the growing sentiments of the whole civilized world, conceived their property to be in danger and foolishly appealed to war, and by skillfull political handling they involved with themselves the whole South on this result of error and prejudice. I believe that some of the Rich and slave-holding are prejudiced to an extent that nothing but death & ruin will ever extinguish, but I hope that as the poorer and industrial classes of the South realize their relative weakness, and their dependence upon the fruits of the earth & good will of their fellow men, they will not only discover the error of their ways & repent of their hasty action, but bless those who have persistently maintained a Constitutional Government strong enough to sustain itself, protect its citizens, and promise peaceful homes to millions yet unborn. . . .
>
> To those who submit to the Rightful Laws & authority of their State & National Government promise all gentleness and forbearance, but to the petulant and persistent secessionist, why death or banishment is a mercy, and the quicker he or she is disposed of the better.[50]

Two famous incidents in Georgia reveal Sherman's attitude toward aristocrats who occupied the highest levels of southern society, and dedicated themselves to maintaining the slave culture. Howell Cobb was a major general in the Confederate army, a former secretary of the treasury under President Buchanan, and a southern legislator who had incited war for years. Upon Sherman's approach, this self-styled defender of the South did not fight; he ran for his life, abandoning his plantation and the townspeople to the devils from the North. When Sherman learned that this "patriot" had run, he took what food he needed, then ordered that the rest be distributed to those left behind. And then, with the simple instructions to "spare nothing," he ordered the place burned to the ground. "That night huge bonfires consumed the fence-rails, kept our soldiers warm, and the teamsters and men, as well as the slaves, carried off an immense quantity of corn."[51] For that southern leader, war and individual ruin were now synonymous. The cowardice and failure that his life represented would be ineradicably etched onto the minds of everyone looking on; *his* claims to virtue were hollow, *his* cause was a sham, and now *his* support was gone. The locals crowded around Sherman and abandoned their former masters.

On November 23 Sherman came to the town of Milledgeville, the seat of the Georgia legislature, and the second of four state capitals he would occupy. Not a shot was fired; the mayor surrendered the city to Captain William Duncan, and the two men shared a glass of wine; there was no looting of the city.[52] In the preceding weeks, the Confederate legislators and military officers had exhorted the people to defend Georgia to the last man, leaving a scorched earth before the Union army. Prisoners were released with promises of freedom if they fought; teen-aged cadets from the military school were armed.[53] Pending his arrival, the Georgia legislature exhorted citizens to "die freemen rather than live slaves." Newspapers had been full of exhortations to destroy the resources Sherman would need:

> Corinth, Mississippi, November 18, 1864
>
> *To the People of Georgia:*
>
> Arise for the defense of your native soil! Rally around your patriotic governor and gallant soldiers! Obstruct and destroy all the roads in Sherman's front, flank and rear, and his army will soon starve in your midst. Be confident. Be resolute. Trust in an overruling Providence, and success will soon crown your efforts. I hasten you to join in the defense of your homes and your firesides.
>
> [Confederate General] P. G. T. Beauregard[54]

Sherman found that the "brave and patriotic" governor had run for his life, pausing only to strip the "Governor's Mansion"—Sherman showed his disdain by placing its name in quotations—of carpets, curtains, and furniture, while leaving behind public records, arms, and ammunition.[55] Such courage! This was the man who had urged southerners to fight to the death in a scorched-earth policy against the tyranny of the North, but who refused to confront Sherman personally.

This point has often been lost on students of history: it was the southern leaders who ordered a policy of systematically burning the South, in order to starve Sherman's army. The *Albany Patriot*, a southern paper, told its readers that "it behoves [*sic*] us to contribute our whole energies to the same end."[56] Another editorial enjoined that if Georgians do not "come down upon Sherman like wolves and sheepfold, they are unworthy of themselves. Fight him in the front, fight him on the flank, fight him in the rear. Remove everything valuable from his path and throw every obstruction in his way."[57] The consequences of destroying their own lands and attacking Sherman's juggernaut would have been awful for civilians. But this is what the southern leaders were demanding—while the leaders themselves ran off.

At Milledgeville, a small group of defenders retreated over a bridge that Sherman easily secured, and he occupied the mansion. People from the area surrounded him with appeals for help, which Sherman answered by supplying food—withheld by the former occupants—from the farms. Sherman's army systematically destroyed the arsenal along with the rail depot and every public building that could be used for hostile purposes, sparing private property that could not be used on behalf of the Confederacy. Sherman exempted, for instance, several mills and thousands of bales of cotton, now assured that they would not be used for the rebel cause. Among the population, any residual desire to resist the North evaporated. Word of his march went ahead of his army and prepared the ground for his bloodless entry into Savannah.

When Sherman entered the town of Sandersville, he saw rebel cavalry burning supplies in a field. He ordered buildings in the area burned, and as the smoke wafted through town he told the citizens that if they made any attempt to burn supplies on their route, he would execute his orders "for the general devastation." The burning stopped. Sherman later observed that, "with this exception, and one or two minor cases near Savannah, the people did not destroy food, for they saw clearly that it would be ruin to themselves."[58] In his study of Union policy toward civilians, historian Mark Grimsley understood that "the extent to which houses and towns were burned . . . turns out to be much exaggerated."[59] The exaggeration, of course, is part of the psychological image of the "Attila from the West" that disarmed his opponents and made it possible to avoid the carnage of sieges and battles. But there were few if any murders on the march, and only a handful of rapes—all the sources agree on this. It was the destruction of *property* that brought shame to the South, and its shame demanded that the bringer of shame be a figure larger than life, cruel and rapacious.

Upon reaching Savannah, he demanded the surrender of the garrison under the southern commander General William Hardee: "Should you entertain the proposition, I am prepared to grant liberal terms to the inhabitants and garrison; but should I be forced to resort to assault, or the slower and surer process of starvation, then I shall feel justified in resorting to the harshest measures, and shall make little effort to restrain my army—burning to avenge the national wrong which they attach to Savannah and other large cities which have been so prominent in dragging our country into civil war."[60]

Hardee responded with a refusal, also taunting Sherman with the claim that *he* (Hardee) had to this point operated according to the rules of civilized warfare—presumably by fortifying himself in a city full of civilians. Yet, on December 21, the Confederates ran northward through routes that Sherman had not yet secured; Hardee's garrison left behind heavy guns, stores, and cotton—and a population with no desire to fight. The majority of Georgians had never been solidly in favor of secession—in many places the secessionists were in the distinct minority and had imposed their will by force—and Sherman was their liberator from southern elites with whom they had never had common cause.[61] Many Georgians had wanted him to make South Carolina, the hotbed of the rebellion, pay for this war. Sherman allowed the population of Savannah—some twenty thousand—

to stay in their homes, and his stay at times may have had less the temper of a military occupation than of a celebration.

In six weeks, Sherman cut through Georgia, destroyed the rail lines supplying the Confederate armies, ruined the plantations and motivations of those who thought they had the right to own human beings like stockyard animals, and took the earthen Fort McAllister—which had withstood two years of sieges—in fifteen minutes. He captured Savannah, its arms and its goods, without firing a shot.[62] Sherman sent a telegraph to Lincoln on December 22, 1864: "I beg to present you, as a Christmas gift, the city of Savannah, with 150 heavy guns and plenty of ammunition, and also about 25,000 bales of cotton."[63] The southern garrison at Fort Sumter, South Carolina, had held off northern sieges for nearly four years. Thirty-five hundred tons of artillery ordnance had reduced its walls to rubble but had not brought its defenders to surrender. When the garrison heard that Sherman was on the way, the soldiers abandoned the fort without a fight. The North erupted in celebration, for all it had known for almost two months was southern press reports of Sherman's starvation amid a burning Georgia countryside.[64] This may sound like hagiography, but it remains factually accurate.

In a letter to General Halleck on Christmas Eve, 1864, Sherman had expressed his understanding of why his march into Georgia cut through the propaganda of southern newspapers: "I attach more importance to these deep incisions into the enemy's country, because this war differs from European wars in this particular: we are fighting not only hostile armies, but a hostile people, and must make old and young, rich and poor, feel the hard hand of war, as well as their organized armies. I know that this recent movement of mine through Georgia has had a wonderful effect in this respect. Thousands who had been deceived by their lying newspapers to believe that we were being whipped all the time now realize the truth, and have no appetite for a repetition of the same experience."[65]

Casualty figures are the bottom line: as recorded in Sherman's *Memoirs*, for the Georgia campaign, of the 62,204 men that left Atlanta, Assistant Adjutant General L. M. Dayton reported 103 killed, 278 missing, 428 wounded; total 809. This was in an age when thousands of young men fell in frontal assaults, and when hospitals were filled with the screams of the dying and piles of amputated limbs. There had been some hundred thousand casualties at the battles of Gettysburg, Second Bull Run, and Antietam. Grant may have lost more than seven thousand men at Cold Harbor alone. In his preliminary report, sent to Secretary of War Stanton after the capture of Fort McAllister, Sherman reported that he had not lost a single wagon on the trip and that "our teams are in far better condition than when we started."[66] This was an army that got healthier—and more highly motivated—as it marched through hostile territory. In the last year of the war, the safest place in the South was in Sherman's army. Thousands of locals voted with their feet as they tried to follow him.

"I do not like to boast," he said, "but I believe this army has a confidence in itself that makes it almost invincible."[67]

Before the capture of Savannah, Grant had ordered Sherman to transfer his infantry via ship to Virginia. He prepared to comply. But after Savannah, Grant changed his orders: "Without waiting further directions, then, you may make your preparations to start on your northern expedition without delay.... I will leave out all suggestions about the route you should take."[68] Sherman cut north into the Carolinas, again deceiving his enemy by leaving his objectives ambiguous. He bypassed the places they expected him to attack, especially Charleston and Augusta—the former he called a "dead cock in the pit."[69] Like the Sicilian legions that Scipio had enlisted for the attack on Africa, Sherman's troops were burning to make their real enemy suffer for loosing this war on their homeland. Public buildings in Columbia were burned, and to this day many in South Carolina consider Sherman the greatest devil who ever lived.[70]

As he set off for the Carolinas—with 60,079 men, out of the 62,204 he had begun with—Sherman also observed the morale of the civilian population of the South. He wondered why his enemy had allowed him to take so easy possession of alluvial land in southern South Carolina. The reason, he surmised, was the "terrible energy" that his army had displayed in the earlier campaigns, an energy that his army was now prepared to direct at South Carolina, "the cause of all our troubles."

Sherman's men were veterans of such western campaigns as Tennessee and Vicksburg; they were not "northerners" of the same stripe that dominated Grant's Army of the Potomac. (To Sherman, a "Yankee" was a New Englander.) Like Aurelian's veterans of the Gothic campaigns, his men were not used to retreating—a skill that McClellan had taught the Army of the Potomac—they were rather used to advancing. Their reputations marched ahead of their boots, preparing the locals for their arrival, leaving the decision to fight up to the enemy, but leaving no doubt of its outcome should they decide poorly. Sherman wrote of many friends in Charleston that he would have liked to protect, but he placed his men and mission first. "I would not restrain the army lest its vigor and energy should be impaired."[71] As to the people of the South, although some in South Carolina might have been prepared to fight with greater courage than the Georgians, "It was to me manifest that the soldiers and the people of the south entertained an undue fear of our Western men, and, like children, they had invented such ghostlike stories of our prowess in Georgia, that they were scared by their own inventions. Still, this was a power, and I intended to utilize it."[72]

As he again marched, his enemy was psychologically groomed for his approach. His arrival was preceded by a myth of invincibility, a weapon more powerful than bullets. "I observe that the enemy has some respect for my name, for they gave up Pocotaglio [in South Carolina] without a fight when they heard that the attacking force belonged to my army. I will try and keep up that feeling, which is real power."[73] This power spread in two directions: to his enemy, who gave up before the very rumors of his approach, and to his own men, who knew they were winning and would soon be home. In less than three months, America's national nightmare was over. This power is the source of the two narratives about Sherman that emerged after the war: that of a destroyer and a violator of southern society, and that of soldier who efficiently met the goals of the Union.[74]

As to the charge that he was a warmonger who gloried in fighting, one fact defines his campaign: *his army did not have to fight a single major battle after the siege of Atlanta*. Bentonville, North Carolina, would be the worst—with casualties of just over four thousand for both sides combined—but even then Sherman called off an assault in order to complete the campaign. He was too busy winning the war to waste his men fighting battles.

Sherman's Moral Perspective

Sherman's writings about the purposes and ends of war, and his attitude toward noncombatants, are powerful statements about the moral aspects of his military policy, including the moral purpose of the conflict as a whole. As far as he was concerned, the southern states—by attacking Union forts—had rejected constitutional deliberation and demanded that political issues be resolved by force of arms. Sherman used the methods of their choosing—guns and bayonets—and then demanded that southern secessionists renounce them and return to the authority of the Constitution. To that end, he directed his forces against the property of the southern plantation class. In what follows, it is important to remember that no one has accused Sherman of actually using his forces to kill civilians—only to destroy the property and resources vital to the southern war effort. It is also vital to remember the actual effects of his victory: an end to slavery in the South, and the restoration of the constitutional authority of the Union that made emancipation possible.

Sherman spent much time explaining his attitude toward his enemies, and to the war as a whole, in letters and in his memoirs, most remarkably in a startling correspondence with Confederate general John Bell Hood, who defended Atlanta. The mayor of Atlanta and the city council had chimed in, offering their reasons why Sherman should consider their needs over the requirements of his army. Sherman's answers cut to the heart of the moral issues the general faced in fighting a war—and demonstrate Sherman's engagement with the broader aspects of his mission.

The issue came to a head with Sherman's need to fortify Atlanta and to remove all opposition in his rear before moving on to the sea. He left wounded soldiers in Atlanta but did not intend to distribute forces behind him to protect supply lines as he marched. He needed his entire army with him when he hit the coast. He also needed complete use of the single railroad supply line back to Chattanooga, to gather resources for the march. Thus, he made plans to move the population of Atlanta—a total of some ninety-six hundred persons—out of the city, either north or south as they wished. He would supply transportation to drop-off points under a two-day truce.

He presented this plan to the Union War Department, through General Halleck, the liaison between the Union army and the government, on September 4, 1864:

> I propose to remove all the inhabitants of Atlanta, sending those committed to our cause to the Rear & the Rebel families to the front. I will allow no trade, manufactories or any citizens there at all, so that we will have the entire use of the railroad back and also such corn & forage as may be reached by our troops.
>
> If the people raise a howl against my barbarity and cruelty I will answer that war is war, and not popularity seeking. If they want peace, then they and their relatives must stop the war.[75]

In his later report to General Halleck, he gave five reasons for the movement: to use houses for military purposes; to shorten lines of defense; to hold Atlanta, which was fortified and stubbornly defended, and "we have a right to it"; to avoid starvation; and to prevent inhabitants from corresponding with enemies. "These are my reasons; and, if satisfactory to the Government of the United States, it makes no difference whether it please General Hood and *his* people or not."[76] Following Sherman's proposal of a truce to effect the removal of the inhabitants, General Hood replied, accepting Sherman's plan as a matter of necessity. He then lectured Sherman on the morality of his actions:

> And now, sir, permit me to say that the unprecedented measure you propose transcends, in studied and ingenious cruelty, all acts ever brought to my attention in the dark history of war.
>
> In the name of God and humanity, I protest, believing that you will find that you are expelling from their homes and firesides the wives and children of a brave people.[77]

Sherman responded, detailing the numerous cases in which Confederate armies, on the approach to Georgia, had also removed civilians from battle zones. He then demolished Hood's mini-sermon with satire: "I say it is a kindness to those families of Atlanta to remove them now at once from the scenes that women and children should not be exposed to, and the 'brave people' should scorn to commit their wives and children to the rude barbarians who thus as you say violate the Laws of War, as illustrated in the pages of its dark History."[78]

Hood had also castigated Sherman for trying to "justify your shelling Atlanta without notice." Sherman, according to Hood, was bound to alert his enemy of his intentions to attack. Hood made this claim after Sherman had spent months chasing the Confederate army from Vicksburg to Chattanooga to Atlanta—as if Sherman's arrival were some kind of sneak attack—and after Hood had retreated into civilian areas behind military defenses. Sherman would hear none of it. He had ordered his men to cannon immediately any buildings from which they are fired upon, without hailing them first. This could result in the deaths of civilians—a situation that Hood had created, by retreating into the city. Tactically, Sherman wrote that Hood "defended Atlanta at a line so close to the town that every cannon-shot, and many musket-shots from our Line of investment that overshot their mark went into the habitations of women and children." This was a tactical stretch on Sherman's part, but it remains true that Hood had control of the town and had placed civilians too close to his defensive lines. Sherman also expressed a deeper, moral point: that Hood controlled the areas where civilians were at risk, and Hood was to blame for leaving civilians in those areas:

> I was not bound by the laws of war to give notice of the shelling of Atlanta, [quoting Hood's words] "a fortified town, with magazines, arsenals, foundries, and public stores"; You were bound to take notice. See the books.
>
> This is the conclusion of our correspondence, which I did not begin, and terminate with satisfaction.[79]

"See the books" was a personal kick at Hood, who had graduated near the bottom of his class at West Point; Sherman was far better versed in military theory and history. Hood could not deny that he knew of Sherman's approach. More broadly, a general is responsible for securing the areas under his control. If Hood wanted to protect civilians, let him move his troops away from those civilians, or let him move the civilians—as Sherman now proposed. Sherman recognized that true noncombatants should not be left to starve—and his actions during the march have generally born this out. But true noncombatants were not responsible for the war and had no desire to continue it. They were not enemy sympathizers, who would pass information or supplies to his foes.

The plan for the evacuation of Atlanta did not send civilians to gas chambers or prison camps, but rather to areas away from the fighting, in an orderly movement under a truce, with provisions for their survival. But the story of the evacuation has grown beyond reality; Sherman ordered some 3,000 people moved, and only 1,644 actually went. It does remain curious that while southern officers and newspapers were representing Sherman as a rapacious killer of all that moves, southern leaders entrusted their own families to his care.[80] Generals William J. Hardee and others directed their families in Savannah to seek the care of Sherman, who did what he could—while Hood had wanted the civilians in Atlanta to remain under Sherman's control and not to be moved to Hood's own areas. Nevertheless, the idea that disarmed the South—the cruelness of Sherman, the "Attila of the North," who burned all in his path—has become his reputation into the present day. When Sherman approached Savannah, and the Confederates ran north, leaving him the city without a fight, perhaps it was the rumors about Atlanta that spared the population of Savannah the cruelty of a siege.[81]

As far as Sherman was concerned, the fundamental point remained that all blame for the war lay with the South. Southern states had rejected debate in favor of attack, had whipped up their own people into a fury against the Union, and had fired on the Union flag. For this they deserved all the curses a people can pour out: "You who in the midst of peace and prosperity have plunged a nation into War, dark and cruel war ... talk thus to the marines but not to me who have seen these things and who will this day make as much sacrifice for the peace and honor of the South, as the best born southerner among you. If we be enemies let us be men, and fight it out as we propose to do, and not deal in such hypocritical appeals to God and humanity."[82]

The criticisms that Sherman has faced over the years—of cruelty to civilians, shelling and burning civilian targets, and his refusal to enlist black troops as soldiers—began during his siege of Atlanta. The mayor, James M. Calhoun, and two members of the city council complained against his evacuation order in a letter: "Many poor women are in an advanced state of pregnancy ... some say 'I have such a one sick at my house; who will wait on them while I am gone?' ... what has this *helpless* people done, that they should be driven from their homes, to wander strangers and outcasts, and exiles, and to subsist on charity?"

But no one was in a position to tell Sherman that war is cruel. He responded famously to their complaints, in a chain of reasoning that warrants rereading:

> I have read it carefully, and give full credit to your statements of the distress that will be occasioned, and yet shall not revoke my orders, because they were not designed to meet the humanities of the case, but to prepare for the future struggles in which millions of good people outside of Atlanta have a deep interest. We must have peace, not only in Atlanta, but in all America. To secure this, we must stop the war that now desolates our once happy and favored country. To stop war, we must defeat the rebel armies which are arrayed against the laws and Constitution that all must respect and obey. To defeat those armies, we must prepare the way to reach them in their recesses, provided with the arms and instruments which enable us to accomplish our purpose.... The use of Atlanta for warlike purposes is inconsistent with its character as a home for families. There will be no manufactures, commerce, or agriculture here, for the maintenance of families, and sooner or later want will compel the inhabitants to go. Why not go now, when all the arrangements are complete for the transfer, in-

> stead of waiting till the plunging shot of competing armies will renew the scenes of the past month? . . .
>
> You cannot qualify war in harsher terms than I will. War is cruelty, and you cannot refine it; and those who brought war into our country deserve all the curses and maledictions a people can pour out. I know I had no hand in making this war, and I know that I will make more sacrifice to-day than any of you to secure peace. But you cannot have peace and a division too. If the United States submits to a division now, it will not stop, but will go on until we reap the fate of Mexico, which is eternal war . . . Once admit the Union, once more acknowledge the authority of the national Government, and, instead of devoting your houses and streets and roads to the dread uses of war, I and this army become at once your protectors and supporters, shielding you from danger, let it come from what quarter it may.[83]

These familiar passages cut to the heart of Sherman's attitude toward an enemy that had started a war that his command now charged him to end: h*e accepted no guilt for a war that was not of his making.* This sense of rightness allowed him to prosecute the war to its conclusion quickly, with his force directed at the true source of southern power rather than merely at military positions dependent upon that power. He saved the lives of his own soldiers, refusing to sacrifice them to arbitrary rules, dictated by an incompetent enemy, who impudently claimed a higher moral position by virtue of his defeat, invoked "God and humanity" only after his own bullets failed, and hid behind crying, unarmed, pregnant women.

In his analysis based on Augustinian just-war theory, Stout applies the categorical rules of proportionality and against attacking civilians, in a way that separates moral evaluation of the war from its ends. Stout concludes that Sherman's "cause was just and indeed holy, but the conduct profane." Hood did use "human shields," writes Stout, which put civilians into danger—but this does not, Stout maintains, excuse Sherman's goal-directed actions to follow and destroy the Confederate army. But how then could Sherman have acted morally, according to these rules? He would have had to be willing to sacrifice his own men —and thus his "just" and "holy" cause—in order to save those who were supporting Hood, and whom Hood was hiding behind. Sherman would have had to sever means from ends, and to elevate Hood's terms of engagement over his own goals, in order to follow rules exploited by his enemy. How many soldiers might have died, had Sherman accepted this? What might the consequences have been, had he not taken Atlanta quickly, and had Lincoln had lost the election?

In practice, these rules would have empowered the proslavery side, and would have condemned its opponents. The injunction against attacking Confederate positions because civilians were in the area—and then against removing civilians under a flag of truce—was used as a moral weapon by General Hood. Hood was attempting to create a moral barrier between himself and his stronger, more astute, foe. But the strength of this barrier would depend upon Sherman's acceptance of the terms established by his enemy. Had Sherman accepted this, the side with the just cause in the war would have voluntarily accepted a weaker position before a slave-owning enemy. Sherman would have had to bend his efforts toward protecting the lives and property of those supporting the enemies of the Union—and to use his own men as means to that end, rather than to the end of victory. One might give pause and ask why this should be considered moral—especially given that Sherman had proposed moving civilians out of the war-torn city under a truce.[84] Hood's attempt failed because Sherman refused to accept guilt for the risk posed to civilians that was required to end the war.

This was not total destruction, and not gratuitous attacks on unarmed noncombatants, but rather war that reached into every recess of southern society, that exempted no pocket of resources from its grip, and that allowed no enemy to go on thinking that victory was possible. If this evacuation order hurt those privileged with aristocratic birth and wealth in the South, Sherman's attitude was, so be it; a painless war is a contradiction in terms. As General Lee had said, looking upon the carnage at Fredericksburg: "It is well that war is so terrible. We should grow too fond of it." In Sherman's mind, it was vital that the people of Atlanta experience directly the nature of the war they had been advocating; this was the only way to make them reject it. Once again, the key to this mission was not murder, but the destruction of property, in order to make clear that the national bloodletting would soon end with the reunification of the nation.

Sherman also understood that his job had a scope wider than the particular task he was facing. His job was not to ensure the comforts of the particular people of Atlanta, but rather to make possible "peace, not only in Atlanta, but in all America." Despite his personal racist attitudes, he showed the same wide-ranging understanding of the question of slaves. Accused of cruelty against blacks by political opportunists in the North, he observed: "The idea that such men should have been permitted to hang around Mr. Lincoln, to torture his life by means of suspicions of the officers who were toiling with the single purpose to bring the war to a successful end, and thereby liberate *all* slaves, is a fair illustration of the influences that poison a political capital."[85]

It is true that Sherman did not enlist African Americans into his army. He also opposed emancipation, until Lincoln announced his Emancipation Proclamation in September 1862.[86] Nor did he allow many of the families of slaves to accompany him—but this was war, and he was protecting his men and refusing to take on more than he could feed. He asked one old man to explain to his people that they could not accompany the army, because this would slow up the work of freeing them.[87] The famous debacle at Ebeneezer Creek—where Union general Jefferson C. Davis pulled up a bridge and stranded slaves, some of whom drowned—demonstrates the savagery of the cavalry under Confederate general Joseph Wheeler. It was Wheeler's cavalry—not Sherman's army—that was the danger to the noncombatants. Sherman stands accused of failing to rescue those civilians from southerners, a strange charge given the nature and purpose of his march, and the nature of his enemy.

Many former slaves saw Sherman as their savior and said so. Questioned by Secretary of War Stanton, without Sherman present, the Reverend Garrison Frazier of the Baptist Church said, "We looked upon General Sherman, prior to his arrival, as a man, in the providence of God, specially set apart to accomplish this work, and we unanimously felt inexpressible gratitude to him, looking upon him as a man who should be honored for the faithful performance of his duty."[88]

Sherman's stated aim was to create a peace on proper—that is, constitutional—terms. No long-term solution to the rebellion was possible until the authority of the Constitution had been restored over all the states. To fail to have established those terms in order to follow rules divorced from context would not have been pacific. It is not a kindness to accept a peace today that turns tomorrow's children into soldiers—a point that Thomas Paine knew well, as he had once looked beyond his own revolutionary trials: "Let us," he wrote two generations earlier, "not leave the next generation to be cutting throats."[89]

To end the motivations for war, there would have to be more than mere defeat of southern forces; the southern commanders would have to admit the defeat in a formal surrender, and the confederacy would have to cease to exist. There is a difference between the southern defeat and its surrender. The *defeat* was a fact: their ability to prevail was destroyed past the point of rebuilding. The *surrender* was the public decision of the political and military leadership, and ultimately of the civilians, to accept the fact of defeat. The surrender was their recognition of the reality of the situation, an admission of impotence, the collapse of all hope for victory, and the permanent renunciation of aggression.

The Union and Confederate officers all recognized a certain code and an underlying civility between enemies that includes an honorable surrender. An honorable surrender is an honest admission that one has been bested in war by a superior force, that one renounces further fighting and trusts the enemy commander to do the right thing. For a surrender to be objectively proper, the fact of defeat must be true, so that the recognition of military defeat is grounded in the facts.

The northern victory placed the Union back on the path of expanding the scope of liberty—this time without the contradiction of slavery in the Constitution. Before their surrender, southern leaders would not even consider a negotiated end to the war that did not include recognition of the south's "right" to keep slaves. As late as February 1865, Confederate president Jefferson Davis refused to meet with President Lincoln without prior recognition of confederate independance. Southern leaders, and those who supported them, had to be defeated and forced into complete, sincere surrender before they would abandon this position. After the surrender at Appomattox, Lee told Grant that the emancipation of slaves was not likely to be a problem, because few in the South would want to restore slavery if they could.[90]

Sherman saw such surrender as having broad political and social effects. Following his movement northward through the Carolinas, and after the surrender of Lee's army at Appomattox on April 9, 1865, Sherman negotiated the surrender of Confederate general Joseph E. Johnston and an army larger than that of Lee. The surrender that Sherman obtained included civil aspects that went beyond his authority as a general and were rejected by the Union government. He was ordered to recommence hostilities and to demand an unconditional surrender. Sherman followed orders, and Johnston surrendered unconditionally.

Sherman wrote that he had wanted to prevent the dispersal of southern troops into guerrilla bands under unsavory commanders. A humorous picture of independent citizen squads in the South, running into the woods to do battle with the enemy, was written by Samuel Clemens.[91] But there is a serious truth behind the humor. Had the southern officers not surrendered their armies, and bound their men to give up the fight, the result could have been as Sherman feared: "I now apprehend that the Rebel Armies will disperse, and instead of dealing with six or seven states, we will have to deal with numberless bands of desperados headed by men such as Mosby, Forrest, Red Jackson, & others who Know not, and care not for danger or its consequences."[92]

Sherman saw the military as better capable of demanding and receiving the enemy's surrender than the politicians; "I perceive the politicians are determined to drive the confederates into guerrilla bands, a thing more to be feared than organized war," he wrote to his wife in April 28, 1865.[93] Once the surrender was received, he thought that retributions should stop and that the Union should accept the South back in. "Men who are now so fierce and who would have the Army of the Potomac violate my truce, and attack our enemy disheartened discomfited & surrendered will sooner or later find foes to face of a different metal."[94] The demonstration of overwhelming force against the South did not end every manifestation of southern aristocratic racist culture—in the 1950s some politicians were still singing the "states' rights" tune as an excuse to deny education to African Americans —but it did emasculate the threat of a terrorist war that existed up to the end of the fighting.

Why did the war drag on for so long? This is a question hotly debated, with a complex set of answers. The South was clearly inferior in depth of resources, but it nevertheless stalemated the war almost to the point of a negotiated settlement with the North—a distinct possibility had Lincoln not been reelected. I suggest two basic reasons, relating to means and ends, that contributed to the delay in ending the rebellion. In terms of means, the North took years to find a general who could formulate and command an integrated strategy against the South and had the will to pursue it. At Vicksburg, Grant brought a unified order to the western armies, focused them on a single goal, grappled with the problems of logistics, and separated the South from western resources. Under Lincoln's authority, he then extended this integration into an overall strategic plan.

But it is in their understanding of the ends of the war that Grant and Sherman turned the course of the conflict as a whole. McClellan had the organizational ability and the resources but was not able to command with the fearsome contempt for a foe that such a position demands. Only a clear demonstration of overwhelming force against the economic and ideological center of southern aristocratic culture extinguished the fire of the war as a whole. It was Sherman's march that directed the moral, political, and military aspects of the struggle against the South itself. This chapter may sound overly encomiastic, but it remains true that today, more than 140 years later, we live under the peace established by Lincoln, Grant, and Sherman—a peace that has made all further progress possible. The affirmation of liberty for all has, since then, been slow, arduous, and painful, but it has not unleashed the insatiate dogs of war. Few times in history has a victory led to such long-term success.

Evaluating Sherman

Any evaluation of William Tecumseh Sherman must begin by recognizing the thousands who would have died had Sherman accepted the "awful arithmetic" of the pitched battle, and who would have remained enslaved under a negotiated separation of the South. Any evaluation of Sherman must project the sanguinary consequences of calls by southern leaders for civilians to stand with hoes against hardened soldiers with cannons as they burned their own fields—and what would have happened had he left any hope in the minds of southerners that an uprising could work. Where would the civil rights movement of later generations have been, had Sherman left the system of slavery untouched in a desire not to harm the social fabric of the South? Sherman did not destroy the South; he saved it—not its slaveholding culture and its facades and plantation porches, but its capacity to achieve its proper, uniquely American, future—with a shock that "allow[ed] the Union and peace to settle once more over your old homes."[95] Those motivated to know how peace can be established and preserved ought to step back from prior assumptions and consider that Sherman's acceptance of Johnston's surrender marks the conclusive end of the only war between the American states since the founding of the Union.

Sherman's burning of Atlanta—an action against property and not its population—not only kept the North on track to win the war but was a demonstration that exposed the failure of the southern leadership. This demonstration put the civilians into a peculiar position: they could obey their leaders' calls to resist and thus die—or they could admit the failure of those leaders and give up. Like the eastern campaign of Aurelian, however, Sherman's success depended on rumors of overwhelming strength and his will to use it. Such rumors have since magnified, over time, into a view of his campaign as a rampage of rape, killing, and terror—this despite general agreement that there were few if any gratuitous murders, and perhaps six rapes, during the march. Sherman's reputation today—like that of Aurelian—is an extension of the myth created in his own day. But it is important to separate fact from fiction, and to recognize that overwhelming force, used in a morally legitimate cause and directed against property, served to free the slaves, end hostilities, and return the United States to constitutional authority.

The rumors that began during the march have formed the backbone of Sherman's later reputation.[96] After the war, some claimed the march was simple and unchallenging—derision similar to what Scipio faced from Fabius after smashing Carthaginian control of Spain.[97] Others claimed that he combined brilliant tactics with terrorism against innocent civilians; in these views, his burning of Atlanta had little to do with winning the war, but much to do with the rise of the Ku Klux Klan, and with feelings of animosity that have lasted into the present age. Others have bought into a southern narrative "in which cultural values and traditions were sacrificed for mere results"—as if the emancipation of four million chattel slaves was a "mere result" that was not worth the traditions of those who wished to buy and sell them, and as if moral judgments should not be concerned with such results.[98] Victor Davis Hanson has attacked this view head-on, noting that the "usual ledger" of condemnations—rapist, terrorist, burner of cities—reached their heights in claims that Sherman was actually wrong to free the slaves because of the negative economic effects on the slave owners.[99] This charge continues, even though President Lincoln had expressed his willingness to provide for "compensated emancipation," which he used in the District of Columbia.[100]

But should Sherman have allowed slave owners to keep their plantations, while they defended the ownership of human beings in a war against the Union? Former president Bill Clinton, in a speech to Georgetown University, answered yes, illustrating views that have become mainstream today. Sherman, Clinton said, practiced a "mild form of terrorism" that had "nothing whatever to do with winning the Civil War" but was a "story told for a hundred years." During his youth, "people were making excuses for unconscionable behavior [continued segregation] by talking about what Sherman had done a hundred years earlier." But who should get the blame for the segregation in the south a century later: the southerners who created such segregation by legalizing the racial superiority expressed by Alexander Stephens, or Sherman, who demonstrated the failure of the southern social system? Who is the real terrorist: those who terrify an entire class of persons into submission and chase down escapees with hunting dogs, or those who terrify the slave owners by burning their mansions? Where would the civil rights movement be today, had Lincoln, Grant, Sherman, and others not ended the institution of slavery?[101] It is those who continue to excuse unconscionable behavior who deserve public censure, not those who ended it.

Such views end up condoning a double moral inversion. First, they elevate property over lives, not only taking the focus off the slave owners but also legitimating the "awful arithmetic" of Grant and Lee in order to protect farmhouses. But, second, such views imply that southern society was legitimate to pursue its course and that Sherman was wrong to challenge that society. Terrorism against such people does not work, said President Clinton—ignoring the fact that Sherman was successful and that Clinton grew up in a free American state because Grant, Sherman and their comrades had kept the Union together and prevented the spread of slavery.

The central issue in the Civil War remains an alternative: the promotion and expansion of liberty, or the preservation of statism and slavery. Sherman won a victory that established the proper moral and political context needed to smash the view that had hamstrung the ancients: that some people are inferior to others, by nature, necessity, fortune, or ethnic identity, and that their proper position is permanent enforced servitude. Sherman's focus on the southern social support for the war allowed the majority in the South who were not in favor of secession to be counted—an act of empowerment that would bear full fruit in later generations. The later affirmation of individual rights for all in the civil rights movement would have been inconceivable had the South continued to sanction the ownership of human beings. For Sherman, the moral truly was to the physical as three to one.

Chapter 6

"The Balm for a Guilty Conscience"

British Appeasement and the Prelude to World War II, AD 1919–1939

The Unstable Armistice of World War I

On September 1, 1939, twenty years and nine months after the armistice that ended World War I, millions of Germans obeyed their leader's call for another grand slaughter of national aggrandizement. "I demand not that my generals understand my orders, but that they obey them!" screamed Hitler, who got what he wanted.[1] The attack on Poland was the climax of twenty years of diplomacy, economic transfers, and treaties in which most European leaders had worked fervently to avoid violence. They fell prostrate before Hitler's "Lightning War." The deepest reasons why so many Germans joined the armies of the Nazis, hailed their leader, and followed his orders cannot be found in economic stagnation, political dissatisfaction, or bad feelings about the last war. Such factors affected many nations that did not attack. In essence, the Nazis tapped into a certain climate of ideas, which empowered them to harness feelings of national superiority to a willingness to sacrifice for the State, the Race, and the Leader. The power of these ideas mobilized the mass support that the Nazis enjoyed among "Hitler's Willing Executioners."[2]

But another force, outside of Germany, also pushed the world toward blitzkrieg and Auschwitz. This force was also a set of ideas, held in the minds of Germany's opponents, which restrained them from confronting Hitler when they could. In the mid-1930s, British politicians in particular were constrained not by the incapacity to act but by a lack of will. Certain moral ideals—which had risen to the cultural forefront after the horrendous experience of World War I—prevented British leaders from recognizing the fundamental contradiction between their own policy goals and those pursued by Germany. Many British leaders accepted the basic legitimacy of German claims to territory and national resurgence and were thus unwilling to oppose those claims in a principled way. The central question of this chapter is why—Why did they adopt this attitude toward Germany?

The appeasement of Germany in the late 1930s has become a synonym for weakness, focused on a single man, Prime Minister Neville Chamberlain, who claimed "peace in our time" by handing over Czechoslovakia to Hitler in September 1938. But this demonization of the prime minister can be misleading. Chamberlain's appeasement, far from being a new plan by a weak man to deal with an emergency, was the culmination of a long policy of negotiation and compromise that had much support in Britain. British leaders accepted the legitimacy of Hitler's basic territorial goals and his "right" to rearm, and then argued over the best means to give to Hitler that which he could not have taken by force. To understand how and why European leaders became de facto allies of Germany as it rekindled the fires of war, we must unpack the moral ideas by which Hitler disarmed his opponents.

The road to World War II began on November 11, 1918, with the armistice that ended World War I. German leaders had sued for peace in October 1918, but no Allied soldier was in Germany to force an end to the war. The war ended with a cease-fire agreement between the Allied powers (primarily the United States, England, and France) and what was left of the Axis powers (most important here, Germany). Three major empires had been destroyed (Hapsburg Austria-Hungary, czarist Russia, and Ottoman Turkey), vast areas of France and Belgium were ruined, European colonial possessions across the globe were in disarray, Russia was in civil war, and millions of civilians were destitute. But German troops marched home from foreign soil "unvanquished from the field," to an unconquered Germany.

The American general John Pershing and others had wanted to make the Allied victory unimpeachable. "We should," Pershing wrote, "continue the offensive [into Germany] until we compel unconditional surrender." The German commander, General Erich von Ludendorff, told his government that "an immediate armistice" was needed "to avoid catastrophe"; the previous day, he had said that if he were in command of the Allies, he would "attack even harder."[3] But the Allies were constrained by U.S. president Woodrow Wilson to accept a negotiated "peace without victors," an agreement between equals rather than a victory over the vanquished. This was the first time that the American government explicitly renounced victory as the object of a war. When Allied leaders balked, Wilson threatened to begin separate negotiations with the Germans, and the Allies acquiesced. The price paid by the Germans was the elimination of the Prussian monarchy and the kaiser, the loss of certain German territories, deep reductions in military capacities, and years of economic reparations. The price paid by the French and Belgians was continued dependency on Britain and the United States for protection against an unrepentant Germany. Marshal Ferdinand Foch, general of the Allied armies in World War I, expressed the views of many when he said presciently, "This is not peace. It is an armistice for twenty years."[4]

The armistice was never intended to solve the crushing problems of the postwar world. Europe was to be rebuilt through a series of treaties, especially the Treaty of Versailles with Germany, written and adopted at the Paris Peace Conference of 1919.[5] Despite the terms of the armistice, it was a conference between victors; the German delegates arrived, briefcases in hand, but were shuffled off ignominiously and excluded. The Germans were handed an imposed settlement—a *diktat*—that they never accepted as legitimate. The treaty mandated a new deliberative body, the League of Nations, to which individual nations were supposed to subordinate their major foreign policy decisions. In the two decades that followed, international discussions were dominated by the consequences of the war, which threatened to explode into even more gruesome carnage—a prospect that technology made unthinkable.

To preserve the peace, the new world was to be built on a new set of ideals. In October 1918, the Germans had agreed to negotiate in terms set out in President Wilson's so-called Fourteen Points, his definitive foreign policy statement of ideals. Wilson said that what America wanted

> is nothing peculiar to ourselves. It is that the world be made fit and safe to live in; and particularly that it be made safe for every peace-loving nation which, like our own, wishes to live its own life, determine its own institutions, be assured of justice and fair dealing by the other peoples of the world as against force and selfish aggression ... the only possible program, as we see it, is this:
>
> I. Open covenants of peace, openly arrived at, after which there shall be no private international understandings of any kind but diplomacy shall proceed always frankly and in the public view....
>
> XIV. A general association of nations must be formed under specific covenants for the purpose of affording mutual guarantees of political independence and territorial integrity to great and small states alike....
>
> An evident principle runs through the whole program I have outlined. It is the principle of justice to all peoples and nationalities, and their right to live on equal terms of liberty and safety with one another, whether they be strong or weak.[6]

Wilson's ideas broke with the international system of alliances created in the previous century. In his view, the pursuit of "national interests" through separate treaties, and the reliance on a balance of power as a deterrent to war, had failed. Wilson's ideals were based on a worldview in which nations were equal parts of a global whole, collective decision making was centered in a League of Nations, and peace was based on cooperation between equals rather than enforcement by victors. Powerful nations were to be constrained by the need for a consensus before action. All nations would be equally free to determine their national destinies. No nations—even those that started the last war—should be held below others. With these new ideals—*equality, national self-determination,* and *collective security*—World War I would truly be the war to end all wars, and the result could be perpetual peace.

Wilson's stress on relationships between equals within an international whole was an outgrowth of deeper philosophical ideas that Wilson—a former president of Princeton University with a Ph.D. in political science—had received from a long line of intellectual predecessors. Similar ideas had passed from the seventeenth-century philosopher Thomas Hobbes to the late eighteenth-century German philosopher Immanuel Kant and beyond. In 1651, following the English Civil Wars, Hobbes had written in his *Leviathan* that men in a "state of nature"—without laws, standards of justice, or a political authority to enforce justice—had no way to deal with one another except force. Man's natural condition was a war of each against all, unless and until a great sovereign government imposes its will over all. The sovereign defines and enforces justice over people who must obey, because the only alternative is anarchy and perpetual war.

Kant's essay *Perpetual Peace*, written in 1795, developed these ideas in an international context. In Kant's view, the sovereign republics of the world exist in just such a lawless state of nature, a "state of War" without standards of justice and institutions to resolve their disputes. Each is unrestrained, and each preys upon all, because there is no single international authority. To achieve perpetual peace, nations must subordinate their foreign policy actions to such an authority. This "league of peace" (*foedus pacificum*) would seek "to make an end of all wars forever": "For states in relation to each other, there cannot be any reasonable way out of the lawless condition which entails only war except that they, like individual men, should give up their lawless (savage) freedom, adjust themselves to the constraints of public law, and thus establish a continuously growing state consisting of various nations [*civitas gentium*], which will ultimately include all the nations of the world."[7]

The League of Nations put the Kantian ideal into practice, in a new world order that eschewed the competitive nationalism of the previous century. Wilson's ideals—equality, national self-determination, and collective security—were the League's foundation, were derivable from Kant's ideas, and were highly influential. These ideals promised an effective alternative to the unpredictable actions of unrestrained sovereign nations. They became the moral compass that shaped the decisions of British leaders.

"Equality" as an international ideal demanded "equality of rights" between nations, regardless of their actions in the Great War. To hold one nation below others—to demand, for instance, limits to arms production, military strength, or economic development for some nations but not for all—was to violate a central principle of the new international politics and to invite further conflict. The "continuously growing [international] state" was not to be dominated by any one nation, but managed through deliberations between all nations, each equally sovereign.

"National self-determination" alleged each nation's right to create its own institutions in the form it desires and to chart its own future toward its own ends. Such self-determination was *national* in character and thus sanctioned the "right" of a nation to establish an authoritarian or even dictatorial government. In Kant's theory, this "right" was not unlimited; Kant assumed that every nation would have a republican government and that democracy was "necessarily a form of despotism."[8] But Kant's theory had little relationship to practice; many nations are not republican but rather despotic. In the real world, the notion of "national self-determination" could not be limited to republican governments. Democracy—the expression of a consensus through a public vote—became the litmus test for a nation's true self-determination. Wilson thus vowed to "make the world safe for democracy," which in practice included the sanctioned capacity to vote one's nation—and one's self—into tyranny.[9]

"Collective security" between nations was a practical requirement that followed from the other two. It demanded that sovereign nations delegate major foreign policy decisions to the League of Nations. Nations would resolve rising tensions that had formerly led to war peacefully, by "adjust[ing] themselves to the constraints of public law," a consensus of the League to be achieved through international deliberation. The League would have no enforcement powers—but each nation would subordinate its political sovereignty to the consensus of the League and agree not to act without its sanction. The League promised to replace the failed balance of power between alliances with a new era of cooperation.

Following the armistice, however, others—the French in particular—recognized a plain truth that ran counter to Wilson's idealism: that Germany had launched the war against France through Belgium and that the Germans had not formally repudiated this aggression. Of course, there were two sides to this conflict, and the French had their own past to contend with, but German guns had wreaked the greatest devastation in France, as they had in 1871, and it was in France that the next war would likely be fought. As French president Raymond Poincaré said at the opening of the Paris Peace Conference on January 18, 1919: "In the hope of conquering, first, the hegemony of Europe and next the mastery of the world, the Central Empires [Germany and Austria/Hungary], bound together by a secret plot, found the most abominable pretexts for trying to crush Serbia and force their way to the East. At the same time they disowned the most solemn undertakings in order to crush Belgium and force their way into the heart of France. These are the two unforgettable outrages which opened the way to aggression."[10]

This is rhetoric for sure—but also a factually accurate assessment of how the armies had actually moved. A plenary session of the Paris Peace Conference of January 25, 1919, established the Commission on the Responsibility of the Authors of the War and on Enforcement of Penalties. "On the question of the responsibility of the authors of the war," the commission stated, in its May 6 report, "This responsibility rests first on Germany and Austria, secondly on Turkey and Bulgaria." Its conclusions left no room for anyone to hide:

1. The war was premeditated by the Central Powers together with their Allies, Turkey and Bulgaria, and was the result of acts deliberately committed in order to make it unavoidable.
2. Germany, in agreement with Austria-Hungary, deliberately worked to defeat all the many conciliatory proposals made by the Entente Powers and their repeated efforts to avoid war.[11]

To drive the point home further, the famous "War Guilt" clause of the Treaty of Versailles, article 231, placed economic responsibility for the war squarely on the Germans: "The Allied and Associated

Governments affirm, and Germany accepts, the responsibility of Germany and her allies for causing all the loss and damage to which the Allied and Associated Governments and their nationals have been subjected as a consequence of the war imposed upon them by aggression of Germany and her allies."[12]

In other words, the treaty explicitly recognized guilty parties and held them responsible for the war. For the French and Belgians, this was a practical matter of the greatest importance; history left no doubt that an armed Germany on their border was a matter of life and death. Old men still remembered the German invasion of 1871. "Mark well what I'm telling you," said French leader Georges Clemenceau, "in six months, in a year, five years, ten years, when they like, as they like, the Boches will again invade us."[13] Many French concluded that Germany's capacity to fight had to be permanently and systematically weakened—a task that could also satisfy a desire for vengeance reaching back to 1871. The Treaty of Versailles took large areas of land from Germany, forbade it to rearm, and required it to pay reparations for years into the future. Much of this was driven by the idea that Germany would inevitably return to the pursuit of empire by conquest.[14]

But the treaty contradicted the nonpunitive terms of the armistice, which the Germans had accepted, and which they thought bound their enemies from acting as victors. This became a moral issue: there was a massive contradiction between the ideals promoted by Wilson and the practical need to restrain German power and hold Germany responsible for the damage it had caused. Given this contradiction between theory and practice, the treaty lacked the moral authority to preserve the peace and was often seen as an unjust attempt to punish Germany by denying the Germans the "equality of rights" that Wilson's ideals had promoted for all men.

To the Germans, the war-guilt clause was a moral outrage—even if its goal was to establish financial and not moral responsibility—and they vowed to discredit it. By January 1919 the German Foreign Ministry had established a special War-Guilt Section to manage the propaganda needed to absolve Germany of responsibility for the war.[15] The Germans had already begun to look like victims. Many did not know how the war had started, that their position on the front had been hopeless, or that the terms of the treaty were the responses of people who had been savaged by years of bombardment.[16] For many Germans the armistice had come out of the blue; their leaders had assured them for years that victory was inevitable—until November, 1918, when these same leaders told them to accept a settlement dictated by their enemies.

It is vital to remember that no Allied soldier was in Germany to force the surrender in 1918 and that German militarism was not discredited inside Germany. Fritz Ebert, who became the German chancellor two days before the armistice, famously greeted returning German troops in December as "unvanquished from the field." Ludendorff—who had asked for the armistice in October 1918—condemned the new German government in February 1919, charging that "the political leadership disarmed the unconquered army and delivered over Germany to the destructive will of the enemy.... Thus was perpetrated the crime against the German nation." The "crime" was to strip Germany of its authoritarian government and its capacity to wage war.

The result was widespread acceptance of the *Dolchstoss*—the "stab in the back"—by what Hitler would one day call the "November criminals."[17] In the years after World War I, intellectuals and politicians broke with reality and reconstructed their history: they posited Germany as a victim of a deep betrayal. By the 1920s, the *Dolchstoss* had become a foundation myth: we are in a reduced position, said the mythmakers, because of a criminal conspiracy against our nation. The Paris Peace Conference became an extension of the plot; many Germans came to see themselves as denied "equality of rights" by the Versailles Treaty. The day before the election of July 1932, Chancellor von Papen said, "It is unbearable that fourteen years after the end of the war there is no equality of rights [to military arms] for us." The next day the Nazis received 13.7 million votes. Six years later, in 1938, Hitler spoke of how "for 15 long years we were a spineless and hopeless object of international oppression" until he gave Germany back its greatness.[18]

But just as important, for twenty years many British leaders also accepted the German claims to redress of grievances. At stake were certain territories, largely German-speaking, that had been taken by the treaty; the economic reparations demanded of Germany; and Germany's capacity to rearm. Many British leaders were unable to oppose such claims, because they accepted the moral right of the Germans to demand them. "Equality of rights" and "national self-determination" became moral catchphrases that could not be resisted.

Looking ahead in time, when the World Disarmament Conference convened in 1932, a guiding principle was "the grant to Germany, and to other powers disarmed by Treaty, of equality of rights in a system which would provide security for all nations."[19] When Germany walked out of the conference—and the League—in October 1933, Hitler's stated reason was that the League did not accept Germany's "equality of rights" to military arms. This excuse undercut British opposition; Foreign Minister Sir John Simon told the House of Commons in February 1934 that "Germany's equality of rights [in armaments] could not be resisted."[20] Prime Minister Neville Chamberlain stated more than once that "the principle of self-determination" was what his policy was designed to achieve. "Negotiations [with Germany] could not be resumed except on the basis of considering ways and means to put the principle of self-determination into effect. If we would not accept this basis it means war. Let there be no mistake about that."[21]

During the 1920s and early 1930s, the contradiction between the ideals of the League and the intent to keep Germany weak by enforcing the Versailles Treaty was resolved by systematically obstructing and undercutting the treaty. A concentrated, relentless series of attacks against the treaty ensued, on many levels and within a sea of conflicting rationales. One writer, in 1940, concluded that "in the history of propaganda nothing outrivals the success of the German effort against the Versailles Treaty."[22]

Repudiating the Versailles Treaty

The conflict between the ideals of the postwar world and the practical requirements of the Treaty of Versailles began immediately; the first battle concerned German economic reparations. The "war-guilt" clause of the treaty had mandated reparations but did not specify either the terms of payment or the amount. This highly complex issue dominated international discussions for more than a decade. The French wanted to recoup losses sustained in the war, but British distrust of the French, along with German protests, bolstered the conclusions of many in Britain that reparations were unjust and a threat to peace. Although reparations were in part a vindictive attempt to weaken Germany, many were physical goods required to restore property destroyed or looted by the German army. In addition to coal and timber, they included telegraph poles, livestock, and materials needed to rebuild the library of Louvain, Belgium, burned by German troops.[23]

The economist John Maynard Keynes, a young adviser at the Paris Peace Conference who resigned over the Versailles Treaty, placed the blame for Europe's problems squarely on the treaty, and on Britain, Italy, and France. In his highly influential book, *The Economic Consequences of the Peace* (1920), he wrote:

> For one who spent in Paris the greater part of the six months which succeeded the armistice an occasional visit to London was a strange experience. England still stands outside of Europe. Europe's voiceless tremors do not reach her. Europe is apart and England is not of her flesh and body. But Europe is solid with herself. France, Germany, Italy, Austria and Holland, Russia and Roumania and Poland, throb together, and their structure and civilization are essentially one. They flourished together, they have rocked together in a war, which we, in spite of our enormous contributions and sacrifices (like though in a less degree than America), economically stood outside, and they may fall together. In this lies the destructive significance of the Peace of Paris. If the European Civil War is to end with France and Italy abusing their momentary victorious power to destroy Germany and Austria-Hungary now prostrate, they invite their own destruction also, being so deeply and inextricably intertwined with their victims by hidden psychic and economic bonds...
>
> So far as possible, therefore, it was the policy of France to set the clock back and to undo what, since 1870, the progress of Germany had accomplished.[24]

Keynes's best seller argued that reparations would prevent German economic recovery and undermine the unity of Europe. But Keynes's description of Europe as a single tremulous body, his argument framed in *moral* terms, his focus on the *economic* condition of Europe, and his failure to emphasize the *political* sovereignty of the European nations allowed him to elevate economic factors over political causes and to see a *nationalistic* war, fueled by *ethnic* interests, as a *civil* war. The Versailles Treaty was sublimely mild next to the brutal Pan-German Empire that the Germans under Chancellor Bethmann-Hollweg had planned to impose on Europe, had they won the war—and was no worse than its earlier counterpart in 1871 France, the Treaty of Frankfurt, which took Alsace-Lorraine for Germany and left a German occupation army in France until the latter paid off indemnities.[25]

Keynes's book shifted blame for the postwar mess onto the Allies and fostered a sense of *meaculpism* ("self-guilt") in Britain. As one scholar put it, with Keynes's book, "Meaculpism was born. Doubt flourished; German guilt faded; British guilt spread."[26] As Keynes wrote in his famous attack on the treaty,

> The policy of reducing Germany to a position of servitude for a generation, of degrading the lives of millions of human beings, and of depriving a whole nation of happiness should be abhorrent and detestable,—abhorrent and detestable, even if it were possible, even if it enriched ourselves, even if it did not sow the decay of the whole civilized life of Europe. Some preach it in the name of Justice. In the great events of man's history, in the unwinding of the complex fates of nations Justice is not so simple. And if it were, nations are not authorized, by religion or by natural morals, to visit on the children of their enemies the misdoings of parents or of rulers.[27]

Keynes's "One-Europe" view—representative of the ideas that permeated European intellectual life—and his claim that reparations were an unconscionable burden on Germany motivated many British to correct the alleged inequities imposed upon Germany, most of all by restraining the French from enforcing the treaty. "Largely on the basis of Keynes's indictment, opinion in the 1920s and 1930s held France as responsible for the instability and unrest in Europe" and turned the peace into a "gambit in the traditional Franco-German rivalry."[28] While the Germans obstructed reparations payments to the point of destroying their own currency, the British backed a series of concessions to the Germans. British politicians became de facto allies of Germany against France. As one historian put it, "By becoming the leading advocate of appeasement, Britain could redress the balance of injustice [against Germany].... Appeasement was the balm for a guilty conscience."[29]

Thanks largely to historian Sally Marks, we now know that from the outset claims that Germany was brutalized by reparations were grossly inaccurate.[30] The public was intentionally deceived by leaders on both sides. In 1921 the London Schedule of Payments publicized the reparations at 132 billion marks. This told the public that harsh terms had been imposed, which satisfied those—especially the French—who sought retribution from Germany, while strengthening the sense of injustice among many Germans. Behind the scenes, however, complex bonds were established that lessened the actual burden on Germany.[31] Germany's real liability came to fifty billion marks, which was less than it had offered to pay and well within its means. Reparations were publicized at nearly three times more than the allies ever intended to collect—and over six times more than they did collect—which created a "myth of reparations" that had serious consequences for popular understanding of the peace.

An important component of the myth was the claim that reparations destroyed the German economy. From 1919 to 1923 Germany suffered the worst economic inflation in history. The German mark collapsed, to the point that it took billions to buy a loaf of bread. Prices rose so fast that German restaurants wanted to be paid before their patrons ate; the money would lose value during the meal. Many people came to think that reparations caused the inflation, which contributed heavily to German dissatisfaction, the rise of radical parties, and ultimately to Hitler. But is it true that reparations destroyed the German economy?

The facts are otherwise. The inflation in Germany that began in 1919 was not coextensive with reparations payments. The German cash payment of a billion marks, in the summer of 1921, was the last payment until 1924. The inflation raged most destructively at the time when no payments were made. Germany paid nothing during the hyperinflation of 1923, and when payments were highest, in the late 1920s, there was little inflation. All told, between 1919 and 1931 Germany paid about twenty billion marks in reparations, about half of which was in physical goods.

Even this figure fails to account for the cash flow of foreign investment, from the U.S. to Germany. The Germans transferred some 2 percent of their national income as reparations to the allies from 1919 to 1931, while direct foreign subsidies to Germany and defaults on foreign investments amounted to 5.3 percent of that total national income.[32] But the myth of ruinous reparations was strengthened because the foreign money went largely into private industry, whereas reparations payments were taken from the German people by taxation or the printing of money—another split between public proclamations and reality. All in all, there was a flow of cash into Germany. "In the end," wrote Sally Marks, "the victors paid the bills."[33]

The true cause of the inflation was the German government itself. After the war, with the value of the German mark about half of its 1914 worth, the government printed billions of marks in order to pay debts to finance the slaughter. The resulting price rises were caused by the massive increase in the supply of paper money. By the mid-1920s, the Germans had brought their printing presses under control, and with a new currency—the Rentenmark, replaced in 1924 with the Reichsmark—the inflation quickly subsided. By the late 1920s, the dependence of the German economy on American loans was total; and in 1928 Chancellor Gustav Stresemann warned the Germans that they were living on "borrowed money."[34] Far from being a victim of rapacious reparations, the Germans were hugely beholden to overseas aid. The American stock market collapsed in October 1929, and the German economy followed suit when the Americans began to call in loans. Recovery was underway by 1932—at which time the Nazis became the largest party in the national legislature, the Reichstag.

Why did the Germans go to such great lengths to obstruct the reparations payments? Why did they not just negotiate a lower figure, make the payments, and lessen tensions while rebuilding? The simple fact is that the reparations were never seen as an economic issue in Germany. They were considered to be a political insult against the German nation.[35] To obstruct the Treaty of Versailles and the payments became an assertion of German national will. Many German leaders were willing to see their own economy collapse in order to avoid the political and moral affront of making payments to France and others. They could then blame vengeful French miscreants for the economic consequences—and further absolve themselves of responsibility for the war.

In Britain the reparations were not seen primarily as an economic issue either—they were a *moral* issue. Given the disinformation, and absent an explanation for the inflation, many British accepted unearned blame for the German economic collapse. France was increasingly isolated in its attempts to hold Germany in check, while British and American leaders pushed for concessions to assist Germany in its recovery. A good example of this is the British reaction to the French occupation of the Ruhr River valley, an industrial area in Germany.

On January 9, 1923, the international Reparations Commission found Germany in default on material transfers to France. The French had real devastation on their soil, and to secure the resources they needed, they moved a small force into the valley, with French, Belgian, and Italian engineers, to produce the coal that constituted the reparations.[36] The Germans responded with a strike that lasted until September. The German inflation, which reached its climax in this year, was falsely blamed on the French occupation. Sympathy for the Germans grew; and even though British and American leaders knew that the Germans were "deliberately ruining" their own currency, they felt compelled to protect the German economy.[37] British leaders proposed a four-year moratorium on German payments, financed by American loans, to restore Germany. These actions effectively placed the British on the side of the Germans and in opposition to the French. A flood of voices amplified Keynes's view that France was attempting to "seize, even in part, what Germany was compelled to drop," and under intense international pressure, the French withdrew in August 1925.

Misunderstandings about the reparations have persisted to the present day, leading, for example, to the idea that World War II could have been prevented had reparations not created the raging inflation that prostrated Germany. But even if this were true, economic distress does not explain the war. If it did, poor Austria and Hungary, bereft of empire, would have attacked, and poor China would have invaded prosperous Japan. Despite the Great Depression and Europe's failure to pay war debts, America did not wage war against Europe in the 1930s, and there was no talk in England of a preemptive war against Germany. Italy, a victor in World War I, attacked Abyssinia and supported Hitler. Austria and Hungary, vanquished in war, were overrun by Hitler. World War II was started by the leaders of statist governments—institutions with unlimited military and police powers—that rose over populations who venerated such authority. Statism—any system in which the individual is subordinated to the unlimited power of the state—was the indispensable precondition of the carnage. Stated concisely, "Statism *needs* war; a free country does not."[38]

What made war inevitable if not opposed—in both Europe and the Pacific—were two militant ideologies that had deep roots and enjoyed popular support. For years before World War II, German leaders linked the existence of Germany with expansion into a greater Aryan nation. "Room; they must make room. The western and southern slavs—or we! ... Only by growth can a people save itself," said one pre-1914 voice, who already grasped the essentials of Nazi foreign policy. Kaiser Wilhelm II saw a "battle of Germans against the Russo-Gauls for their very existence." Another said, "We must become land-hungry, must acquire new regions for settlement."[39] Twenty years later, on November 5, 1937, such ideas translated into Hitler's private statement to his top staff, his "last will and testament" in which he made clear that expansion by force for *Lebensraum* (living room) was his intent and that Czechoslovakia had to die so that Germans did not live under inferior Slavs. Many generals were shocked—but they did not arrest or kill the Leader. They obeyed.[40]

As history would show, there was an irreconcilable conflict between the policy goals of the British and those of the Germans, because they governed under fundamentally different principles. Individual freedom—even if imperfectly protected—is not compatible with militant racial authoritarianism, and to treat them as equals allowed killers to determine the future of a continent. While the Germans of the 1930s were swept up in a rising tide of nationalism that aimed to return Germany to greatness, the British, plagued by self-doubt and guilt over Germany's economic plight, were morally and intellectually disarmed from taking the measures needed to prevent Germany from regaining territory and rearming. The final result was to constrain the defenders and unleash the aggressors.

"Collective Security" and the False Idol of Consensus

A highly compressed selection of a few events from 1919 to 1939 suggests that the ideals of national equality and national self-determination, accompanied by fear of an even greater war, helped guide British leaders through a chaotic mix of international problems. Many British politicians thought that a key to peace in Europe was to lower tensions by mediating and correcting injustices, of which the Versailles Treaty was the worst. But far from demanding moral rigidity, the ideals of equality and national self-determination required leaders to grant legitimacy to both sides of an issue, using the method of pragmatic negotiation. The appeasement that Hitler leveraged so masterfully was an application of this method.

British policymakers were concerned with an increasingly vulnerable global empire, within which European squabbles could look much less threatening than a crisis in Asia or India. Britain's trade and foreign investments had fallen drastically since World War I and during the 1920s. Britain was saddled with foreign colonies taken from Germany—a financial drain that could not be defended militarily. Britain had to choose its battles carefully; it was unable to project power across its empire and was economically unprepared for another long war. Further, many British did not see Germany as the major threat to their empire; Russia and Japan were arguably more worrisome. It was important, many thought, to avoid entanglements in European defensive pacts—a Britain so occupied could lead the Japanese to take advantage of British weakness in the Pacific, which could cause violence in the Middle East. One way to lower tensions and avoid trouble, many people believed, was to pressure France to make concessions to Germany.

The Ruhr River crisis in January of 1923 is again a case in point. The occupation might not have been necessary had the British and Americans supported French and Belgian demands that Germany comply with the treaty. But Britain abstained from the vote to declare Germany in default because, in the analysis of Sally Marks, to admit that Germany was in violation of the treaty would have required the British and Americans to do something, and they did not want to act.[41] French occupation took the blame for the German economic distress, and international pressure rose to reverse the injustice. Britain tried to walk a middle path. To support France against Germany would sanction an outrage that could lead to no good result, but to repudiate payments outright was politically impossible. So the British "denounced the occupation as immoral and illegal, but rendered it feasible by permitting France to mount it on British-controlled railways in the Rhineland."[42] Once again, public pronouncements contradicted actions.

In August 1923, Gustav Stresemann became chancellor of Germany, and with Germany's money destroyed, he lobbied to reorganize the reparations. British leaders knew that German leaders had destroyed their own currency to obstruct reparations, yet they evaded this fact and pushed for concessions. The result was the Dawes Plan of April 1924. A London conference in July and August 1924 arranged its details, including a moratorium on German payments, an effective end to all enforcement of the Versailles Treaty, and the removal of French forces from the Ruhr Valley within a year. The French were isolated and pressured to restrain their attempts to enforce the treaty, even while the French franc approached collapse.[43] French premier Poincaré was forced from office, and his replacement, Edouard Herriot, could not withstand the intense international pressures. While France retreated under deafening criticism, Germany received loans and foreign investments more than equal to the payments, many backed by American guarantees. Yet the shame of a *diktat* remained for the German people, for their finances were to be administered under a board appointed by their enemies in a war that was supposed to have ended five years earlier. By restraining the French, the British had become de facto allies and financiers of the Germans—who were again seen as victims.

The Versailles Treaty was under withering attack but had not yet been abrogated. In 1925 the treaty was formally modified by the Locarno Agreements, an ambitious attempt to bring order to the wreckage that the treaty had become.[44] These agreements—initiated by Germany, openly negotiated, and voluntarily accepted—were the most far-reaching steps taken since 1919. Advocates claimed that these agreements would bring Europe together, by setting up open international protocols that would supersede balances of power and individual obligations between nations. Although purporting to strengthen the ability of the allies to respond to threats, these agreements in fact subordinated the defense needs of nations such as France to an international consensus. This is a relevant passage, from Article 5:

> Where one of the powers referred to in Art. 8, without committing a violation of Art. 2 of the present treaty or a breach of Arts. 42 or 43 of the treaty of Versailles, refuses to submit a dispute to peaceful settlement or to comply with an arbitral or judicial decision, the other party shall bring the matter before the Council of the League of Nations, and the Council shall propose what steps shall be taken; the High Contracting Parties shall comply with these proposals.[45]

Such agreements rely on reactive, consensual processes of attaining permissions in order to avoid war. They require a nation to act forthrightly against aggression, but only after the League of Nations sanctioned the action. This subordinated the decisions of sovereign nations to a form of international democracy—which is the literal meaning of Kant's demand that nations submit to the "constraints of public law." That a sovereign nation *should* enforce the terms of a treaty when another nation threatens, alone and preemptively if necessary, was explicitly rejected.

Such treaties can be effective only if the signatories share fundamental moral principles that affirm the value of freedom and individual rights and that reject the use of force to promote national aggrandizement. But governments that accept those principles did not plunge Europe into war. Absent such principles, the practical result handcuffed nations seeking security to a long process of international deliberation, while potential aggressors could gain time to rearm. Any projection of power by a defensive nation would likely be a violation of the treaty, whereas a program of rearmament by a rogue nation could be construed as a legitimate pursuit of "equality" and "national self-determination"—which is exactly how Hitler was viewed, ten years later.

The Locarno Agreements were greeted with elation by public figures, especially Stresemann, a sponsor.[46] But German leaders were still pursuing the return of Germany as a world power, and they used the agreements to achieve that end in the West. Meanwhile, they passionately resisted such agreements in the East, isolating smaller nations between the USSR and Germany in order to settle longstanding claims. The biggest was East Prussia, a German-speaking area that was divided from Germany by the "Polish corridor," a strip of land that allowed Poland access to the sea through the port of Danzig (today, Gdansk). No German leader accepted this separation of ethnic Germans from Germany, or a minority German position under a Slavic government.[47] The German-Russian agreement of 1922, signed at Rapallo, April 16, 1922, absolved both parties of mutual economic reparations—further evidence that Germany was not economically oppressed—and the Treaty of Berlin of April 1926 freed Germany from Soviet interference in a European war.[48] Such agreements suggest that Hitler's 1939 pact with Stalin was no change in German policy. Germany and Russia could ally if it suited their needs—which should have challenged the idea that fascism and communism are necessarily enemies.[49]

By the late 1920s, Germany was deeply dependent on American aid, which deepened the consequences of the stock market crash in 1929. In 1930 the allies of Versailles adopted a new attempt to rectify matters with Germany, the so-called Young Plan, named after American financier Owen D. Young. It was another concession to Germany that again cut reparations payments, to some twenty-six billion marks, to be paid over fifty-eight and a half years. Economic turmoil led to a moratorium in fiscal year 1931–32; the turmoil was increased by a flight of capital after Hitler's strong showing in the German elections.[50] At a conference in Lausanne in 1932, it was clear to everyone that a resumption of payments by Germany was impossible.

The Young Plan did achieve one concrete result: the French were pressured into a pullout from the Rhineland, the contested area between France and Germany that both the Versailles Treaty and the Locarno Agreements had forbid Germany from militarizing. To create an essential buffer on their eastern border, the French began work in earnest on the Maginot Line. A German occupation of the Rhineland could be prevented now only by an enforced decision of the League of Nations or by the unilateral action of a single nation willing to endure global condemnation. Although Germany was still prohibited from rearming, opposition to it was crumbling. The will of the Allies to resist had become a defensive mentality; their political institutions had been subordinated to the League of Nations; and nearly everyone agreed that Germany was entitled to an "equality of claims" and "national self-determination."[51]

One clue to the political climate in Europe can be seen in the "moderate" German leader Gustav Stresemann. Chancellor for four months in 1923 and foreign secretary under four governments, he was known in 1919 as "a violent supporter of the policy of annexation, a fanatic for unrestricted submarine warfare." His famous statement that "not the second of October, when Germany's decision to request an armistice was made, but the ninth of November [the date of the abdication of Kaiser Wilhelm II], was the death day of Germany's greatness in the world" expresses his view of Germany's authoritarian nature. He defended the policies of the kaiser at Weimar on February 6, 1919. To bolster Germany's position he guided Germany into the Dawes Plan and the Locarno Agreements, but he refused an eastern Locarno agreement in part because he rejected the separation of East

Prussia from Germany. His prestige rose enormously: he shared the Nobel Peace Prize in 1926. One of his proudest moments, which he spoke of in his Nobel Prize acceptance speech, was the withdrawal of the French from the Ruhr valley in 1925. This was exceeded only by the French evacuation of the Rhineland, immediately before his death. He was a German statesman of the first rank—but his commitment to a powerful German state was as unwavering as that of General von Ludendorff.[52]

Between 1928 and 1933, the German economy bottomed out, and the Weimar Republic collapsed. Successors to Stresemann, such as Heinrich Brüning, often painted themselves as "moderate" alternatives to radical parties. As Donald Kagan notes, Brüning was committed to austere economic policies designed to allow Germany to "resist any external pressure and be in a position to exploit the world economic crisis in order to bring pressure to bear on the remaining Powers." His goals—which he shared with Hitler—were to shake off reparations, the Young Plan, and arms limitations.[53] Brüning would allow his countrymen to undergo any hardship for Germany's resurrection as a world power. Brüning claimed in his memoirs that he had wanted Hitler to become a sharper, more "extreme," alternative to him in foreign policy, so that he [Brüning] might gain concessions and thus strengthen Germany. He got half of his wish.

The result of such maneuvering was that the more consistently nationalistic party—the Nazis—rose in strength, and Hitler was soon named chancellor. French withdrawals, British concessions, American loans, Russian alliances, and years of diplomacy did not prevent this rise—the overall commitment of the German people to a greater German state was resurgent. As an opponent of any international restraints on German rearmaments, Hitler was not unique, only more consistent, and he spent the first few years working with other German leaders to attain their common aims. There is a huge controversy over this point, but the incessant themes exhorted by every German leader between 1918 and 1933 were to repudiate the Versailles Treaty, to achieve a capacity for national mobilization, and to return Germany to national greatness.[54] Hitler's attempts to undermine the treaty were true to the spirit of German policy, and the public reacted as it had been taught since 1918. Allied leaders either agreed with these goals or evaded their consequences.

Hitler's "Peace Speech" of May 17, 1933, proclaimed his peaceful intentions with a great show of passion and pledged Germany's renunciation of offensive weapons. But Hitler founded this promise on a demand for "equality in arms," a claim that the *Times* of London called "irrefutable."[55] In Britain, proposals to oppose Germany would be greeted by storms of protest and demands that the money be used for socialist programs at home rather than for arms. Public praise would follow any suggestion of conciliation and appeasement of Germany. On October 14, 1933, Germany—denied "equality of arms"—withdrew from the League of Nations and the World Disarmament Conference. England soon began separate negotiations with Hitler over how much he would be allowed to rearm.

The Wages of Unearned Guilt

It will be worthwhile to pause and consider how certain ideas rose to the forefront in British discourse and conditioned British responses to these developments. Keynes's *Economic Consequences of the Peace* helped to elevate a sense of meaculpism into a central factor in British foreign policy. Historian Walter Rock has noted "a curious and perhaps unique trait in her [Britain's] national character—indulgence in a guilt complex with regard to her history."[56] At the end of World War I, the French, devastated by the German invasion and under no doubts about who had invaded whom, were less inclined to such guilt. But it was expressed amid a vibrant give-and-take among journalists, intellectuals, and politicians. Despite the complexities of this discourse, sympathy for Germany and against Versailles was powerful and was founded on certain moral views.

The *Sunday Times*, in a series of articles written by Herbert Sidebotham (alias "Scrutator"), set out to denigrate attempts to defang Germany by undercutting Britain's sense of moral self-value. He stated the egalitarian ideal plainly: "Perhaps the first and most welcome change of mind is to disabuse ourselves of the idea that we and our friends are better morally than other nations [for] in the domain of public morals all nations of western Europe are much on the same level."[57]

Various news sources had followed this basic approach, taking the idea that Germany and England were morally equal to its logical conclusions. By 1938 the following equation of dictatorship with capitalism—misnamed "rule by the rich"—appeared in the *Tablet*, a weekly under Catholic control: "The difference [between the western democracies and the fascist states] is so very much less than it appears in the papers.... What matters for the soul of man is not whether there are elections and plebiscites, but whether there is among people an essential respect for men as men. *Plutocracy and Authoritarianism have both their confessions to make*."[58]

Such ideas were abetted in England by newspaper editorials that stressed the validity of German territorial claims, which many people saw as legitimate despite the character of the regime. The *Times*—the most important paper in Britain, which "had gained the reputation of an official spokesman for the British government"[59]—had firsthand evidence of Nazi crimes, provided by veteran correspondents with years of experience in Germany. But the editors interpreted this evidence against their view that great injustices had been done to Germany since 1919 and that Germany was right to demand redress. The editor from 1922 to 1941, Geoffrey Dawson, and his assistant, Robert McGowan Barrington-Ward, saw attempts to prevent Germany from regaining control of disputed territories as vicious attacks. In a letter —written after Hitler had destroyed Czechoslovakia—Barrington-Ward asked: "Would you have fought, and would Canada have fought, to prevent the re-occupation of the Rhineland, the *Anschluss*, or the union of the Sudetenland with the Reich? Did we give the Germans at any time the hope and opportunity of accomplishing their aims peacefully?"[60] The "aims" he cites approvingly were the absorption of the Rhineland, Austria, and Czechoslovakia.

Taking the idea of *meaculpism* to its logical conclusion, the *Times* editors came to blame the British themselves for the very rise of the Nazi party. "Nothing has come my way to convince me that what was inexpedient and immoral then," wrote Barrington-Ward on the Versailles Treaty, "has suddenly become both moral and expedient merely because we brought the Nazis to power in Germany."[61]

Given such basic evaluations of Germany's position in relation to Britain, such commentators presented the news in a way that was intended not to incite readers into forming conclusions hostile to a policy of compromise with German leaders. Even with all the facts, it would have been difficult for many of the liberal-minded British to grasp the evil rising in Germany. The *Times* reports made it nearly impossible. As a result, the many calls by the *Times* and other papers for the British to rearm had little effect; even for those upholding the need for British rearmament, the basic reason for the rearmament had been obscured.[62]

Although it is not clear that Dawson intentionally twisted the facts, he admitted that he shaped the stories for reasons other than the truth—even when the *Times* had every reason to be very proud of its reporting. Following the bombing of the Spanish town of Guernica on April 26, 1937, the *Times* was the first to state the truth, that German airplanes on loan to the Spanish had committed the atrocity. The criticism from Germany of the *Times* story was much louder than criticism of the raid. Dawson wrote to a *Times* correspondent in Germany on May 23, 1937: "I do my utmost, night after night, to keep out of the paper anything that might hurt their [German] sensibilities. . . . I can really think of nothing that has been printed now for many months past to which they could possibly take exception as unfair comment."[63]

Dawson wrote correctly that the "essential accuracy" of the story had not been disputed. He was therefore confused about the vituperance—he was trying to gain "reasonable relations with Germany" and had taken pains not to offend German "sensibilities." He did not "rub in or harp on" this butchery. He had printed the facts. Why were the Germans taking this so badly? The answer is obvious, although Dawson did not see it: the very statement of the truth is what offended the Germans. Dawson was trying to establish a standard of "fairness" and "reasonable relations" with a dictatorial police state that bombed undefended towns as an exercise in military training and then lied about it.[64] The *Times* wielded an enormous influence on British leaders, and they, too, wanted to deal with Germany as a nation under a rational government, while evidence grew that this was not the case.

Pacifist movements in Britain had a hand in the resulting paralysis. On June 27, 1935, the results of a "Peace Ballot" were announced in England.[65] Some eleven million people voted on questions relating to the League of Nations and disarmament. Of course, not everyone voted, but so many votes could not be ignored. The outcome roundly validated British support for the League of Nations and elevated international institutions over military preparedness by England. The results included:

1. Should Britain remain a member of the League of Nations?
   Yes: 1,090,387 No: 355,888
2. Are you in favor of an all-round reduction in armaments by international agreement?
   Yes: 10,470,489 No: 862,775
3. Are you in favor of an all-round abolition of national military and naval aircraft by international agreement?
   Yes: 9,533,358 No: 1,689,786
4. Should the manufacture and sale of armaments for private profit be prohibited by international agreement?
   Yes: 10,417,329 No: 775,415
5. Do you consider that, if a nation insists on attacking another, the other nations should combine to compel it to stop by:
   (a) economic and nonmilitary measures?
   Yes: 10,027,608 No: 635,074
   (b) if necessary, military measures?
   Yes: 6,784,368 No: 2,351,981

If the ballot was an accurate measure of British public opinion, the English people had little willingness to oppose Hitler preemptively, and any politician who proposed to act was liable to lose his job. The resultant climate of opinion absolved British politicians from acting on the facts. Conservatives often opposed rearmament on pragmatic grounds; many thought that socialism was inevitable, and to oppose it political suicide. Neville Chamberlain, Conservative chancellor of the exchequer in 1932, agreed with Keynes that economic problems were worse than the threat of war; he opted for public welfare over military defense, and as chancellor and prime minister he opposed strong increases in armament spending, even after Munich.[66] In response to criticism from Churchill, Conservative prime minister Stanley Baldwin defended himself, saying that, had he demanded rearmament in 1933, "I cannot think of anything that would have made the loss of election from my point of view more certain."[67] England began to rearm more earnestly after 1936 but, to avoid criticism, the government did it without the confidence of open public disclosure. It would have taken extraordinary courage, integrity, and foresight for a politician to buck the tide of opinion that the poll represents. It would likely be one's last act in government.

In evaluating their response to the rise of Hitler, it must be stressed again that British leaders were charged with protecting an empire of global reach. Military disarmament in Britain had been debated in the 1920s, and British leaders knew that their empire was vulnerable at any number of places. It was not Hitler's actions in Europe that led to a serious consideration of rearmament but rather Japan's attack on Manchuria in 1931. In December 1933 the British cabinet's Disarmament Committee had written that the use of force to stop Germany was "unthinkable."[68] In 1936 the first lord of the admiralty warned against Japanese threats to Far Eastern territories.[69] The British underestimated their own forces for years and overestimated Germany's. German propaganda was very successful in masking Germany's economic and military weakness into 1939.[70] Meanwhile, German leaders knew that they were militarily no match for Britain in a protracted war; as late as March 1938, the German army experienced serious breakdowns when it moved into Austria. A confrontation with Hitler in the mid-1930s would have been dangerous, but it could have rallied allies and motivated German generals to reverse Hitler's ambitions.

There were, of course, many people in Britain who wanted to build the necessary forces and oppose Hitler, but they were not controlling events. While the Western allies still had greater military capabilities than Germany—and had strong and specific reasons to ally against Hitler—there was simply no will to oppose Germany. Conversely, the Germans had the will—which meant that they would soon have the capabilities.

Two aspects of this situation formed a crippling combination. First, the false idea that Germany was unjustly oppressed, combined with fear of a new war, fueled a sense of guilt in the Allies and prevented them from restraining Germany, all the while support for Hitler inside the Reich increased. Second, the fact that Germany was *not* crippled allowed it to rearm, while Britain did so ineffectively. Hitler's aggressive ideology demanded that the Germans rise up and end this unconscionable slighting. The paralysis of the Allies allowed them to do so. The *Manchester Guardian* made the consequences clear in October, 1939: because of the "surrender at Munich," the world "has reverted to the system under which justice is at the mercy of power."[71]

The British people wanted peaceful English socialism at home; they got brutal German National Socialism across the world.

The Climax of British Appeasement

After attaining power, Hitler set out to get all he could through the acquiescence of his foes. But he also demonstrated that he would back down when confronted directly. His sole foreign policy defeat, before 1942, was the premature attempt to take over Austria by a coup d'état on July 25, 1934. Nazis and sympathizers had engaged in a terrorist war against the Austrian government and manufactured attacks on the German population that served as pretexts to protect the "equality of rights" to "national self-determination" of the local Germans from Austrian "oppression."

Hitler's actual role in this remains unclear, but when Nazi henchmen shot Austrian Chancellor Dollfuss in the throat, Hitler prepared a press release claiming victory.[72] The Austrian government acted quickly: Austrians rose up against the Nazis, assassins were hanged, and the Italian leader Mussolini mobilized four divisions at the Brenner Pass to protect Austria from German troops. At midnight, a new release claimed regret for the murder and called it an internal Austrian matter. For Hitler it was too early. He brought Nazi organizations under control and stopped the campaign of terror. But he also held to his commitment, on page one of *Mein Kampf*, to the absorption of Austria as a nonnegotiable matter of principle across the span of his life. It was beyond the scope of *his* compromises ever to renounce that claim. As he said in a speech of April 28, 1939, regarding the actions of the "criminals of Versailles," who had split Austria from Germany: "I have always regarded the elimination of this state of affairs as the highest and most holy task of my life.... I should have sinned against my call by Providence had I failed in my own endeavor to lead my native country and my German people of Ostmark back to the Reich and, thus, to the community of German people."[73]

It is a tragedy that the British did not take Hitler's goals seriously, grasp their implications beyond claims to "self-determination," and formulate a policy to oppose them. When Mussolini approached England and France for an agreement to support Austria, the English refused and the French issued a mild statement. Drawing a line at Austria, which England could have defended with French and Czech assistance, and supporting the French in the Rhineland in 1936, might have prevented the encirclement of Czechoslovakia in 1938 and made the later pact with Poland, which England could not defend, unnecessary. In 1934, however, Austria remained free precisely because one nation—Italy—had acted without waiting for an international consensus. Hitler would resolve that problem over the next few years by making a firm friend out of Mussolini.

In March 1935 a group of British civil servants drew up a paper on the need to rearm; the cabinet softened its anti-German rhetoric before publication, and it had little effect.[74] But on March 17, 1935, Hitler—who had pulled out of the League of Nations and the Disarmament Conference in 1933 on the pretext of "equality of rights"—confirmed the accuracy of the report by repudiating the Versailles Treaty provisions against rearmament. The move was greeted with stupendous celebrations in Germany. Hitler took care; on May 21 he made a powerful antiwar speech, calculated to encourage the British to proceed with negotiations over arms limits. Hitler shouted passionately that Germany repudiated war: "No! National Socialist Germany wants peace because of its fundamental convictions.... Germany needs peace and desires peace.... Whoever lights the torch of war in Europe can wish for nothing but chaos." He applied this to every area of contention: French frontiers are guaranteed, claims to Alsace-Lorraine are renounced, Poland is "the home of a great and nationally conscious people," and Germany has no ambitions in Austria. He implied that Germany would return to the League when it disavowed the injustices of the Versailles Treaty, but promised to uphold the military portions of the treaty and the Locarno Agreements voluntarily until then.[75]

The *Times* of London again took the bait enthusiastically: "It is to be hoped that the speech will be taken everywhere as a sincere and well-considered utterance meaning precisely what it says." To have its readers be "taken" this way was no doubt the goal of the *Times*, but those readers had no way of knowing that the explicit policy of the editor—who had accepted the validity of German claims—was to avoid offending the "sensibilities" of the Germans. Hitler said that Germany was ready to agree to any level of disarmament, including limits to his navy at 35 percent of British forces, which he was willing to negotiate directly with Britain. The British did precisely what the *Times*—and Hitler—hoped; they accepted his right to equality of claims and pressed ahead with separate negotiations over the size of Germany's armed forces. In other words, they conceded the principle that Hitler had a right to rearm while arguing over the amount.[76] The proposed limits allowed German factories to be retooled for military purposes—a years-long process—which was all Hitler wanted for the moment. By compromising over the amount, the British conceded everything—the principle behind his right to arms, and the means to their practical development.

More demonstrations of weakness were coming his way. In October 1935 Mussolini moved his navy through the Suez Canal and attacked Ethiopia; the British did not even close the canal. They appealed to the League of Nations, which passed economic sanctions after weeks of debate. British leaders were paralyzed by the need for consensus; by disavowing their right to independent self-defense against clear aggression, they subordinated their judgments and constitutional responsibilities to an international committee. The League became both an excuse for inaction by opponents of Hitler and Mussolini and a demonstration to Hitler and Mussolini that this excuse could be used to their advantage.

Hitler observed the world's reaction to this, as well as to the Spanish Civil War, by sizing up his enemies and drawing the conclusion that he would face no serious opposition if he moved carefully. By this point, he must have seen in his enemies a *policy* of appeasement, a *policy* of neither stating the reality of any situation openly nor acting as it demanded. Inside Britain, a complex set of political and economic concerns tempered the political climate, but the net result left Hitler free to act. His own military and economic capacities were as yet decisively inferior to those around him—but he understood that the moral is to the physical as three to one and that his foes would not use their power if he cloaked his goals in the moral language of "national self-determination" and "equality of rights" and did not provoke them too early.

On March 7, 1936, he raised a trial balloon; he occupied the Rhineland with a force of some thirty-six thousand police and light-armed soldiers.[77] This rash and potentially fatal move was a direct violation of the Locarno Agreements, but Hitler used the Franco-Soviet Pact, an attempt by France to obtain security guarantees that had been signed a few days earlier, as a pretext. But to Hitler's mind—as his own statements would later affirm—the French withdrawal of 1930 was a much more important indicator of what the Allies would actually do: nothing.

In the Rhineland the Germans were exposed and could have offered only a limited military response to French resistance. At Nuremberg, General Jodl said that "the French covering army could have blown us to pieces." Hitler himself later said, "If the French had marched into the Rhineland we would have had to withdraw with our tails between our legs" because even a "moderate resistance" would have been impossible.[78] The orders to withdraw in the event of opposition, issued by War Minister Werner von Blomberg, meant that, at minimum, German units that had crossed the Rhine were to retreat.[79] The Czechs, Romanians, and Poles offered to support France—a hundred divisions could have moved on Germany—and the German air staff had reported that it would be unable to prevail in a war with France and Czechoslovakia. But Hitler understood the psychology of his enemies far better than his generals did. Preoccupied with the Italian invasion of Ethiopia, the British and French did not oppose what many saw as a legitimate German movement.

The Rhineland occupation served Hitler's need to cement his power inside Germany through a vote of confidence from the German people. The risk he took was so great that territorial aggrandizement would not have justified it at that time. Victor Klemperer—a businessman, journalist, and a Jew who lived through the Nazi horrors—wrote in his diary about Hitler's Reichstag speech of March 8: "Three months ago I would have been convinced there would be war the same evening. Today, vox populi (my butcher): 'They [the French and British] won't risk anything.' General conviction, and ours too, that everything will remain quiet.... His [Hitler's] position is secured for an indefinite period."

On March 23, Klemperer wrote of how the Rhineland occupation had affected the minds of people previously lukewarm in their support of Hitler:

> It will be a tremendous triumph for the government. It will receive millions of votes for "peace and freedom." It will not need to fake a single vote. Internal policies are forgotten.—Exemplum: Martha Weichmann, who visited us recently, previously democratic. Now "Nothing has impressed me as much as re-armament and marching into the Rhineland." ... It all impresses the foreign powers and, despite a condemnation from the League of Nations and the proposal of a supranational police authority for the Rhine zone, will also be a stupendous victory for Hitler. He flies from place to place and gives triumphal speeches. The whole thing is called an 'election campaign.'"[80]

An election campaign indeed: on March 29 Hitler held a plebiscite; the votes overwhelmingly affirmed his policies. Generals who doubted the Führer's genius fell silent or were purged. The occupation of the Rhineland was one of Hitler's most important domestic policy successes.

Inside England, Sir Anthony Eden, British secretary of state for foreign affairs, sized up the situation in a speech to the House of Commons on March 26, 1936. He observed that Hitler, purporting to be aggrieved at the recent Franco-Soviet Pact—an agreement pursued by France because it was unable to gain security guarantees from England—had multiple opportunities to bring Germany's grievances before the world. In Eden's words, Germany's objectives were proper; it was the method that was wrong: "I believe it to be the judgment of this country that even those in this country who think that Germany has a strong case deprecate the fact that she has chosen to present it by force and not by reason."[81]

Eden then offered three proposals by which "international law was to be vindicated." Germany was to submit to arbitration, suspend fortifying the Rhineland, and agree to an international force in the area. (This last was the "supranational police authority" in Klemperer's diaries.) Eden goes on to make it clear that he does not intend to act forcefully himself: "I must make it plain that these proposals have always been proposals. They are not an ultimatum, much less a *diktat* [a standard charge leveled against the Versailles Treaty]. If an international force were the difficulty, and if the German government could offer some other constructive proposals to take its place, His Majesty's Government will be quite ready to go to the other Powers interested and try to secure agreement upon them."

In other words, Eden, later known as an opponent of appeasement, now thought it out of bounds not only to demand that Germany fulfill a treaty needed to protect France and Belgium from a repetition of an invasion twenty-two years earlier but also to uphold an agreement—Locarno—that Germany had voluntarily signed. Eden could only make proposals, no longer to the League as a prerequisite to action, but now to Germany as a prerequisite to action. He was now subordinating British interests not to an international consensus but to Germany, *because* Germany had moved its forces in defiance of two treaties. Further, because Germany had a "strong case," the British government was willing to become Hitler's advocate in bringing

Germany's demands to the other allies and to pressure them to accept those demands.

That anyone might have actually thought these proposals "harsh" is a telling comment upon Eden's age, but it was an age in which people said that "I suppose Jerry can do what he likes in his own backyard, can't he?" Labour Party members generally opposed sanctions against Germany, and the Conservative member of Parliament Harold Nicolson said, "The country will not stand for anything that makes for war. On all sides one hears sympathy for Germany. It is all very tragic and sad." The British secretary of war told the German ambassador that the British people would not fight for the Rhineland.[82] In his speech, Eden told Hitler, with all the clarity of a direct personal guarantee, that England would not act alone: "Let us make our position on that absolutely clear. We accept no obligations beyond those shared by the League except the obligations which devolve on us from Locarno."[83] Once again, England could do nothing without international sanction, but Hitler was free to move.

Eden then stated his goals in terms that would come to characterize the age:

> "Our objectives in all this are three-fold—first, to avert the danger of war, second, to create conditions in which negotiations can take place and third, to bring about the success of those negotiations so that they may strengthen collective security, further Germany's return to the League and, in a happier atmosphere, allow those larger negotiations on economic matters and on matters of armament which are indispensable to the appeasement of Europe to take place. I assure the House that it is the appeasement of Europe as a whole that we have constantly before us."[84]

We will return to the issue of appeasement; suffice it here to say that peace was not Hitler's aim, a point that could have been understood at the time. Eden's objectives, on the other hand, were to achieve peace through negotiations, peace being understood as a situation in which people negotiate and achieve the "happier atmosphere" needed for collective security. Without solid objectives for Britain, Eden had reconfirmed his commitment to assist Hitler in achieving Hitler's objectives. Hitler, on the other hand—who destroyed the very meaning of agreements by violating a treaty that Germany had requested, negotiated, and signed—had removed a layer of French military defenses, ratcheted up the commitment of the German people to his rule, and boosted his industry and manpower, all the while confirming a policy of appeasement in his enemies.

Hitler had plenty of evidence for this assessment. Following the resignation of Anthony Eden as foreign minister in February 1938, the *Times* printed the remarks of the Conservative Lord Lothian, from the House of Lords, subtitled "Moral Failure of the League." The cause of the rise of the German totalitarian state was shifted onto attempts by the British and Allied governments to enforce the treaty:

> The failure of the League was in large measure a moral failure; it was a failure because they had put peace first and justice second; whereas justice was the only condition on which peace could rest.
>
> The treatment which had been accorded to Germany in the years after the war ... [which] rendered quite untenable the thesis that Germany alone was responsible for the War was the fundamental reason for the troubles we were in....
>
> We heard a great deal of the violation by Herr Hitler of the treaty because he returned his own troops to his own frontier. We heard much less to-day of the violation with which the French army, with the acquiescence of this country, crossed the frontier to annihilate German industry and in fact produced the Nazi Party. There were very few people left in the world who did not deplore the failure to come to terms with Germany long ago. The result was that we had a large totalitarian Germany, very formidable and very dangerous, and it was the existence of that Germany which governed the whole international situation to-day, because without it neither the Far Eastern situation nor the Mediterranean situation would be difficult to handle.... [The orthodox League view, that no changes be made to treaties without consent, meant] the duty of this country to uphold that view by economic sanctions, and in the last resort by war all over the world.[85]

The terms of the "moral failure" are clear: the unwillingness of League members to allow Germany to gain its allegedly just objectives. "Moral failure" meant the failure to recognize German claims to the

Rhineland—"his own frontier"—despite the Locarno Agreements. Justice would require absolving Germany from producing the material goods needed to replace mass destruction in France—"the annihilation of German industry"—even though loans had kept German industry alive. Most of all, Germany was not held responsible for the last war, and was not recognized as the primary cause of problems, even though "dangerous" and "totalitarian" Germany "governed the whole international situation." As a result, Britain was now said to be committed to "war all over the world." Ultimately, Lothian dismisses this as an "Anti-Fascist Crusade" rather than the "traditional attitude of the League as an instrument for bringing nations together." To bring dangerous totalitarian nations together with Britain was Lothian's form of justice—an aim essentially shared with Eden.[86] There were rejoinders, of course, from those who urged Britain to maintain its commitment to the League and to remain loyal to the League's decisions—but they did not control British policy.[87]

By March 1938, Hitler had achieved his "most holy task" by annexing Austria. On February 12, exactly a month before German troops moved, Hitler summoned Austrian chancellor Kurt von Schuschnigg to Berchtesgaden and stunned him with a naked threat of invasion. Austria was preparing a defense against Germany, and Hitler wanted this stopped. Hitler made Austria's position unmistakably clear, all the while demonstrating the disastrous consequences of failing to oppose the German entry into the Rhineland two years earlier: "Don't believe that anyone in the world will hinder me in my decisions! Italy? I am quite clear with Mussolini: with Italy I am on the closest possible terms. England? England will not lift a finger for Austria.... And France? Well, two years ago when we marched into the Rhineland with a handful of battalions—at that moment I risked a great deal. If France had marched then, we should have been forced to withdraw.... But for France it is now too late!"[88]

The Austro-German agreement of July 11, 1936, had pledged the German government to recognize the "full sovereignty of the Federate State of Austria" and made direct reference to Hitler's speech of May 21, 1935. But Austria had promised to follow a policy "corresponding to the fact that Austria regards herself as a German State."[89] To make Austria's German essence explicit was perhaps all Hitler wanted at the time. In a Reichstag speech on February 20, 1938, Hitler affirmed his commitment to "protection" for Germans outside Germany, a matter that superseded any agreement to respect Austria's "inner political order," which he saw as permeated with communists. On February 24 Schuschnigg affirmed his commitment to preserve Austria as a nation. His surprise call for a plebiscite, to be held March 13, infuriated Hitler, who knew that support in Austria for reunification with Germany was limited and not strong in Vienna. On March 11 Schuschnigg, isolated, a strongman not strong enough to oppose the Führer, resigned under threat of invasion, and German troops moved in. The "absorption" (*Anschluss*) of Austria was complete. The world saw crowds cheering Hitler's arrival—as if a choice were possible to them and events were not carefully staged. This time there was no opposition from Italy; Hitler had made firm friends with Mussolini.[90]

Many British papers, having largely accepted the legitimacy of Germany's claim, assumed that Austria was a done deal.[91] Criticism was reserved for the way it was done, which was bound to "encourage [German] confidence in the methods that achieved it."[92] The *Times* was afraid it would move the British government and people to "instinctive resentment and condemnation" and threaten the policy of appeasement—as if instinct governed human affairs, and appeasement was an unchallengeable policy—yet a headline in the same issue read "The Rape of Austria."[93] As to British officials, one said, "Thank goodness Austria's out of the way," and Prime Minister Chamberlain said, "At any rate the question was now out of the way."[94] After years of tension, they were content to have Hitler controlling events for them. But with an ally gone and Czechoslovakia surrounded on three sides, the stakes had been raised dramatically.

According to a widely accepted interpretation, however, Operation Otto, the meticulously planned German drive on roads to Vienna, in good weather and without an enemy in sight, virtually broke down en route, and Hitler was furious at the shortcomings in his military machine.[95] In 1942 he complained that "we saw over eighty tanks immobilized on the side of the road—and what an easy road it was!"[96] According to Heinz Guderian, a German officer on the drive, the drive stalled when "the German war machine was held up on my orders for the reception of Hitler, and for no other reason." Yet this offers no compelling reason to change the basic conclusion, that the power of the German forces was overstated in the minds of many British leaders. Guderian continues: "Defects in heavy artillery could not have appeared since we did not possess any heavy artillery.... At the time we possessed only light tanks. Heavy tanks were as non-existent as heavy artillery, and therefore could not have been loaded onto railway trucks."[97]

Whether or not the British and French could have easily prevented the German advance by force, and whatever the facts are of the German movement, it is beyond doubt that their acquiescence to the moral propriety of German demands had greatly abetted Hitler's rise. Any chance to present the German people with a failure for Hitler was lost.

Flush with this success, Hitler wasted no time in bringing a new crisis to the forefront: his demand for the Sudetenland (Southland) in Czechoslovakia. Czechoslovakia—the nation of the Czechs and Slovaks—was an amalgam of various nationalities, including some three million Sudeten Germans, the most vocal of whom chafed at being marginalized under "inferior" Slavs. The campaign for Czechoslovakia bore all the marks of Hitler's style: terror campaigns, run by his stooges in a front political party; inciting violence as a pretense to protect Germans; and making claims to moral legitimacy based on the "self-determination" of the German-speaking population and their "right to equality." Chamberlain told the cabinet that, if Hitler demanded a popular vote, "it would be difficult for the democratic countries to go to war to prevent the Sudeten Germans from saying what form of government they wanted." Given the ideal of "national self-determination," a brutal dictatorship is merely one "form" of government among many, and legitimate if a foreign dictator asserts that there is support for it.[98]

Many British were disarmed by these ideas and failed to grasp the meaning of the racial collectivism behind such claims. Certainly the British were not ready for war—but would allowing the Czechs to fall make the situation any better? Rather than standing firm and recognizing that the loss of Czechoslovakia would place all of southern

Europe under the Nazi thumb, they made moral arguments upholding the Germans' right to nationalistic self-determination. Britain, which had first pressured the French to compromise with Germany, now pressured the Czechs to give up their defenses without a fight. France had a mutual defense agreement with Czechoslovakia and with Poland, but would act only with British support. All parties were hoping that the Sudetenland was Hitler's last demand—as if Poland and East Prussia did not exist. The British consistently underestimated their own forces while overestimating those of Germany, assuming "a best-case analysis of the German situation with a worst-case analysis of Britain's" in order to avoid action.[99] Meanwhile, the Czech government was increasingly isolated and vulnerable to collapse.

Inside Germany, Hitler again faced opposition from his generals. Pessimistic military reports were not a monopoly of the British; many generals in Germany knew they could not easily prevail in a war with the Czechs, who had a large army and strong defenses. The Germans would overcome the Czechs if the war were short, but the cost could have collapsed the German economy, and Hitler would not have gained the undamaged equipment of a major army. Hitler was not, however, banking on Germany's military power; he rather expected the Western allies not to use their own. Had England and France issued an ultimatum and followed it up with mobilization, a coup by the German generals might, perhaps, have removed Hitler from power. This was probably the last time that Germany could have been stopped without war.

Geoffrey Dawson, editor of the *Times*, maintained his editorial stance in favor of granting terms to Hitler. He claimed that it "had found favour in some quarters" for the Czech government to allow the Sudetenland to secede—a phrase that claimed authority for his opinions, as if they were reflecting government policy. The *Times* stated famously:

> It might be worthwhile for the Czechoslovak Government to consider whether they should exclude altogether the project, which has found favour in some quarters, of making Czechoslovakia a more homogeneous State by the secession of that fringe of alien populations who

> are contiguous to the nation with which they are united by race. In any case the wishes of the population concerned would seem to be a decisively important element in any solution that can hope to be regarded as permanent, and the advantages to Czechoslovakia of becoming a homogeneous State might conceivably outweigh the obvious disadvantages of losing the Sudeten German districts of the borderland.[100]

The Sudeten German Party, a Nazi front, had stimulated the pretenses needed for Hitler's intervention, but Dawson noted that by September 5 the Czech government had addressed the Sudeten concerns; the problem now "offers no reason for war." He hoped that this would be noted at an upcoming Nuremberg rally—as if the Nazis scrupulously examined the facts in order to be accurate at public rallies. What the Czechs and the Germans both knew was that Czech defensive fortifications were in the Sudeten areas and that this proposal would have left the "homogeneous state" defenseless.

Like Dawson, Prime Minister Neville Chamberlain was also at a loss to explain Hitler's actions. On September 8 he told his colleagues: "I keep racking my brains to try and devise some means of averting catastrophe"—although he had probably decided weeks earlier to simply ask Hitler what he wanted.[101] Chamberlain was not a weak man; he had extensive experience in government and could be very demanding on his staff, even though he had been prime minister for only a year. But his approach to foreign policy issues combined pragmatism with a concern for justice, equality, and rights to national self-determination. In the words of one historian, Chamberlain and his supporters "believed that questions of strategic interest and war were not the first priority of statesmen. They believed that morality in terms of righting past wrongs and of supporting minority rights, national aspirations, and the like should receive special consideration in the determination of foreign policy."[102]

Chamberlain was acting on several other premises. First, he was acting on the "worst case" judgments of his staff.[103] German propaganda had successfully exaggerated the strength of Germany's forces, and Chamberlain feared German bombing of London, as the French feared for Paris. He had decided, in advance of meeting Hitler, not to fight for the Czechs, and he created arguments to support that decision. Second, stressing Britain's military weakness, which had deepened while the budget was under his sway as chancellor of the exchequer, he justified his decision by saying that "no democratic state ought to make a threat of war unless it was both ready to carry it out and prepared to do so."[104] He never really considered the power that leadership by Britain could have wielded in bringing allies into the fight—nor did he consider the costs of allowing Czechoslovakia to go without a fight.[105] One must wonder whether England would ever be strong enough in his mind. Third, he believed in negotiations, the give-and-take between opponents over particular points of dispute, which led him to size up the Hitler-of-the-moment rather than recognize the broader reasons why Europe was facing this particular crisis.

Most of all, though, Chamberlain thought that a problem such as Czechoslovakia was a fundamental injustice to the minority German-speakers and that Hitler was right to demand it back. National self-determination was a right, and it applied to the Sudeten Germans as well. That Hitler was a dictator did not, to Chamberlain, make the issue any less legitimate—as long as Hitler did not use force. Chamberlain wanted to be moral, and by accepting the legitimacy of Hitler's demands, he was disarmed from taking a principled stand against the demands themselves. All that was left was to negotiate the means to accomplish them—which is exactly what he did.

On September 15 Chamberlain got onto a plane for the first time, flew to Berchtesgaden, and had a cordial meeting with the Führer. Hitler approached Chamberlain with a disarming graciousness that masked an intransigent attitude. When the discussion turned to the Czechs, Hitler said that the talks could continue if the British accepted the principle of self-determination.[106] Chamberlain agreed and told Hitler privately that it mattered not to him whether the Sudeten Czechs remained part of Czechoslovakia or not.[107] The two men were agreed; what was left was to argue over the amounts and the details.

The effect of this meeting on the Czechs was devastating; Hitler's proxy in the Sudetenland, Konrad Henlein, had fled to Germany with his Friekorps military squads, and the Czechs were getting the crisis under control—until Chamberlain cut the rug from under them by legitimating Hitler's demands.[108] The central importance of Chamberlain's meeting must be strongly emphasized: the brutal effects on the Czechs began not only with the content of the discussions but in his very agreement to speak with Hitler. This strengthened the Führer's hand, motivated potential allies to back out, emboldened Nazi supporters inside Czechoslovakia, and destroyed the Czech resistance.

Chamberlain came home, convinced the British government to accept Hitler's demand as "self-determination" (not cession), and told the French ministers on September 18 that "violent action could be avoided," he said, "[only by] certain measures of self-determination ... the German Government would be willing to discuss the ways and means of putting the principle into effect in an orderly fashion."[109] The French balked at self-determination—recognizing that this would split the Balkans into myriad pieces—but accepted "cession" as a principle to limit the issue to the Sudeten areas. The next day, Chamberlain presented the cabinet with cession, as if it were the idea of the French rather than Hitler's. Chamberlain was manipulating language to convince his own government and Britain's strongest European ally to accept Hitler's demands. With the word games concluded, the Czech government was forced to accept the terms of its own disembowelment.

Chamberlain returned to Germany on September 22, this time to a stormy meeting with a madman who first demanded to occupy the area in two days, then on September 28, and then on October 1. This confused Chamberlain; Hitler was not acting as a reasonable man should. But he was acting like a man in unbending pursuit of goals he had held for two decades, which he had published in books and screamed in speeches, and who knew that he could get what he wanted from compromises made willingly by opponents who accepted the moral legitimacy of his objectives.[110] Chamberlain evaluated Hitler without holding in mind the full context of why he was even there—the origins of the crisis in Nazi terror, the meaning of the German threats and the rearmament, the consequences for Britain and

Europe, and the defenseless position of eastern Europe should the Czechs lose the Sudetenland.

Back in London, Parliament, after a bitter split, rejected these demands, even as British public opinion had hardened against Hitler.[111] Churchill records his view of Parliament's attitude toward the Sudetenland Czechs: "Some Ministers found consolation in such phrases as 'the rights of self-determination,' 'the claims of a national minority to just treatment'; and even the mood appeared of 'championing the small man against the Czech bully.'"[112]

While the Czechs struggled to keep control of their nation, and while the clock ticked on a German ultimatum set to expire on 2:00 PM, September 28, Chamberlain offered the following public and private communiqués. Publicly, he addressed the British people on September 27: "How horrible, fantastic, incredible, it is that we should be digging trenches and trying on gas-masks here because of a quarrel in a faraway country between people of whom we know nothing! … war is a fearful thing, and we must be very clear, before we embark on it, that it is really the great issues that are at stake."[113]

In case this was not clear enough to Hitler—that there was no "great issue" at stake—Chamberlain then sent a message to Hitler offering his personal assistance to Hitler in attaining his goals:[114] "After reading your letter, I felt certain that you can get all essentials without war, and without delay. I am ready to come to Berlin myself at once to discuss arrangements for transfer [of the Sudetanland to Germany] with you and representatives of the Czech Government, together with representatives of France and Italy if you desire. I feel convinced that we could reach agreement in a week."[115]

Agendas of meetings between world leaders are very important, and Chamberlain has stated his purpose clearly: to discuss the means of granting to Hitler everything he could not have taken by force—as Eden had earlier offered to promote Hitler's goals in the Rhineland. Chamberlain's letter guaranteed to Hitler that, although Hitler might "distrust" the Czechs, Britain and France had the "power … to see that promises are carried out." After a twenty-four-hour extension of the ultimatum, the third meeting took place in Munich on September 29. This was a four-power conference between Britain, France, Italy, and Germany. At 2:00 AM on September 30, dead tired after twelve hours of talks, Chamberlain—now willing to act outside the League of Nations—accepted Hitler's ultimatum. The Czechs, with the undamaged resources of their army, passed into the German Reich. The French government approved the result overwhelmingly—it was absolved of the responsibility to act. Tens of thousands of people sent thanks to Chamberlain for keeping their children out of a European bloodbath. Energy surged through the Reich with another infusion of material resources and psychological encouragement. In a few months, Czechoslovakia ceased to exist—and Hitler, who had decided that Chamberlain "will be too cowardly to attack," pushed ahead with war.[116]

Marshal Keitel was asked at the Nuremburg trials for the German generals' opinion of the Munich agreement. Keitel answered: "We were extraordinarily happy that it had not come to a military operation, because throughout the time of preparation we had always been of the opinion that our means of attack against the frontier fortifications of Czechoslovakia were insufficient. From a purely military point of view we lacked the means for an attack which involved the piercing of the frontier fortifications.... I believe I may say that as a result this greatly increased Hitler's prestige among the generals."[117]

Of course, Keitel would have obeyed had his Führer commanded —but his assessment of the situation was accurate. Hitler's actual strength in 1938 was a fraction of that facing him. He had five divisions on the west facing France, against fifty-six French divisions that could be called up within five days.[118] The equipment on both sides was in appalling shape, mostly of World War I vintage—but a long war would have found Germany deeply vulnerable to shortages of raw materials and an economic crisis.[119] Neither French nor British leaders actually debated the consequences of allowing Hitler to take Czechoslovakia and its army without a fight.[120] Neither seriously pursued an alliance of European powers, which might yet have turned the population against Hitler. A war of the Czechs and the French against Germany, with British support, would not have been easy and might have expanded horribly—but the end would not have been in doubt, and the cost in lives would have been a fraction of those destroyed in World War II.[121]

One important reason why the British were willing to hand over the spoils of war without a war was moral: for twenty years, Germany, having plunged Europe into war twice in two generations, had taken on the image of a weak and victimized party trying to regain its legitimate "equality of rights."[122] The British had no love for Hitler—but they thought that the demands of the German were, on some level, just. The utter injustice of Hitler's claim had come to flower in the very idea that the Sudetenland Germans—represented by stooges under the Führer's control—were victims of the "Czech bully." But the moral ideal of equality—expressed in the view that we are each part of a greater whole—is never unwilling to sacrifice a part for the preservation of the whole. The insidiousness of the idea was expressed by Churchill: "England and France said 'both the French and British governments recognize how great is the sacrifice thus required of Czechoslovakia.'" As philosopher Leonard Peikoff put it, in his philosophical study of Nazi Germany, "No one can claim that they did not sacrifice enough individuals."[123]

Chamberlain claimed "peace in our time"—a phrase he regretted almost immediately—but "our time" ended immediately for the Czechs, and a few months later for everyone else.

Causation and Appeasement

This highly selective and abbreviated account of Britain's relationship with Germany has focused on certain key ideas that conditioned the thinking of political leaders who faced enormous decisions. Given these ideas, "Hitler and his demands were like a funnel into which British attitudes on every question from armaments to xenophobia were poured; what emerged from the funnel was the single policy of appeasement."[124] Many British leaders, fearful of a new war, facing problems of awesome complexity, and having accepted Germany's claim to equality of arms and national self-determination, became, intentions notwithstanding, Hitler's de facto allies. The political legitimacy, energy of achievement, and physical resources that Hitler took from every success increased the commitment of the German people to follow their Führer into war. One single projection of power into Europe by France or Britain between 1933 and 1938—just one direct setback for Hitler—might have put an end to his career. This was the time that the issue at stake was most clear—and that Hitler was most vulnerable inside Germany.

The homestretch of the road to Munich began in 1919, when the Allied powers claimed the fruits of victory—the right to dictate Germany's basic course into the future—without achieving and proclaiming victory. The biggest of those fruits was the destruction of the German-Prussian monarchy, a fundamental change in Germany's political organization that was imposed from without, but which lacked the context of a defeat to legitimate it in Germany. Germans were asked to accept the status of a defeated power, in a war in which victory and defeat had been explicitly disavowed. The Allies contradicted their own aims and then inverted cause and effect: they wanted the effects of victory without achieving it. They thus failed to discredit and destroy the central motivations of the German leadership and its support among the population: the desire to aggrandize the state by achieving hegemony over Europe. With this cause still in place, the Allies then tried to disrupt the physical factors that made the war possible: German territories, industry, and armies. The first deadly inversion was to elevate the means to fight over the motivations to fight, and to try to limit the former without challenging the latter.

The attempt failed because the Germans took it as an affront to their nation rather than as a necessary step to preventing a war. This set in place a spiral of hostility toward the treaty and the Allies—especially the French—that was quiescent for certain periods of the German democracy of the 1920s, but it never entirely went away. With blame for their economic situation placed on the treaty instead of their own military aggression, and with a deep sense of duty to the state focused on the need to regain its status, Hitler was able to unlock the energy of the German people and harness them in sacrificial service to his war.

The Allies accepted the basic territorial, economic, and military armament goals that Hitler set and were unable to state clearly why those goals were invalid. When the Chamberlain government adopted a moralistic approach to foreign policy, it adopted ideals promoted by Wilson as its moral compass. But placing those ideals into the service of the nation as an end in itself—*national* equality of rights, and *national* self-determination—the protection of the rights of the individual was unacknowledged and obscured. A nation that voted itself into a brutal tyranny and murdered whole classes of its own citizens was given equal status with nations for whom such atrocities were repugnant. Allied leaders accepted not only the basic goals that Hitler claimed as his national destiny but the nationalist framework within which those goals were understood. This was the second inversion—the elevation of the nation as a thing with "rights" over individuals.

The result was a series of obfuscations, and of contradictions between public utterances and truth. It was to be a "peace without victors"—which the victors would enforce in a treaty that the vanquished did not write. The League of Nations was to be a forum for all nations and not an alliance—which Germany was at first not allowed to join, and which worked to enforce the treaty against Germany. Germany was to accept "war guilt"—while the Allies accepted that the guilt was really not true. Germany was to pay "reparations" collected from taxation—while foreign loans flowed into private German industries.

To ensure peace in Europe became the British obsession—but what did the British mean by "peace"? Peace came to mean a state of affairs where open deliberation reigns—but when a tyrant arose who openly challenged this state of affairs, broke treaties his nation had asked for and signed voluntarily, and presented an antihuman dictatorship that eschewed all promises, a return to peace meant to entice him to enter the very framework he was destroying. Hitler's pretense at entering that framework disarmed British leaders and led them to act in virtual alliance with him. The British approach was appeasement—an attempt to avoid a crisis for the moment by granting an enemy's demands, in the hope that peace will, somehow, follow.

The primary meaning of appeasement, according to the *Oxford English Dictionary*, is wider than any specific method: "the action or process of appeasing, pacification, satisfaction."[125] Eden used "appeasement" in this original sense when he told Parliament that "it is the appeasement of Europe as a whole that we have constantly before us." By appeasement, Eden meant the same thing that Winston Churchill had in mind when he said, in 1921, that "the aim is to get an appeasement of the fearful hatreds and antagonisms which exist in Europe and to enable the world to settle down. I have no other object in view."[126] In 1929 Churchill wrote, "I should be prepared to make peace with Soviet Russia on the best terms available to appease the general situation."

This wide sense of "appeasement" leaves open the specific *method* by which peace can be established. In this wide sense, W. T. Sherman was a master appeaser, ending the Civil War and reestablishing constitutional authority by burning Atlanta and collapsing the southern will to war—while McClellan's actions brought continued war and bloodshed.

But British leaders in the 1930s—unable to refute Hitler's claims to equality of rights and national self-determination—fell back on a tradition of pragmatic negotiation and compromise, a means to dealing with others that worked when they shared rational goals. This exemplifies the narrower sense of appeasement, which is a specific method of trying to achieve peace—the satisfaction of the desires of an opponent. Concerning appeasement in this narrower sense, the *Oxford English Dictionary* further states: "Freely used in political contexts in the 20th century, and since 1938 often used disparagingly with allusion to the attempts to conciliation by concession made by Mr. Neville Chamberlain, the British Prime Minister, before the outbreak of war with Germany in 1939; by extension, any such policy of pacification by concession to an enemy."

This narrow sense of "appeasement" came to mean the grant of demands to an enemy *because* he has made a credible threat of force, in the hope this will end further threats. Because appeasement is a means, it must be evaluated against the ends being sought. Chamberlain's own goal in negotiating with Hitler was to create peace through diplomacy by bringing Hitler into the European framework of negotiations. Chamberlain failed to grasp that Hitler was already destroying that system and that Chamberlain would abet the destruction if he rewarded Hitler for doing so.

The key here is to recognize the essential incompatibility between British policy goals and those of the Germans. Freedom under a decent government is not equal to life in a racist dictatorship. No nation has a right to "national self-determination," if this means the destruction of the rights of its citizens, and their subordination to the state. By accepting Hitler's goals as legitimate, Eden, Chamberlain, and others became his allies in achieving those goals. Because England had nothing to gain from Hitler, every negotiation began with *his* aims in mind and focused on the means by which he could achieve *his* goals. Any compromise or concession to Hitler aided him in attaining his goals. With Hitler's every success, the German will to fight increased—in Hitler's mind, in the commitment of his generals to the cause, and in popular support for his leadership.[127]

Neville Chamberlain's statement to the House of Commons, in December 1937 was that "only ... by a real understanding and effort to meet other's needs" could peace be maintained. But the deadly flaw here is to forget that a "need" exists only in relation to a purpose. The nature of the purpose is vital to assessing the legitimacy of the need. Foreign Secretary Eden does advocate appeasement of Germany in his 1936 speech, but not specifically in the passage in which he calls for the "appeasement" of Europe. His appeasement rather exists in the granting of any legitimate status to Hitler's Germany—and engaging in any discourse with Hitler—based on Hitler's having marched into the Rhineland. It would not be lost on Hitler that he was receiving these proposals *because* he had acted aggressively. The Czech position collapsed not when Chamberlain made a bad deal with Hitler, but rather earlier, when it became clear that Chamberlain was going to make *any* deal with him.

Hitler's stature rose as millions of German speakers came under his control, and his opponents retreated. Seeing the Führer prevail, opponents of Hitler inside Germany fell silent—he had succeeded again.

Hitler understood that his major weapons were not the military but the uncertainty of his enemies. He faced no opposition because his enemies thought they *should not* oppose him. The root cause was a *desire* in the minds of the Allies that Hitler not be evil and that their moral standards should apply to him. "Germany will refuse that fundamental consent to the rule of law until she obtains equality of rights and treatment from the League," said one respondent to Anthony Eden in Parliament; "Germany is already breaking the shackles of Versailles, and we ought to have struck them off before now."[128]

The conclusion is chilling: at every step up to 1936—from the Ruhr Crisis of 1922, to the French withdrawal from the Rhineland in 1930, to Hitler's aborted moves against Austria in 1934, to his repudiation of the Versailles Treaty and Mussolini's attack on Ethiopia in 1935, to Germany's remilitarization of the Rhineland in 1936, and the occupation of Austria in the spring of 1938—Germany could have been stopped, and a world war avoided, by British leadership to unite the Czechs, Poles, Austrians, and French. Epaminondas of Thebes did it against the Spartans—who retreated in fear—but British leaders did not provide such leadership for countries that could have been emboldened to defend themselves. Once those allies were gone, and Germany's strength had massively increased, they committed their nation to the defense of a country that was solidly beyond their range: Poland.

In 1936 a country with no army to speak of challenged countries with superior forces, and attained a vast military superiority within three years, because its opponents were intellectually confused, morally disarmed, and devoid of clear objectives. Hitler projected moral power into the center of the British decision making by calling on the ideals that British leaders had accepted as unquestionable. By late 1938 the British people knew what was up—the change in public opinion was seen almost overnight—and Chamberlain had to take his policy of appeasement underground. It gives one pause to consider that the British people might have supported a policy of strength against the Germans earlier, had it been forthrightly presented to them.

The failure was resolved only when European and American allies opened their eyes, stopped evading the nature of the Nazi threat, and repudiated the method of appeasement. Unconditional surrender ended Hitler's war and destroyed the motivations for a new war. The territory and weapons that Germany controls in our own day are no threat—because, since 1945, the will to war in Europe has been eviscerated. The big lesson here is the need to face facts, to understand the ideas that motivated the actors, and to act as the facts and motivations demand. British leaders evaded the nature of Germany's rise for more than a decade, while the strategic balance shifted from one in which Germany had the will to act but not the capacity, to one in which its capacity exceeded that of its opponents. The will—a militant ideology—plus the capacity made war inescapable.

Basic Chronology of the Prelude to World War II in Europe

1870–1871	Franco-Prussian War
1899–1902	Boer War
1904–1905	Russo-Japanese War
1907	August: Anglo-Russian Entente, sets off Triple Alliance
	Second Hague Peace Conference
1908–1909	London Naval Conference
1911–1912	Tripolitan War
1912–1913	First Balkan War
1913	Second Balkan War: Treaty of Bucharest
1914	June: Archduke Francis Ferdinand assassinated in Sarajevo, Bosnia
	The Guns of August, 1914
1917	Russian Revolution; Armistice of Brest Litovsk
1917–1922	Russian Civil War
1918	January: Wilson's Fourteen Points
	November: Armistice
1919	January: Versailles Peace Conference convenes
1920	Russo-Polish War
1920–1922	Greco-Turkish War
1919	Weimar Republic formed
1919–1922	German inflation at its height
1920	League of Nations established
1921	August: U.S. Peace Treaty with Germany
1922	Rise of Fascism in Italy
1921–1922	Washington Conference sets regulations on warships
1922	Treaty of Rapallo; mutual Russian-German recognition
1923	January: French occupation of the Ruhr valley
	November: Hitler's Munich Beer Hall Putsch
1924	Dawes Plan
1925	April: Hindenburg elected president of Germany
	June: Geneva Protocol against chemical weapons
	August: French evacuate Ruhr valley; France starts Maginot Line
	October: Final Protocol, Conference of Locarno
1926	April: Treaty of Berlin: Soviet Union and Germany
	Germany admitted to League of Nations
1927	September: President Hindenburg repudiates German responsibility for WWI
1928	August: Kellogg-Briand Pact renounces aggressive war

1929	Young Plan to reorganize German finances; collapse of U.S. stock market
1930	Allies evacuate the Rhineland; London Naval Conference sets limits to naval arms
1931	December: Allies withdraw from the Saar, Austria
	Japan invades Manchuria
1933	January: Hitler made chancellor
1932–1934	World Disarmament Conference in Geneva
	October 1933: Germany walks out of League of Nations and the conference
1933	Austria: Parliamentary government suspended; Nazi Party made illegal
1934	Hitler threatens Austria; Dolfuss shot in attempted Nazi coup
1935	January: 90 percent vote in Saar (Austria) to reunite with Germany
	March: Hitler denounces disarmament terms of Versailles
	June: Anglo-German Naval Agreement (distances France from England)
	October: Italy attacks Ethiopia
	December: Five-Power Naval Conference
1936	January: Anglo-Italian agreement provides safe passage in Mediterranean
	Spanish Civil War (to 1939)
	February: Franco-Russian Pact ratified
	March 6: British overtures to Hitler for an air pact
	March 7: Germans reoccupy the Rhineland; Hitler denounces Locarno pacts
	October: Italian-German Agreement on Austria; start of Rome-Berlin Axis
	November: German-Japanese Comintern Pact
1937	March: Italian-Yugoslav pact secures borders
	November: Hitler, in secret, announces goals of *Lebensraum* by war
1938	March: German invasion of Austria
	September: German-Czech crisis; Munich Agreement
1939	German annexation of Bohemia and Moravia; violation of Munich agreement
	British Guarantees to Poland
	Italian conquest of Albania
	August: German nonaggression pact with Russia
	September 1: German invasion of Poland

Chapter 7

"Gifts from Heaven"

The American Victory over Japan, AD 1945

The Great Reversal

Between 1889 and 1931, a nation of seventy million people systematically implanted, into their minds and their society, an ideology of sacrifice to an emperor-god. These ideas soon metastasized into a continental war, launched first against Manchuria in 1931, then against China in 1937. In 1941 a coordinated campaign of attacks was launched against the U.S. fleet at Pearl Harbor, as well as the Philippines, Hong Kong, Malaya, Indonesia, and the islands of Guam, Wake, and Midway. By 1942 the war had reached the Aleutian Islands, New Guinea, and Burma—and it threatened Australia, India, and the west coast of America. The seemingly invincible Japanese Empire of the Rising Sun controlled one-seventh of the globe.

By the summer of 1945, however, the Japanese had lost it all. Surrounded by an impregnable armada, they lay prostrate before merciless American bombers. The best of their youth had killed themselves in suicidal battle. Their fleet was sunk. More than sixty cities had been firebombed. Two had been atom-bombed. More than six million troops and civilians were largely cut off in Asia. They were militarily defeated and psychologically shattered, and they faced the possibility of a famine that could kill millions.

Rather than starvation, however, something entirely different blossomed in Japan. Over the next five years, under unprecedented foreign occupation, and with zeal as great as that with which they had once armed for battle, the Japanese reformed their nation. They purged their schools of religious and military indoctrination, ended a feudal agricultural system, emancipated women, abandoned aggressive warfare, and adopted a new constitution. Imperial subjects became citizens, "divine" decrees were replaced with rights-respecting laws, rulers became administrators, feudal cartels became corporations; propaganda organs became newspapers; women achieved suffrage, and students learned the principles of self-reliance and self-government. Hiroshima, formerly a military headquarters, became a world center for nonviolence, and the grandchildren of those who once lined up feverishly for war now marched passionately for peace. War as a national policy gave way to peace as a national policy.

This was the great reversal: an entire nation changing its basic direction, its people rearranging their institutions as they rearranged their minds, and emerging from the rubble of defeat as a first-rank economic producer. What stood between the suicide battalions of July 1945 and the new Constitution of November 1946 was an overwhelming military drive against Japan's social and political center, an intransigent demand for unconditional surrender, the incineration of more than 250,000 Japanese citizens by napalm and nuclear attack, and military occupation. The result, for the three generations to follow, has been peace of a kind seldom seen in history: not universal peace, but an institutionalized, peaceful relationship between former enemies, based on mutual respect for rights and productive trade.

Why did such a fiery, violent victory lead to such a beneficent result? The story turns on the moral nature of the Japanese defeat and rejuvenation, but it begins with the cultural background to the Japanese attacks, and the intransigent American drive to victory.

The Cultural Background to the Japanese "Social Pathology"

The rise of a malignant state—or, in contrast, the creation of a government that protects the rights of its people—does not happen in a social vacuum. Japanese scholar Takemae Eiji posed the right question:

"What social pathology propelled Japan on its course of aggression, leading to defeat and occupation by foreign armies?"[1] This pathology —what Tsurumi Kazuko called "socialization for death"—can be found in virulent sacrificial ideas that the Japanese deliberately inculcated in themselves and their children over the decades preceding the war.[2]

The historical starting point for Japan's relations with modern Western nations was in 1853, when Commodore Matthew Perry sailed into Japan and presented a letter from President Millard Fillmore demanding that Japanese ports be opened to U.S. trade. Over the next decade the shoguns—feudal lords of a sort selected by noble families —signed treaties with the Americans, Russians, French, and British, dramatically expanding Japan's trade with the West. The Japanese grappled with foreign influences that could conflict with their traditional values. The split between the desires of many for Western technology in a world dominated by European powers and the wishes of many others to maintain their traditional values against the encroachment of Western ideas led the Japanese to graft the products of Western culture onto their society. They remained loyal to traditional warrior ideals, powerful clan ties, coercive economic cartels, a land system akin to sharecropping, and strong family and state power over individuals—but now armed with the technology of their European competitors.

In the Meiji Restoration of 1868, dissident samurai warriors drove out the shogun and elevated an imperial dynasty into preeminence. An imperial rescript (or decree) of February 1870 declared the emperor to be a living god, and his throne to be a holy office.[3] The emperor had been largely powerless in the past, perhaps even a hostage taken by leading families to maintain their power. Now he was refashioned into the nominal head of the nation, the divine anchor for its military rulers, and the living embodiment of the "Yamato race" and the *kokutai*, the Japanese "national essence." In the following decades, Japanese leaders created a mythology in which the imperial throne had been occupied from time immemorial by a direct line of succession from the sun goddess Amaterasu. This national mythology was deliberately connected to Shinto, "a cluster of beliefs and customs of the Japanese people centering on the *kami*, a term which designates spiritual entities, forces or qualities that are believed to exist everywhere, in man and in nature."[4] Japanese leaders used Shinto to legitimate the political sovereignty of the emperor, nationalizing Shinto shrines and directing rituals in those shrines to venerate him and to obey his wish.[5]

The emperor, the people, and the land were unified in the cult of the holy *kokutai*—Japan's "national polity," or "fundamental structure." Kokutai was a central concept that linked the spiritualism of Shinto with a transcendent conception of the nation, akin to its Platonic form, that was connected to Shinto mythology and promulgated through state control of Shinto shrines. The Meiji Constitution of 1889 codified this mythology into law: "The Empire of Japan shall be reigned over and governed by a line of Emperors unbroken for ages eternal"; "The Imperial Throne shall be succeeded by Imperial male descendants, according to the provisions of the Imperial House Law"; "The Emperor is sacred and inviolable." The connection to the military was central: "The Emperor has the supreme command of the Army and Navy"; "The Emperor declares war, makes peace, and concludes treaties"; "Japanese subjects are amenable to service in the Army or Navy, according to the provisions of law."[6]

The Meiji Constitution was the product of Japanese investigations into Western constitutional forms. Japanese leaders ultimately rejected English and American options in favor of Prussian-based authoritarianism, a conception of the political state with no room for individual rights, no limits to the power of government, and no citizens, only subjects whose minds and bodies were subordinated to the embodiment of the kokutai, the emperor. The Japanese people learned that "the way of the subject is to be loyal to the Emperor in disregard of self, thereby supporting the Imperial throne coextensive with the Heavens and with the Earth."[7] The imperial "wish" became law, disseminated through rescripts and written into each subject's mind as a moral absolute and an unquestioned divine commandment.

But the emperor was not the direct ruler of the nation, for he was conceived as a god, and gods do not often interact with men. The emperor reigned, he did not rule, and he did so in aloof isolation. The Meiji Constitution established a legislature, the Diet, which served at the behest of the emperor. It also reconstituted the ancient relationship between the warriors and the emperor and placed the authority of the emperor behind a warrior class, which grew in power over decades. The emperor's wish became the means to legitimate the decisions of a military clique—who saw themselves as the guardians of the nation.

The key to the power that the emperor and the military held over Japanese culture was the system of government schools. In 1890 the emperor issued one of the most important documents in modern Japanese history, the Imperial Rescript on Education. The purpose of the rescript was to "counteract ... interest on the part of the masses in things Western and a corresponding neglect of Japan's traditional culture."[8] This decree bound every child to a system of education that focused on worshipful obedience to the emperor, sacrifice to the nation, warrior virtues, and military training. School principals would recite the rescript before their students in a rigid ritual, and mispronunciation of a single word could end in suicide.[9]

Every Japanese child went through this indoctrination. From the moment he could speak, veneration of the imperial throne and duty to the state were his highest ideals. To learn grammar, he wrote out, in longhand, lesson books extolling service to the emperor. He learned *kodo*, the imperial way, and was told that morality was *on*, his obligation to his emperor and his parents. It was drummed into him that the emperor was the embodiment of the kokutai, and that the nation was greater than any of the individuals who composed it. He learned that spreading the emperor's dominion was honor, and that defeat was dishonor, which was worse than death. He memorized the emperor's words—especially the Rescript on Education—and recited its tenets before an image of the emperor, the source and standard for moral judgments.[10] He dreamed of fighting for the emperor.

The educational focal point for the students was the imperial portraits. In a position of worship, and with an attitude of penitent obedience, children faced the portraits, which were ceremonially uncovered while they recited the Educational Rescript. The emperor's image took on the status of a religious icon in their minds. Honor guards watched over the portraits, and teachers died trying to save them from fires. After the sinking of the battleship *Yamato* in 1945, a surviving sailor reported that a comrade had locked himself and the imperial portrait in his cabin as the ship went down, in order to protect the emperor's image.[11]

The essence of this indoctrination was the submission—indeed, the subsumption—of the individual to the imperial will and the state. As a result, fanatical military ideals resonated with people on the street—educated people—who were trained from their earliest days in "blind submission to the Throne and the Imperial state."[12] Historian John Dower relates how an imperial subject, upon hearing that the emperor was to speak on the radio for the first time, knelt and repeated the words of her youth as they rose in her mind: "Should any emergency arise, offer yourselves courageously to the State."[13] This was the moral law she had deeply absorbed. She did not want to die—but she and millions of others were ready to do so if the emperor wished it.[14]

The result was the inculcation of theological militarism into Japanese culture, a synthesis of selfless devotion to the nation, the race, and the emperor with long-standing military ideals in a rigid social structure. The response of individual Japanese to this "socialization for death" took many forms, ranging from militant fanaticism to weary resignation, from undying imperial loyalty to a willingness to ridicule the emperor or even replace him after the war. Not every Japanese civilian was a mindless fanatic, but none could fully escape the ideas drilled into him from childhood under the state system of education. Those who rose to determine the basic direction of the nation's policies were the most extreme—that is, the most principled—of those fanatics. Any doubts people might have had were drowned in the tsunami of their indoctrination and the cultural environment it created. If all else failed, and someone forgot his position as a subject by questioning either the emperor or the actions of the military, the state security forces—charged with maintaining ideological conformity—would take direct action.

With this political, social, and educational system in place, and under military leadership that was motivated to attain a place of dominance in Asia, the Japanese set out to create an empire, "as befitted their destiny as a superior race."[15] This was to be most of all a *moral* war, a holy defense of the traditional values of the kokutai and the emperor, to be fought to the death.

The Attacks Begin

In 1894–1895, a war on mainland Asia left Japan with control over parts of China. But this was followed by a humiliating retreat from Liaotung Peninsula in Manchuria, a retreat not forced by the military success of Japan's foes but under diplomatic pressure from European nations. Many Japanese were highly motivated to regain what they believed had been unjustly taken from them in 1895; the 1904–1905 Russo-Japanese War allowed them to do so. A negotiated peace, the so-called Treaty of Portsmouth, brokered by U.S. president Theodore Roosevelt (for which he won the Nobel Prize), somewhat affirmed the Japanese position. But it also required the Japanese to give up territorial claims in Asia that many thought they should keep. Once again, the peace did not satisfy those Japanese who saw a moral right to control the Asian mainland and oppose the spread of "things Western" in Asia.

After World War I, many European leaders saw Japan—a victor at the Paris Peace Conference—as a competitor that would cut into their share of the "Chinese Melon."[16] Some Japanese concluded that European colonization in the Far East would pose a moral danger to their values, and they renewed their commitment to take control of Asia. The factors were in place for the most radical military officers to solidify their control over policy—but they did not all agree on the best way to serve the emperor. Factions formed within the army over which policies would best protect the Japanese "national essence." Some officers favored an outright invasion of Asia, whereas others favored slower or more indirect action. Everyone agreed about the ultimate aims, however; they disagreed only about the best methods to achieve them. In this climate of ideas, the most radical (i.e., consistent) side won.

The invasion of Manchuria in 1931 by the Kwantung army on Asia—acting without orders from the government in Tokyo—broke the treaty that had ended the Russo-Japanese War in 1905.[17] This ominous action demonstrated the ability of the army to resist political control and act independently, should its officers decide that they better served the nation than the government did. In October 1931 a conspiracy of officers rose against the government in Tokyo and in support of the army. The minister of war in Tokyo sent a radio communication to the army, excerpted here:

1. The Kwantung Army is to refrain from any new project such as becoming independent from the Imperial army and seizing control of Manchuria and Mongolia.
2. The general situation is developing according to the intentions of the army, so you may be completely reassured.[18]

The commander of the Kwantung army replied that he had acted as he had "for the country." Because the different factions agreed on the aims—the aggrandizement of the nation and the emperor—there was much sympathy for the attack, and no effective opposition. Critics collapsed into awe, and political leaders were unable to demand the withdrawal of the army, because its commanders purported to act in service to the emperor and in the interest of the holy kokutai.

Some military officers opposed aggression as a means to aggrandize the state; they offered the last serious challenge to a policy of war. But, sympathetic to the overall aims of their more radical comrades, they were unable to end the aggression in Manchuria. In February 1936 a rebellion by these officers in Tokyo led army leaders to issue a statement, in the name of the emperor, that included: "Your action has been recognized as motivated by your sincere feelings to seek manifestation of the national essence [*kokutai*]." The statement continued: "The present state of manifestation of *kokutai* is such that we feel unbearably awed."[19] The leaders were "awed" before the nationalistic feelings of the rebels—just as the rebels had been awed before the feelings of those officers advocating aggressive war in Manchuria. Everybody was in awe of anybody who shared his feelings for the "national essence." The end result for Japan in the 1930s was that those who advocated the naked essence of their indoctrination—military conquest—rose to control the government. They took the country to war.

In 1937 Japan launched its war against China, this time with the full approval of the Tokyo government. By this point, plans for a wider war were underway. At each step, the supremacy of the emperor, and the duties of his subjects to him, were reinforced through propaganda and imperial decrees. In March 1937 the Education Ministry released the "Cardinal Principles of the National Polity" (*kokutai no hongi*), which reaffirmed the central position of the 1890 Rescript on Education in Japanese life. The emperor was now referred to as a "deity incarnate." In July 1941 the government published the *Shinmin no Michi* (The Way of the Subject), which reaffirmed the national mythology of the emperor's descent from the sun goddess Amaterasu and defined the ruling "national polity" (kokutai) as a "theocracy" in which "the way of the subject is to be loyal to the Emperor in disregard of self, thereby supporting the Imperial Throne coextensively with the Heavens and the Earth." This handbook for subjecthood denounced "individualism, liberalism, utilitarianism, and imperialism" as threats to the Japanese virtues of filial piety and sacrifice to the emperor.[20]

As he went forth to serve His Majesty, every Japanese soldier carried his duties in the form of the *Senjinkun* or Field Service Code, a guide for fighting the emperor's war: "The battlefield is where the Imperial Army, acting under the Imperial command, displays its true character, conquering whenever it attacks, winning whenever it engages in combat, in order to spread the Imperial Way far and wide so that the enemy may look up in awe to the august virtues of His Majesty."[21]

That true character was revealed on December 13, 1937, when the emperor's holy warriors rampaged against Nanking, China. The "Rape of Nanking" may have killed 300,000 Chinese—nearly double the deaths caused by the atomic bombings of Hiroshima and Nagasaki combined.[22] Thousands of women were gang-raped and forced into military prostitution. Thousands of Chinese civilians were herded and machine-gunned, used for bayonet practice, buried alive, doused with gasoline and burned, or decapitated with swords before smiling Japanese troops. The Japanese media covered the killing contests; the *Japan Advertiser* ran pictures of two officers who competed to see who would be the first to kill one hundred men with a sword.[23]

In sum, World War II in the Pacific was launched by a nation whose highest ideals had been made violently hostile to human life. A morality of death as a national policy was leading Japan to the brink of national suicide.[24]

The December 7, 1941, attack on the American fleet at Pearl Harbor was only one prong of a coordinated campaign in Asia and across the Pacific Ocean. The next day, the Japanese emperor issued an imperial rescript on the war—another "divine wish" designed to reemphasize the position of each Japanese person as an imperial subject. The emperor again demanded their subservience; the population bowed their heads and prepared to sacrifice and die. By mid-1942 the American command had retreated from the Philippines to Australia, and thousands left in the Philippines were brutalized in the Bataan Death March. Filipinos were subjected to a nightmare of occupation and jungle warfare. Attacks on Australia and India were looming. The Japanese home islands were far beyond the capacity of American forces to reach effectively.

To roll back and end Japan's drive for empire would require more than military action—serious changes would have to be made *inside* Japan. Before such changes could be made, however, Japan would have to be thoroughly defeated militarily.

The American Drive to Victory

At the start of World War II, Americans were woefully unprepared for war because they did not desire war. In the wake of World War I, they wanted to return to "normalcy," to pursue prosperity through private enterprise. Whether or not President Franklin Roosevelt was eager to shift the American public's focus from his failed New Deal to the war with Japan, he recognized the nature of the Japanese attacks in his request for a declaration of war:

> Yesterday, December 7, 1941—a date which will live in infamy—the United States of America was suddenly and deliberately attacked by naval and air forces of the Empire of Japan. . . .

> Always will we remember the character of the onslaught against us. No matter how long it may take us to overcome this premeditated invasion, the American people, in their righteous might, will win through to absolute victory.
>
> I believe that I interpret the will of the Congress and of the people when I assert that we will not only defend ourselves to the uttermost but will make it very certain that this form of treachery shall never again endanger us.
>
> Hostilities exist. There is no blinking at the fact that our people, our territory and our interests are in grave danger.
>
> With confidence in our armed forces, with the unbounded determination of our people, we will gain the inevitable triumph. So help us God.
>
> I ask that the Congress declare that since the unprovoked and dastardly attack by Japan on Sunday, December 7, 1941, a state of war has existed between the United States and the Japanese Empire.[25]

The president's statement was rhetorically loaded, intended to motivate action against Japan. In the words of Emily Rosenberg, the purpose of "such a speech was . . . to mobilize. It necessarily aims to simplify, to flatten complexity and reduce ambivalence."[26] But the statement was powerful precisely because it was an accurate identification of the essence of the situation. Since the attack, FDR said, "a state of war *has existed* between the United States and the Japanese Empire." Roosevelt wanted Americans to focus their efforts toward making "very certain that this form of treachery shall never again endanger us." His words were directed outward against Japan, not inward at American losses; the entirety of his report about the damage was contained in three short sentences.

Roosevelt thus established a goal-directed posture for American policy, focusing American discourse on the goal of "absolute victory," or "the inevitable triumph." Its achievement required four key steps:

1. Identify the enemy.
2. Make a decision to defeat that enemy.
3. Define victory.
4. Commit and reaffirm, over time, to achieve that victory.

Roosevelt's speech, and the declaration of war, accomplished the first two of these steps—the cognitive act of identifying the enemy, and the political decision to act on that identification. The remaining steps were more complex.

"Achieving victory" would have been insufficient as a goal, for it would have failed to identify the basic nature of "victory." To achieve victory, one must first know what victory *is*. Does victory consist in pushing an enemy army back to its own soil? Holding peace talks or democratic elections? What specifically would constitute victory? The essential answer was: the total and permanent destruction of Japan's will and capacity to fight. What this meant had its own set of problems, but, again in essence, it required the permanent reversal of the Japanese decision and commitment to fight, demonstrated in action by their leadership, army, and population at large, and the repudiation of the ideology behind the war. The meaning for Japan was integrated into two words: unconditional surrender.[27]

The commitment to achieve the unconditional surrender of Japan was reiterated and renewed over years of struggle and tens of thousands of casualties. Roosevelt reaffirmed this commitment in January 1943, at the Casablanca Conference with British prime minister Winston Churchill. As reiterated by a critic of unconditional surrender, Ann Armstrong, Roosevelt connected the idea of a permanent disarmament of Germany, Italy, and Japan to a change in their philosophies, demonstrating his awareness that the causes of wars are in the ideas that motivate people to build and use weapons:

> The President and the Prime Minister, after a complete survey of the world situation, are more than ever determined that peace can come to the world only by a total elimination of German and Japanese war power. This involves the simple formula of placing the objective of this war in terms of an unconditional surrender by Germany, Italy and Japan. Unconditional surrender means not the destruction of the German populace, nor of the Italian and Japanese populace, but does mean *the destruction of a philosophy* in Germany, Italy and Japan which is based on the conquest and subjugation of other peoples.[28]

While still at Casablanca, Roosevelt again stated the "simple formula," in front of some fifty reporters: "I think we have all had it in our hearts and heads before, but I don't think that it has ever been put down on paper by the Prime Minister and myself, and that is the determination that peace can come to the world only by the total elimination of German, Japanese and Italian war power. . . . The elimination of German, Japanese and Italian war power means the unconditional surrender of Germany, Italy and Japan."[29]

The demand for unconditional surrender was restated in November 1943, in Cairo, where Roosevelt, Churchill, and Generalissimo Chiang Kai-shek of China issued a statement that committed the allies "to persevere in the serious and prolonged operations necessary to procure the unconditional surrender of Japan."[30]

We return later to the function of "unconditional surrender." Suffice it here to note that Roosevelt, speaking to Secretary of War Stimson in reference to Germany, said, "The German people as a whole must have it driven into them that the whole nation had been engaged in an unlawful conspiracy." He had also said, at an earlier press conference, that "practically all Germans deny the fact that they surrendered during the last war, but this time they are going to know it. And so are the Japs."[31] Roosevelt, as an assistant secretary of the navy under Woodrow Wilson during the First World War, had seen the consequences of the failure to defeat Germany in 1918. He also recognized that the German and Japanese people had to be convinced beyond doubt of their defeat.

"Victory," and the drive to unconditional surrender, became an integrating idea for millions of people. Americans planted Victory Gardens and flashed the Victory Sign. Military officers became advocates of total victory.[32] The Office of War Information, a propaganda arm of the U.S. government, kept the idea of unconditional surrender alive. A June 1945 pamphlet stated: "Only Unconditional Surrender can lead to the smashing of militaristic hopes and ambitions" in Japan.[33] According to a survey at the time, Americans thought by a margin of nine to one that Japan must be "completely beaten." By the time Harry S Truman took office in April 1945, he might have been impeached had he tried to deviate from this formula. The war in the Pacific was a clash between two of the most powerful nations in the world, driven by two distinctly different moral visions.

The Japanese Decision to Surrender

The months after Pearl Harbor were some of the darkest moments in American history. The destruction of the U.S. Air Force at Manila, General Douglas MacArthur's retreat from the Philippines, and the Bataan Death March remain some of the saddest. At that time, the Americans did not have the capacity to end the Japanese onslaught. Even two years after Pearl Harbor, the Americans had advanced only some two hundred miles up from the south. One journalist commented that, at that rate, they would get to Tokyo by 1960.[34]

But, early difficulties aside, the Americans *were* on their way to Tokyo, and events soon demonstrated that their drive toward victory was very different from expanding the reign of His Majesty. While the Japanese were bound to rituals and seeking death to honor the emperor, the Americans were focused on winning the war and returning home alive. For the Americans, who valued their lives, defeat in battle one day meant they could still fight for their lives the next; for many Japanese, a defeat in battle ended in ritual suicide. Throughout the war, American forces grew in experience while the best of the Japanese—especially their pilots—died at their own hands.[35] The Americans also greatly outproduced the Japanese in weaponry—Japanese economic cartels could not compete with American free enterprise.

By mid-1944 the war had turned unalterably against Japan. In July 1944 the Americans took Saipan and the Marianas, the government of Tōjō Hideki fell, and Japan's management of the war was reorganized under the six-member Supreme Council for the Direction of the War, or the "Big Six." The Big Six could approach the emperor for his sanction only after they had reached unanimous agreement. The military controlled three of the positions, and it could deadlock the council or force the removal of the prime minister at any time.

Absent a specific decision to end the war, the military continued fighting. The effects of the overwhelming American onslaught on Japan in the last months of the war must be understood with reference to the ideas that motivated the Japanese leadership—and the paralysis in Japanese political decision making that developed before the dropping of the atom bombs.

In the last months of 1944, General Douglas MacArthur swept up from Australia and around Japanese troops on New Guinea and returned to the Philippines. In February and March 1945, Admiral Chester Nimitz moved his navy toward Japan from the east and took the island of Iwo Jima in some of the most brutal fighting of the war. These movements placed Japanese cities within the reach of American bombers. On April 4, 1945, the U.S. commands were unified into the U.S. Army Forces in the Pacific (USAFPAC), and MacArthur was placed in overall command.

In early 1945 the Americans stepped up their bombing of Japanese cities. But high-altitude bombing remained long, dangerous, and ineffective; the jet stream blew bombs off course and made it impossible to destroy industrial targets with precision from six miles high. A study of British night bombing in Europe showed that only 20 percent of dropped bombs landed within five miles of their targets. Unable to destroy Japanese industries, Air Force commander General Haywood Hansell Jr. was replaced by General Curtis LeMay. On the night of March 9–10, 1945, LeMay took a huge gamble, which many of his officers opposed as excessively risky for American forces. High-altitude bombers flew low over Tokyo, at five thousand feet, overloaded with incendiary bombs to be dropped on closely packed, balsa wood homes. A horrendous firestorm aided by twenty-five-knot winds blowing in from the west killed more than eighty thousand Japanese.

Up to this point, American commanders agreed that Japan could be defeated only with a land invasion. But in April LeMay wrote that "the destruction of Japan's ability to wage war is within the capability of this command."[36] Perhaps it would not be necessary to spill rivers of American blood in a land invasion.

The Japanese had no defense against the American bombing, and they were, in this sense, defeated. But many Japanese leaders remained convinced that the Americans' will to fight would collapse—if the Japanese people could kill enough during a land invasion. Many Japanese officers were obsessed with the idea of a "final battle" against an American invasion, a denouement that would allow them to preserve the Japanese "national essence."

The result was the basic Japanese strategy of January 1945, *Ketsu-Go*—the "Decisive Defensive Plan for the Homeland." When Okinawa fell in June, the civilian and military authorities issued the "Fundamental Policy for the Conduct of the War," backed by an imperial war rescript, which told the nation to fight to the death with no surrender. All that was needed to force a negotiated settlement was for millions of Japanese civilians to throw their bodies at the Americans in a last charge.[37]

The *banzai* charge was of ancient origin. Americans had seen it at Saipan, where the last remnants of a defeated Japanese defensive force drank a final toast, armed themselves with bayonets, pistols, and sharpened sticks, and charged American armored positions. A national banzai charge would be a suicidal "decisive engagement" against an American invasion of Japan—the final spasm of a suicidal ideology. "One hundred million deaths rather than surrender" was a popular saying, especially in the military.[38] For many Japanese leaders, an American invasion—and the deaths of millions of Japanese civilians—was an energizing *hope*. *Kamikaze* pilots related how "the dream and hope always persisted . . . unwarranted as they were."[39] The hope was pure, morbid fantasy—the culmination of the national mythology that they had been injecting into themselves for two generations.

No issue better illustrates the difference between the Japanese and the Americans during World War II than their attitudes toward an American invasion of the Japanese home islands. The Americans dreaded the idea of such an invasion and would have done anything to win without it. Many in the Japanese military leadership wanted such carnage—as it was their final hope of forcing the Americans to accept Japan's existence as an empire. Americans remembered that battles such as Saipan and Okinawa had required them to kill more than 97 percent of the Japanese defenders. Japanese leaders saw this willingness to die as their great strength—especially when multiplied by millions of Japanese civilians. Trapped between their ideals—which demanded no surrender—and the forces converging on them, a suicidal final charge was their only option.

Army Minister Anami Korechika embodied the inability of the Japanese leadership to reconcile its military ideals with the stark reality of defeat. A staunch supporter of traditional *Bushido* warrior ideals, skilled in archery and swordsmanship, Anami also recognized that intransigence by the Japanese in the face of an American land invasion would lead to mass death. He tried to control the most fanatical officers in the Ministry of War—but he also sympathized with them deeply. He believed that the loss of the imperial line would mean the loss of Japan's very identity as a nation—the kokutai—which was worth more than the lives of all of the emperor's subjects. To avoid this, he was willing to commit any evasion, to embrace any apparition of hope, and to sacrifice any number of Japanese civilians. This was true nationalism—the nation "transcended" any individual lives—and the emperor was its divine embodiment. In his mind, if the Japanese people had the will to fight to the death, the nation might be saved.

The Japanese leaders were not the only ones in denial of their failure. The Japanese population was also disconnected from reality by an endless stream of propaganda that consistently misrepresented the military position of the Japanese and exhorted them to sacrifice. After the surrender in 1945, Kodama Yoshio, a political figure in prison awaiting trial as a war criminal, wrote his book *I Was Defeated* as a statement for his trial:

> Although the nation was resigned to the fact that the decisive battle on the Japanese home islands could not be avoided . . . they still thought that the Combined Fleet of the Japanese Navy was undamaged and expected that a deadly blow would be inflicted sometime either by the Japanese Navy or the landbased *Kamikaze* suicide planes upon the enemy's task forces. Neither did the nation know that the Combined Fleet had already been destroyed and neither could they imagine the pitiful picture of rickety Japanese training planes loaded with bombs headed unwaveringly towards an imposing array of enemy dreadnaughts [big-gun battleships].[40]

A "Die for the Emperor" propaganda campaign leveraged the indoctrination to which the Japanese had been exposed since birth. Without a whimper of popular protest, Hiroshima was made the southern command center for the Japanese Second Army Group, which would coordinate the suicidal "decisive operation." In March the government ordered the establishment of Area Special Policing Units, charged with "a seamless fusion of the military, the government, and the people." Officials prepared to call one million civilians on Kyushu alone into active service. There were no uniforms. Officials suspended all schooling beyond the sixth grade, so that children could stand with sharpened sticks against the Americans.[41] Japanese forces on Kyushu swelled to 900,000. There were no marches for peace, no calls for the leadership to end the war, no vigils for the victims of Nanking, no questioning of the emperor's rescripts. The Japanese continued to work at the Mitsubishi Steel and Arms Works in Nagasaki to churn out materials for weapons.

Richard Frank records a report of July 1945 by a U.S. Air Force intelligence officer, who recognized the nature the Japanese public pronunciations about their war effort and the implications for American attitudes toward the Japanese: "The entire population of Japan is a proper Military Target.... THERE ARE NO CIVILIANS IN JAPAN."[42]

The alternative of surrender or death was made explicit by American and British leaders, who offered the Japanese a chance to accept unconditional surrender and end the war. In July 1945 Roosevelt, Churchill, and Soviet dictator Josef Stalin met at Potsdam, an area in Germany south of Berlin and inside the Soviet zone of occupation. The statement of their commitment to unconditional surrender was a thirteen-point ultimatum, the "Proclamation Defining Terms for Japanese Surrender," included in the Potsdam Declaration of July 26, 1945. The ultimatum was signed by three nations—the United States, Great Britain, and China—because Russia had not yet entered the Pacific war. (There was no international coalition to sign it, and the Japanese continued to maintain embassies, including in Russia.) The statement included the following:

> The result of the futile and senseless German resistance to the might of the aroused free peoples of the world stands forth in awful clarity as

> an example to the people of Japan. The might that now converges on Japan is immeasurably greater than that which, when applied to the resisting Nazis, necessarily laid waste to the lands, the industry and the method of life of the whole German people. The full application of our military power, backed by our resolve, will mean the inevitable and complete destruction of the Japanese armed forces and just as inevitably the utter devastation of the Japanese homeland.
>
> The time has come for Japan to decide whether she will continue to be controlled by those self-willed militaristic advisers whose unintelligent calculations have brought the Empire of Japan to the threshold of annihilation, or whether she will follow the path of reason.
>
> Following are our terms. We will not deviate from them. There are no alternatives. We shall brook no delay. . . .
>
> We do not intend that the Japanese shall be enslaved as a race or destroyed as a nation, but stern justice shall be meted out to all war criminals, including those who have visited cruelties upon our prisoners. . . .
>
> We call upon the government of Japan to proclaim now the unconditional surrender of all Japanese armed forces, and to provide proper and adequate assurances of their good faith in such action. The alternative for Japan is prompt and utter destruction.[43]

The ultimatum accurately represented the facts of Japan's situation, it compared that situation to the awful precedent in Germany, and it made the basic demand clear for the Japanese and the world. It gave Japanese leaders a chance to save the lives of their own people by giving in to the armada facing them. It stated the intentions of the victors not to destroy the population of Japan but to bring justice to war criminals. As an ultimatum, it eliminated any possibility of negotiations between the foes. Ultimately, the fact of the defeat would allow American leaders to reject ambiguities—such as the statement of "terms"—and to enforce the surrender.

The Japanese leadership was unable to agree on a formal reply, and when English-language newspapers indicated that the leadership would "ignore" the ultimatum, Truman was bound to unleash the "prompt and utter destruction." "Ignore" was an inadequate translation of the Japanese word *mokusatsu*, which defies translation without context. It could mean "consider" it, or "table" it, or "take it under advisement"—it was not a rejection. But the leadership was unable to make a coherent statement on the ultimatum, and it did not correct the mistranslation or even state that it was formulating a reply.

During the two weeks that followed Potsdam, the Japanese Big Six engaged in a series of meetings. Its dominant concern was to protect the Japanese "national essence" by preserving the imperial system and the position of the military. Many Japanese leaders still thought that they could cause enough casualties during an American invasion to compel an agreement leaving Japan under imperial control. It was a measure of Japanese attitudes that many Japanese read the Potsdam ultimatum as a weakening of American resolve.[44] If the Americans truly had the will to win, they would not ask for their defeated enemy's agreement in the matter.

As the crisis became more inescapable, the evasions of the Japanese leadership became more fervent. Those leaders remained trapped between the irrational demand of their ideals to fight to the death and the glaring fact of their defeat. Unable to fight effectively, but precluded from suing for peace by their ideals, the Big Six were deadlocked between a hard-line faction represented by Army Minister Anami, who would agree to no terms with his enemy, and a faction that wanted to petition for peace, with conditions to preserve the imperial household. Without unanimous agreement, the Big Six could not issue a decision. Absent a specific decision to surrender, frantic preparations for the final battle continued. Meanwhile, the Japanese emperor ordered the preservation of the symbolic imperial regalia—a mirror, a jewel, and a sword. His priority in the face of destruction was on preserving the physical symbols of the imperial household.[45]

Japanese leaders continued to grasp at any straw, however irrational, to find hope of victory. Some were so deluded as to think that the Russians—who had taken control of eastern Europe and were assembling a massive army to sweep across Asia—would actually enter the war on Japan's behalf. Foreign Minister Tōgō Shigenori (not to be confused with former prime minister Tōjō Hideki) instructed Sato Neotake, the ambassador to Russia, to induce the Russians to adopt a vague, undefined "favorable attitude" toward Japan.[46] One section of Sato's answer to Tōgō's impossible assignment summed up the issue concisely: "If the Japanese Empire is really faced with the necessity of terminating the war, we must first of all make up our minds to terminate the war. Unless we make up our own minds, there is absolutely no point in sounding to the views of the Soviet Government."[47] As Sato realized, Japanese leaders were incoherent and unable to communicate intelligible instructions to their representatives.

There were still no voices in Japan arguing openly for peace, and no prominent or effective "Peace Party." In January 1942, when Tōgō stated before the Japanese Diet that it was necessary to work for peace, his remarks raised a storm of protest and were stricken from the minutes. Prime Minister Tōjō Hideki had faced opposition over his aggressive policies, but the fall of his government in July 1944 was not enough to allow open opposition to the war to coalesce. The "Yoshida Anti-War Group" was a loose-knit group of people who wanted a negotiated end of the war—but it remained underground, and "in practical terms the results of its activities were negligible."[48]

In early 1945 the emperor called in the *jūshin*, or former prime ministers, for good wishes and for advice. Okada Keisuke advised the emperor to consider ending the war—but he urged the use of kamikaze attacks to achieve the victory needed for favorable terms.[49] Only Prince Konoe Fumimaro spoke in favor of peace, in a "Memorial to the Throne" that urged the emperor to end the war. His argument was that the war had unleashed radical social forces that threatened to incite a socialist revolution in Japan. "Regrettably, I think that defeat is inevitable. . . . More than defeat itself, what we must be concerned about from the standpoint of preserving the kokutai is the communist revolution which may accompany defeat."[50] Surrendering to the Americans might be the only way to save the traditional aristocratic state.

But Japan was not a country where one could speak this way with impunity. Two months later, two former officials and a journalist who had helped prepare the "Memorial," including former prime minister Yoshida Shigeru, were arrested. Yoshida was imprisoned for forty-five days and forced to apologize: "Not having thought the matter through sufficiently, I slandered the military, and for this I offer my sincere apologies. I hope you will excuse me on this point. Henceforth I shall revise my mental attitude, and desire to cooperate as a subject in the execution of this war."[51]

In April 1945 Admiral Suzuki Kantarō became prime minister. Several commentators have claimed that the emperor ordered Suzuki to make peace, but Suzuki said at his war crimes trial that he "did not receive any direct order from the Emperor" to end the war—only that he had somehow understood that that was what the emperor wanted. Robert Butow notes that, in discussions leading to the formation of the Suzuki government, one adviser said: "Suzuki could not just openly declare he was going to end the war, but (and it was a significant 'but') that was clearly his intention." Suzuki, however, signed statements as prime minister pledging to mobilize the nation for a last charge—so that "the one hundred million" would throw their bodies forward in defense of the emperor.[52] On June 6 Suzuki had supported the "Fundamental Policy"—that every man, woman, and child should fight to the death. The emperor issued a rescript on June 9, 1945, ordering his subjects to "smash the inordinate ambitions of enemy nations" and "achieve the goals of the war."[53]

Some scholars have purported to look into the minds of Japanese leaders, focusing on the Japanese concept of *haragei*, by which a person might think one thing while speaking or acting in a contrary way. This idea split a person's thoughts from actions and implanted deceit very deeply into Japanese decision making. Given such deception—of self, and of others—a person could later claim that he had really wanted peace, but was able to promote it only by advocating war.[54] But it is the actions of these leaders that matter, not their alleged inner thoughts. They actively, repeatedly, and over years promoted the deaths of millions, as a matter of policy. *After* the war they claimed to have wanted to end the war, but they did not say it openly at the time.

In mid-1945 the government continued to disavow any statement that even hinted at an entreaty to peace. When a Japanese military attaché to Sweden insinuated otherwise, the vice chief of the imperial army general staff corrected the statements: "As we have said before, Japan is firmly determined to prosecute the Greater East Asia war to the very end."[55]

While the Japanese reaffirmed their commitment to fight, the United States began to fulfill its promise of "prompt and utter destruction." On August 6, 1945, Hiroshima was obliterated by a uranium fission bomb. Among the dead were officers in the Japanese Second Army Group, who would have commanded the suicide charge in southern Japan. Truman issued a statement that day, subsequently dropped as leaflets, that if Japanese leaders did not accept the ultimatum, "they may expect a rain of ruin from the air, the like of which has never been seen on this earth."[56]

The atomic bomb was an unprecedented shock to the Japanese leadership. Before Hiroshima, General Anami had maintained that the United States had no such bomb. The Japanese Technology Board had advised that, even if Americans had such a bomb, the "unstable" device could not be transported across the Pacific.[57] But once the bomb fell, its existence—and the American will to use it—could not be denied. Still, Anami refused to give up hope for an invasion, and he claimed that the Hiroshima bomb was the only one the Americans had.

On August 7, the emperor may have expressed his desire, in private, to his chief aide, Lord Keeper of the Privy Seal of Japan Kido Kōichi, to bring about an end to the war, to "bow to the inevitable" in order to avoid "another tragedy like this."[58] Yet he chose neither to do nor to say anything to end the war. On August 8 the Russians attacked Japanese positions in Asia. Russia's attack threatened Japan's northern islands with Soviet occupation, which would likely be followed by the communist revolution the Japanese leadership feared. Still, Japan did not surrender.

On the morning of August 9, 1945, at 10:30 AM—three days after Hiroshima—the Big Six again convened and could make no decision. Anami continued to argue against any form of capitulation. He maintained that the Americans had no other atomic bombs, an understandable conclusion for him, because he would have used them immediately. In the interpretation of historian Sadao Asada, Anami became "almost irrational," declaring, in essence: "The appearance of the atomic bomb does not spell the end of war.... We are confident about a decisive homeland battle against American forces.... there will be some chance as long as we keep on fighting for the honor of the Yamato race."[59] In his mind, millions of Japanese men, women, and children could still save the "national essence"—if only they had the will to fight and die.

Word of the bombing of Nagasaki came while the Big Six were in recess. When they reconvened, Anami had revised his position. He had first denied such a bomb and later conceded that one existed. Now he admitted that the United States might have a hundred atomic bombs and could drop three every day. Anami, Suzuki, Tōgō, and their fellows were forced to accept that the Americans were willing and able to remain offshore and bomb Japan into the bedrock. The rumor spread that Tokyo was to be next, on August 12, three days after Nagasaki.

This complete loss of hope was central to Japan's decision to surrender. As long as its leaders saw even a slim chance of preserving their system, they would grasp at that chance. They had hoped desperately for an American land invasion, but that hope dissipated with the incendiary attacks and atomic bombs. Now there would be no chance to preserve their military system: no great battle, no banzai charge, no honor, no "Nation" to live on—only "prompt and utter destruction."

The Big Six met until 10:00 PM—but remained deadlocked. Three of the six were willing to accept the Potsdam declaration, with the proviso that the imperial house be maintained. The other three—including Anami—demanded further conditions to preserve the position of the military. With no decision possible, Prime Minister Suzuki requested a meeting late that evening—in the emperor's presence. The emperor, in full dress uniform, heard Foreign Minister Tōgō argue for the acceptance of Potsdam and Army Minister Anami argue against it. Suzuki then stepped forward and made an unprecedented request: that the emperor make a decision.

In making his "sacred decision," the emperor chided the army for its failure and noted that the terrible bombs would bring only suffering to his people. He accepted that events did not allow the war to continue—and he expressed his wish for Japan to accept the Potsdam Declaration. This open admission of failure was vital to securing the organized surrender of millions of Japanese troops—but even with the emperor's wish, Anami later had to threaten rebellious officers that anyone who attempted to disrupt the surrender "will have to cut me down first."[60]

On August 10, while thousands of U.S. planes bombed Tokyo and other cities, Japanese leaders sent word to the Americans that they would accept the Potsdam ultimatum, with the proviso that the imperial house would remain sovereign over Japan. The next day, August 11, the Americans replied that the emperor would be subordinated to the Supreme Commander. On August 14 the Japanese government accepted the ultimatum and surrendered. This political decision to submit to the will of the victors was communicated to the army and the populace on August 15, when, for the first time, the emperor's recorded voice emanated from radios across Japan. He told his subjects that circumstances had not turned out to their advantage—again blaming the army for failing to prevail—and that they would have to "endure the unendurable." His subjects bowed before the radios and wept—many in relief, they having expected the emperor's broadcast to be a call to fight to their deaths. For millions of Japanese, the meaning of the American victory was liberation from death. Physically, and psychologically, they were given back their lives.[61]

The American decision to retain the emperor has been widely criticized—and with good reason. Emperor Hirohito was aware of war planning at every step; during the fervent war years, he was apprised regularly of Japan's military resources, sometimes seeing multiple drafts of statements and orders.[62] He repeatedly expressed his desire for further military conquest. He ordered his people to prepare for suicidal charges in his name. He could have expressed a wish to end the war earlier and forced the leadership to confront the need for surrender. He did not. He was a war criminal if anyone was, and he deserved to be executed if anyone did.

But the decision was not in conflict with the surrender, which required not complete destruction, but acquiescence to the will of the victors. The Japanese people overwhelmingly wanted to retain their imperial system, and American leaders knew that to demand its destruction would likely result in a communist insurgency and civil war.

The emperor was retained but constitutionally neutered, redefined from a Prussian-style autocrat into something closer to a British-style figurehead. The emperor accepted his new position fully. On September 2, 1945—the day the Japanese surrender was signed—he issued a rescript commanding his subjects "to lay down their arms and faithfully carry out all the provisions of the instrument of surrender."[63] During the occupation, he denied his status as a deity, accepted a new political constitution, and remade himself into a private man, expressing his desire to return to his chosen field of study, marine biology.

As for Army Minister Anami: given the emperor's sacred decision, Anami was trapped between his loyalty to the imperial throne and his sympathy for junior officers who shared his Bushido ideal of fighting to the end. Some officers demanded that the emperor be replaced with someone who might better promote the "national essence" by prosecuting the war. But Anami refused to countenance such a coup, and thereby prevented a potentially devastating military rebellion. The morning after the acceptance of Potsdam, Anami committed the final act toward which his entire life had been oriented: he committed suicide, by disemboweling himself, leaving behind poetic verses of apology to the emperor for his "great crime."[64]

The shock of the air attacks—and the intransigence of the United States in its demands —had made the issue of surrender an either-or proposition for the Japanese. Sixty years of indoctrination had created a cultural straitjacket that could be removed by nothing less than overwhelming power and intransigence. Under the occupation, it blew apart like a threadbare fabric that had restrained its victims by no means other than their unwillingness to cast it aside. Freed from Anami and his suicidal ideology, the Japanese could remake themselves and their country and affirm their desire to live.

The Occupation

With the surrender, the military regime was revealed as a failure. But it was the American military occupation that brought an alternative to the Japanese people. From September 1945 until April 1952, the

Japanese lived under the orders of the Supreme Commander of the Allied Powers (SCAP). Until March 1951, SCAP was General Douglas MacArthur. All of Japan was under American control; there was no division of the country, no "Communist North Japan."[65] During the occupation, not one American soldier was killed by hostile action in Japan. In mid-1945, when 500,000 troops were anticipated for the occupation, MacArthur was criticized for saying that in six months only 200,000 troops would be needed. But he was correct—and that number fell to 102,000 by 1948. As he reported, "In the accomplishment of the extraordinarily difficult and dangerous surrender in Japan, unique in the annals of history, not a shot was necessary, not a drop of Allied blood was shed."[66]

MacArthur's initial orders were found in the "Basic Initial Post Surrender Directive to Supreme Commander for the Allied Powers for the Occupation and Control of Japan," which was issued by the Joint Chiefs of Staff and based on the Potsdam Declaration: "As Supreme Commander for the Allied Powers your mission will be to assure that the surrender is vigorously enforced." The occupation began with open recognition of the status of the Japanese as a *defeated* people. This concept of defeat, driven deeply into the bedrock of the Japanese mind, established an entirely one-sided relationship with the Americans. The occupation was *forced* upon the Japanese; it was not the result of negotiations: "By appropriate means you will make clear to all levels of the Japanese population the fact of their defeat. They must be made to realize that their suffering and defeat have been brought upon them by the lawless and irresponsible aggression of Japan, and that only when militarism has been eliminated from Japanese life and institutions will Japan be admitted to the family of nations."[67]

Attempts by the Japanese to bargain were cut off. When Japanese officials stated to Truman their desire to control their own foreign embassies, he replied that all instructions "will be communicated by the Supreme Commander at appropriate times determined by him." There was to be no appeal to Washington over MacArthur's head. When Japanese officials tried to instruct the Americans as to the best way to occupy Japan—by keeping American troops out of Tokyo, by letting the Japanese disarm themselves, by allowing officers to keep their ceremonial swords, and by dispatching food immediately—President Truman correctly identified this as an attempt by the defeated to bargain with the victors as equals. He issued a clarification to MacArthur, instructing him to "exercise your authority as you deem proper to carry out your mission. Our relations with the Japanese do not rest on a contractual basis, but on unconditional surrender. Since your authority is supreme, you will not entertain any questions on the part of the Japanese as to its scope."[68]

Unconditional surrender was entirely different from an armistice agreement reached by negotiations. Unconditional surrender began with a *demand*; the alternative was surrender or death, not a choice between negotiating points. There was no contract between the Americans and the Japanese, and American leaders would hear no claims to have violated any contract, akin to the claims to mistreatment raised by the Germans after World War I in Europe. As to claims that the Potsdam Declaration placed obligations on the Americans, the American government adopted the position "that the United States had merely issued an ultimatum 'whereby we told the Japanese what we proposed to do' but was not itself bound to do it."[69]

This American dominance over a vanquished people was overwhelmingly accepted within the U.S. government—and those who misunderstood the situation could find themselves sidelined. During one meeting, a high-level State Department official with a long history of service in Japan suggested that "international law" applied to the occupation, which he claimed was a contract between Japan and the United States. The Japanese, he said, had not really surrendered unconditionally given the language of Potsdam—"the following are our terms." Within three days he was replaced with someone who understood the policy.[70]

Part of the meaning of unconditional surrender was that the Americans assumed no responsibility for the welfare of the defeated Japanese. The initial orders to MacArthur made this clear:

> 13. You will not assume any responsibility for the economic rehabilitation of Japan or the strengthening of the Japanese economy. You will make it clear to the Japanese people that:

a. You assume no obligations to maintain, or have maintained, any particular standard of living in Japan, and
b. That the standard of living will depend upon the thoroughness with which Japan rids itself of all militaristic ambitions, redirects the use of its human and natural resources wholly and solely for purposes of peaceful living, administers adequate economic and financial controls, and cooperates with the occupying forces and the governments they represent.[71]

The point made here was blunt: given the starvation in countries that Japan had conquered, the Japanese had no special right to food. The Americans actually charged the Japanese some costs of the occupation.[72] But MacArthur soon became an advocate for American aid, which eventually totaled more than $20 billion. He opposed reparations, as well as the organized looting, demanded by some nations, of industrial machinery.[73] The essential point, however, is that before the surrender, the American military dropped napalm and atom bombs, not food, on Japanese civilians.

The goals of the occupation reached far deeper than economic matters. The special barbarity of the last year of the war had led American and British leaders to realize that serious change in Japan would require a "decisive rupture with the authoritarian past."[74] The Americans intended to permanently end Japan's will to wage war. An American training film for Americans in Japan displayed this purpose in bold letters: "THIS IS JAPAN'S LAST WAR."[75] Americans at home saw the effort on the cover of the *Saturday Evening Post*: "The G.I. Is Civilizing the Jap."[76]

From the Japanese point of view, from the outset there was never any doubt that they had brought this misery on themselves. As John Dower put it, "Because the defeat was so shattering, the surrender so unconditional, the disgrace of the militarists so complete, the misery the 'holy war' had brought home so personal, starting over involved not merely reconstructing buildings but also rethinking what it meant to speak of a good life and good society."[77]

The occupation proceeded on this basis. Civil reforms flew at breakneck speed. On October 4 MacArthur released his "Civil Liberties Directive," which was followed by a flurry of more specific directives.

The imperial secret police were eliminated; schools were reformed; trade unions were empowered to challenge the feudal economic cartels (the *zaibatsu*); feudal tenancy in the countryside was broken, and many Japanese became small landowners.[78] Women were emancipated, given suffrage and employment opportunities, and the first female officers were soon inducted into the Tokyo Police Department.[79] When Prime Minister Higashikuni Naruhiko and his entire cabinet resigned, the Japanese Diet selected a new prime minister, and SCAP told him to accept the directives.

The most important mission of the occupation was the elimination of emperor worship and religious-political indoctrination. To this end, two major reforms were required: Shinto as a state cult had to be eradicated, and schools had to be purged of indoctrination for service to the state. These were the keys to remaking the moral framework that dominated Japanese political life.

Shinto had been made the national religion in 1882 and was legally tied to political practice in the Meiji Constitution of 1889. Whether or not this connection between religion and state was consistent with the historical tenets of Shinto, the Japanese government employed Shinto mythology to build a national religious cult and used Shinto shrines for political purposes. American leaders in 1945 recognized the danger of Shinto as enforced by the Japanese government. A public statement by John Carter Vincent, chief of the Division of Far Eastern Affairs in the State Department, established a policy that distinguished between the private practice of Shintoism and its politicized form. In response to an inquiry by SCAP, Secretary of State James F. Byrnes replied via telegram (see Figure 7.1), quoting Vincent: "Shintoism, insofar as it is a religion of individual Japanese, is not to be interfered with. Shintoism, however, insofar as it is directed by the Japanese government, and is a measure enforced from above by the government, is to be done away with. . . . there will be no place for Shintoism in the schools. Shintoism as a state religion—National Shinto, that is—will go. . . . Our policy on this goes beyond Shinto. . . . The dissemination of Japanese militaristic and ultra-nationalistic ideology in any form will be completely suppressed."[80]

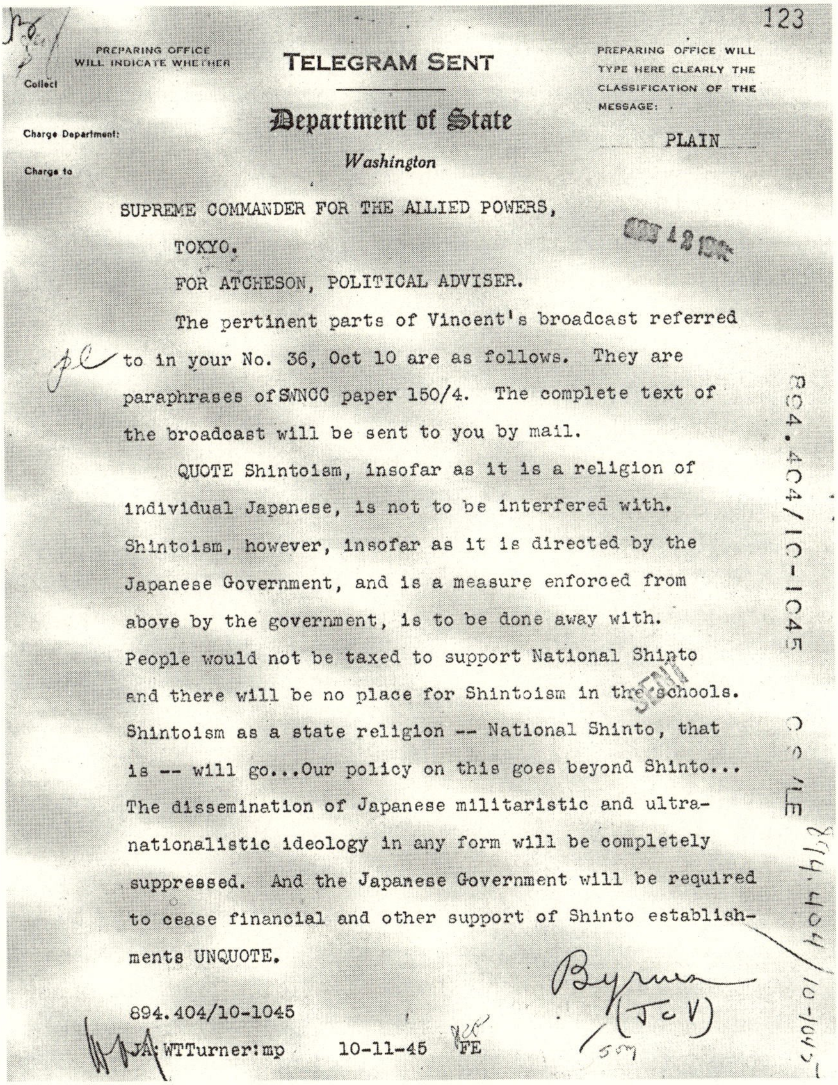

123

PREPARING OFFICE WILL INDICATE WHETHER

Collect

Charge Department:

Charge to

TELEGRAM SENT

Department of State

Washington

PREPARING OFFICE WILL TYPE HERE CLEARLY THE CLASSIFICATION OF THE MESSAGE:

PLAIN

SUPREME COMMANDER FOR THE ALLIED POWERS,

TOKYO.

FOR ATCHESON, POLITICAL ADVISER.

The pertinent parts of Vincent's broadcast referred to in your No. 36, Oct 10 are as follows. They are paraphrases of SWNCC paper 150/4. The complete text of the broadcast will be sent to you by mail.

QUOTE Shintoism, insofar as it is a religion of individual Japanese, is not to be interfered with. Shintoism, however, insofar as it is directed by the Japanese Government, and is a measure enforced from above by the government, is to be done away with. People would not be taxed to support National Shinto and there will be no place for Shintoism in the schools. Shintoism as a state religion -- National Shinto, that is -- will go...Our policy on this goes beyond Shinto... The dissemination of Japanese militaristic and ultra-nationalistic ideology in any form will be completely suppressed. And the Japanese Government will be required to cease financial and other support of Shinto establishments UNQUOTE.

Byrnes
(JCV)

894.404/10-1045

JA:WTTurner:mp 10-11-45 FE

894.404/10-1045

Figure 7.1. Telegram of Secretary of State James F. Byrnes, transmitting the remarks of John Carter Vincent, head of the Office of Far Eastern Affairs, to the Supreme Commander of the Allied Powers in Japan. From W. P. Woodard, *The Allied Occupation of Japan, 1945–1952, and Japanese Religions* (Leiden: Brill, 1972), pl. II.

State-mandated Shinto—the coercion of the Japanese people to follow this mythology and its rituals—was the cardinal means by which the Japanese government was able to motivate the population into suicidal military action.[81] MacArthur's so-called Shinto Directive left the shrines open—a very important issue to many Japanese—but it severed the connection between Shinto and the government. Shinto was reduced from a political mandate to a private matter; this was key to ending the sacrificial, nationalistic mind-set that had infected the Japanese people.

Not only was State Shinto prohibited; nationalistic ideologies as such were out. "The dissemination of Japanese militaristic and ultranationalistic ideology and propaganda in any form will be prohibited and completely suppressed," said the initial orders given to MacArthur. "As soon as practicable educational institutions will be reopened. As rapidly as possible, all teachers who have been active exponents of militant nationalism and aggression and those who continue actively to oppose the purposes of the military occupation will be replaced by acceptable and qualified successors. Japanese military and para-military training and drill in all schools will be forbidden. You will assure that curricula acceptable to you are employed in all schools."[82]

On October 22, 1945, SCAP directed the Japanese to end militaristic teaching, to empower teachers to write their own courses, to reinstate teachers fired for opposing the former government, and to dismiss teachers who refused to cooperate.[83] Directives in October and December 1945 ended courses in "morals" that were platforms for political indoctrination.[84] Textbooks were rewritten. The "school-sponsored or school-directed ceremonies of bowing to the imperial palace ... [and] shouting 'Long live the Emperor'" were ended.[85] In the words of Eiji Takemae, the Americans undertook "an exercise in moral and psychological disarmament balanced by a positive project of institutional reform."[86] Students were taught the importance of challenging dogma and of forming their own judgments.

Acting energetically, Japanese educators excised imperial indoctrination from their classrooms. Before public assemblies, teachers, in tears, apologized for their past activities. The emperor's household instructed schools to eliminate the practice of bowing to the emperor's image, recalled the imperial portraits, and asked that photographs of the emperor be placed alongside others, if school officials cared to hang them at all.[87] In one high school, students boycotted classes and forced their principal to resign; in another, female students denounced their male principal.[88] In 1948 the Educational Rescript of 1890 was rendered null and void by an act of the Japanese Diet.

Young children would be profoundly influenced by the war and its visible effects. There is a powerful photo, in Takemae Eiji's book, of schoolchildren sitting in rows, learning their lessons from a teacher.[89] It was a typical schoolroom scene—except that the children were sitting outside, in the rubble of their bombed-out school. "All this, thanks to the war" was the meaning of their lessons—a repudiation of government propaganda in the form of a clear, concrete demonstration of what their country had done.

The breaking and reconstruction of existing thought structures could be painful. John Dower writes of one Yuri Hajime, who lost his father, brother, and uncle in the war, and who, at age thirteen, was bombed out of his home. Yuri had an immaculately kept language book that he had meticulously copied by hand in school. After the Japanese surrendered, he was told to black out the sections of the book containing the now discredited ideas of the militaristic past. The result left him in tears—but, in Dower's words, it also "left him with a lasting awareness that knowledge could be challenged." The "jumble of contending values" that remained was a step toward finding a sense of personal judgment and freeing himself from the "indoctrinating power of the state."[90]

What were the results of these educational reforms? Theodore Geisel—also known as Dr. Seuss—visited Japan eight years after the surrender and asked students to draw pictures of what they wanted to be when they grew up. The children drew hundreds of pictures of doctors, statesmen, teachers, nurses, and even wrestlers. Only one student wanted to be a soldier—and he wanted to be General MacArthur.[91] The values of these children were already different from those of their parents a decade earlier, when Japanese children dreamed of dying on the battlefield for the emperor.

The importance of the victory here is philosophical—the physical destruction forced the Japanese to destroy and rebuild the concepts by which they understood the world and themselves. Generations of Japanese had formed deeply entrenched mental connections about the nature of the world and their place in it, connections based on the divinity of the "Emperor," the importance of "national essence," and the supremacy of the "Yamato race." The total defeat blasted those connections. The Big Lie that had fueled the Japanese war effort had been blown to pieces, and the truth of the situation now lay bare. One Japanese magazine, in its 1946 New Year's issue, published a photo of the Hiroshima mushroom cloud and titled it "Truth That Emerged from Lies."[92]

The Japanese were reconsidering their long-standing beliefs and coming to see knowledge as the by-product of mental effort rather than the absorption of dogma; virtue as productiveness rather than conquest; their lives and happiness as values to be pursued rather than sacrificed; political leaders as administrators rather than rulers; factories as sources of wealth rather than weapons; and women as equal under the law. The notion of a duty to serve the emperor was replaced with the principles of life, liberty, and pursuit of happiness—written into Article 13 of the new constitution. Mindless obedience was abandoned and replaced with the virtue of rational thinking. Intellectuals began to argue for individualism, recognizing the need for an "autonomous sense of self."[93] These moral reversals were at the center of the Japanese repudiation of war.

John Dower has recorded a range of evidence for this reversal. Rather than veneration, returning soldiers were often treated with contempt.[94] Before the defeat, "thanks to our fighting men" (*heitaisan no okage desu*) conveyed sincere praise for the rise of Japan; after the defeat, the phrase was used sarcastically with respect to the destruction all around them. Military uniforms were called "defeat suits," and surplus military boots, "defeat shoes."[95] Publications changed their titles: *War Technology* magazine was renamed *Peace Industry*; *Wartime Women* became *Ladies' Graphics*; *Wartime Economy* became *Investment Economy*.[96] "Everyone serving for construction" was wartime propaganda used to orient a student's thoughts toward the good of the nation; it was replaced by "construct a nation of peace," and children—including the emperor's son—wrote such phrases as part of their calligraphy lessons.[97] In this moral realignment, many Japanese remembered that they had needed an outside force to bring their freedom into existence. A cartoonist depicted his shackles being cut by U.S. Air Force scissors; he commented: "We must not forget that we did not shed a drop of blood, or raise a sweat, to cut those chains."[98]

Vital to the permanent lifting of the veil of evasion was the creation of a new political constitution for Japan. The story need not be repeated here. The version written by a Japanese committee was deemed "wholly unacceptable" to the Supreme Commander.[99] SCAP became a voice for those Japanese people who wanted constitutional reforms, as against their conservative leadership who wanted window dressing on the Meiji Constitution. In essence, the new constitution corrected the errors of 1889, when individual rights had been rejected in favor of Prussian statism. The key was to turn the emperor into a figurehead within a constitutional government. The emperor did his part; on January 1, 1946, he issued a statement denying that he was an "incarnate deity."

During the following months, there was much haggling, intentional obfuscation, and twisting of verbiage when the English was translated into Japanese. But the result was American in origin, English in its basic organization, and a repudiation of the Meiji authoritarianism. The Japanese Diet adopted it overwhelmingly, and in November 1946 the emperor, in his last official act, ratified it. As the new constitution went into effect on May 3, 1947, a Japanese band played *The Stars and Stripes Forever* in front of the imperial palace—where fanatical subjects of the emperor had once committed suicide. Not until years later did it became public knowledge that MacArthur's staff had written the constitution, but it remains the law of the land in Japan, and in that sense it is a Japanese achievement. The constitution was the climax of the occupation, for it was now impossible to regress into military rule without an open public decision.

The Meaning of the Victory as Unconditional Surrender

Why was America's victory over Japan such a success? One answer lies in the nature of the motive behind Japan's assault and in the nature of the surrender.

All major wars begin with a political decision, and a commitment by the populace, to use horrific violence to achieve certain ends. The political "will to war" is a decision and a commitment to fight for superiority over an enemy.[100] Although it is possible in a brief, small-scale conflict for a leader to act independently of a political decision (for example, a commander may order an attack against an enemy position, as a Japanese commander did in Manchuria in 1931), a war lasting years requires popular support. The ultimate cause of wars is in the ideas—especially moral ideas—that motivate a population to support the war over time. The will to war in Japan—the commitment by the population and the leadership to support the war over years—was founded on the idea that there is something greater than each individual—an emperor-god, a nation, or a race—to which each individual owes his life. Wholesale indoctrination of the Japanese people with this idea is ultimately what gave rise to the horrific aggression they perpetrated on the world. For Japan to lose its will to war, its people would have to repudiate this idea.

Thus, the goal of World War II was to demand that the leadership, and the people, of Japan renounce the decision and commitment to fight and demonstrate that renunciation through long-term action. To achieve this goal, the victors had to defeat the aggressor nation, demand its unconditional surrender, and discredit the ideas that gave rise to its aggression.

It is vital to distinguish between *defeat* and *surrender*.[101] A military defeat is a *fact*, the point at which a nation has no chance of triumphing in war. It is a fact that the Japanese were no longer able to defeat their enemies after mid-1944. But they continued the war because they had not accepted this fact and its implications. Surrender is a *decision* to recognize the fact of defeat, to accept the will of the victors, and to demonstrate such recognition and acceptance in action. It was not enough to defeat Japan's armed forces, for they were the consequence and not the cause of the nation's will to war. Until the underlying causes of the aggression were confronted and discredited, and the nation's population repudiated the basic ideas that motivated it to war, Japan continued to renew its commitment to fight, ultimately by calling on a national suicide charge.

The surrender was a political decision to end the war, and an open, public recognition that the war could not be continued. For the Japanese surrender to be objective, the defeat had to be real, it had to be openly proclaimed by the victors, and it had to be *seen* by the population at large. It is a sad consequence of the war that the Japanese population had to witness the defeat close-up, through horrifying violence. But had their government surrendered before they were devastated by the air war, the surrender could have been seen as a sellout, akin to Germany's acceptance of the armistice in 1918. As it was, several rebellions began within the military; an organized surrender of three million Japanese troops was not a foregone conclusion. But surrender offered the only hope for recovery.

The initial orders to MacArthur following the surrender stressed the need for a proper and open recognition of this fact—that "you will make clear to all levels of the Japanese population the fact of their defeat." These orders were the application of Roosevelt's policy toward Germany; in his comments to Secretary of War Henry Stimson, he emphasized that "it is of the utmost importance that every person in Germany should realize that this time Germany is a defeated nation.... The fact that they are a defeated nation, collectively and individually, must be so impressed upon them that they will hesitate to start any new war."[102]

This is an almost literal reinterpretation of the words of General William T. Sherman: "In accepting war it should be pure & simple as applied to the Belligerents. I would Keep it so, till all traces of the war are effaced ... I would not coax them, or even meet them half-way, but make them so sick of war that Generations would pass away before they would again appeal to it."[103]

Critics of the unconditional surrender policy have claimed that American and British intransigence cost thousands of lives, by extending the war past the point where a negotiated "cessation of hostilities" was possible. But there is an equivocation in this claim, for a "cessation of hostilities" is not the same as a structured, institutionalized peace.

Ann Armstrong has criticized unconditional surrender by citing military officers on both sides of the European conflict who claimed that this demand prevented anti-Nazi factions from rising up against Hitler. Among those officers cited by Armstrong is the German General Alfred Jodl, chief of the Operations Staff of the Armed Forces High Command under Hitler, who said the following after he signed the German surrender on May 7, 1945:

> With this signature, the German people and the armed forces are, for better or worse, delivered into the victors' hands.
>
> In this war, which has lasted more than five years, both have achieved and suffered more than perhaps any other people in the world. In this hour, I can only express the hope that the victor will treat them generously.[104]

That Jodl could still consider the war to be something "achieved" by the German people—that he saw their suffering as worthy of discussion, but not the suffering and slaughter of their many millions of victims—demonstrated his unrepentant attitude, his moral view of the German offensive, and his cognitive break with the facts. Unconditional surrender, as a goal and principle, prevented the Allies from allowing the seeds of the next conflict to remain in place. "No soil would be left from which myths might sprout later that Germany and Japan had not really been defeated."[105] Armstrong implied that this was the case in 1914 when she cited the British military strategist —and fascist supporter of Hitler—John Frederick Charles Fuller, who compared World War II in August 1945 with World War I in 1918, noting that, "whereas in 1918 President Wilson's Fourteen Points offered a fire escape to the beaten Germans, in 1945 [*sic*] President Roosevelt's Unconditional Surrender offered nothing less than total incineration."[106]

One is at a loss whether to begin by noting the results of the 1918 "fire escape" offered to the Germans—a return to war in twenty years—or by asking why the enemy generals of 1945 chose the certainty of continental incineration over an admission of defeat for the regime or even for Germany. Should Jodl have commanded a new German army to rise to further "achievements?" Should another fire escape have been offered to those who had once again set the world ablaze? Armstrong's book was published in 1965—twenty years after the unconditional surrenders in 1945. One wonders what book she would have written, twenty years after the fire escape of 1918?

The same situation existed in Japan, and the necessity of unconditional surrender applied there as well. After the war, former prime minister Yoshida Shigeru called the war a "historic stumble" and an aberration—and he advocated the restoration of conditions in the early Meiji Dynasty. State Department official Dean Acheson knew that the war was not an "aberration" but rather a consequence of Japan's false ideology. His response, that "the economic and social system in Japan which makes for a will to war will be changed so that the will to war will not continue," stated American policy succinctly.[107] Unconditional surrender forced Yoshida to accept the fact that there would be no return to the conditions that had caused the war.

The deeper reasons for the positive consequences of the defeat and the surrender are philosophical, and relate to the mental integrations formed by the Japanese about the nature of the world and their place in it. The "emperor," the "nation," the "national essence," the "Yamato race"—these monolithic abstractions loomed over the Japanese like gods. Breaking the power of the emperor as an "incarnate deity" in the minds of the Japanese was vital to lifting the veil of evasion that had subordinated the minds of the Japanese to authority. To return the Japanese to cognitive contact with reality required an end to the emperor's wish as the source and standard of morality, and an end to a religious myth as the core curriculum in the schools. The power of the air attacks demonstrated the failure of the regime; the occupation connected the failure to the educational system and to Shinto. Seeing the consequences in such a close and horrifying way, the Japanese could then challenge the abstractions, and reintegrate them according to a different standard of moral and intellectual judgment.

The result was a reorientation of the minds of the Japanese, as their existing concepts were smashed and rebuilt. Consider the idea of "war." For Japanese youth in the 1930s—and for the civilians back home—war brought to mind "pleasing the emperor," "honor through sacrifice for the Nation," "venerating one's ancestors," and achieving "renown" among one's fellows. Warriors who committed suicide rather than surrender were considered to be "spiritually pure." Aggressive war was a virtuous activity by which a soldier strove to please his emperor, his nation, his family, and his comrades. The conceptual integration he made was "war" as spreading the reign of His Majesty and the nation, veneration of his ancestors, honor, glory, and goodness.

But what did "war," "honor," "glory," and "goodness" mean in the lands where Japanese soldiers had pursued those ideals? The rapes of Nanking and Manila, the Bataan Death March, the millions looted, enslaved, subjugated, and murdered—for them war meant starvation, suffering, rape, and death. This is the true meaning of military aggression. Wrapped in a veil of ignorance, however, the Japanese back home heard of their glorious advance. They did not know that their fleet was sunk and that their capacity to fight effectively gone—and when they heard of these things, they erected mental apparitions, recited imperial rescripts, and implemented censorship to obscure the truth. They created a "Spartan mirage" for the kokutai, and lived behind the veil.

The firebombings brought the fundamental meaning of war home to Japanese civilians. *This*—point to screaming children, scarred unmercifully—this was what they had been doing to others. Now it has come for them. The bombings concretized the meaning of war and made it impossible to claim that there was goodness in such horror. Accompanied by forthright statements by Allied leaders, the bombings smashed the false intellectual integrations on which the Japanese had been raised. This allowed them to reintegrate the concept "war" into its essentials: blood, smoke, rubble, fear, scars, screaming death. War was now a horror to be rejected, not an ideal to be sought.

The defeat brought the Japanese back to cognitive contact with reality. It broke the connections between sacrifice and glory, death and honor, the emperor's wish and goodness. A new Japanese periodical, *Shinsei*, or *New Life*, wrote that "make-believe and slippery excuses no longer work. The old Japan has been completely defeated. Completely. We must engrave this in our hearts and embark from here on a newborn Japan."[108]

The prose poem "On Decadence" by Sakaguchi Ango—who bore witness to the firebombings—also gave voice to this transformation of values, and its connection to the self-deception that had come from lies: "The heroes of the special attack corps [*kamikaze*] are mere apparitions; human history will begin with those who have become black marketeers. Widows being held up as apostles of virtue are mere apparitions; human history will begin with those who adopt visions of renewal. And finally the Emperor, he too is a mere apparition; a true imperial history will begin with the emperor becoming a common man."[109]

In Sakaguchi's words, the pseudohistory of Japan before its defeat was a series of evasions, myths, and propaganda, which the bombings revealed to be apparitions. The result led to important changes in their orientation toward values. Former warriors could now turn to success in business, widows could forsake mourning in favor of renewal, and the people could value the emperor as a common man. The apparitions that had brought a "social pathology" to Japanese society evaporated, and a morality of death was replaced by a morality of life.

There is a story, repeated by John Dower, that indicates the effects of the victory both on an individual and on Japanese culture. On August 15, 1945, a Japanese businessman, Ogawa Kikumatsu, heard the radio broadcast in which the emperor admitted Japan's defeat at the hands the Americans. Many things might go through one's mind, as one heard one's leader say that one's country had not prevailed and was soon to be subject to an enemy occupying force. Ogawa, his eyes wet with emotion, set to work figuring out how he could make a yen on the situation. He asked himself: what is cheap, easy to mass-produce, small enough to fit in one's back pocket, unknown yesterday but indispensible from this point on?

His answer was an English phrase book. Hundreds of thousands of Americans were on the way, and the Japanese would need help communicating with them. Ogawa sold the idea to a publisher, and in three days or less (so the story goes) he produced a thirty-two-page guide to English. By December, three and a half million copies had been sold.[110] The first three entries were: "Thank you!" "Thank you awfully!" "How do you do?" (They were transliterated phonetically as: *San kyu! San kyu ofuri! Hau dei* [or, *hou dei dou*]!) Perhaps there was no better portent for Japan's future than this: upon hearing of his country's defeat and surrender, Ogawa's first thought was a productive act taken as thousands of Americans prepared to arrive.

Of course, the Japanese experienced pain during the readjustment after 1945—pain caused by the material devastation of the war, which the occupiers made clear was caused by the Japanese pursuit of empire. But they were able to trust the Americans in a way that was never possible for the victims of their own army—another strike against the propaganda of the Japanese government. The Americans took no systematic vengeance against Japan. The return of the Japanese as productive allies and robust economic competitors, standing on their own feet, was of enormous benefit to both sides. When the war with America had begun in 1941, the *Worker Encouragement Press*, a propaganda rag, published a drawing of Japanese people kneeling in worship before the imperial palace. On September 1, 1945, the day before the Instrument of Surrender was signed, the publication, in its new "Guide to Recovery" issue, published the same drawing, with a new title: "We have cried all we can. Let us stand up."[111]

No one can predict the future—but we can consider the past three generations of peaceful coexistence with Japan, and study the actions of those who demanded and achieved the unconditional surrender of their enemy in 1945. The victory united word and deed on moral and cognitive levels, revealed the failed policies of a deranged regime, and allowed the best of the Japanese to rise and to prosper.

Epilogue: The Controversy over the Atomic Bombs

The controversy around the American use of atomic weapons refuses to die, despite the fact that on the morning of August 9 the Japanese were not willing to surrender but by late evening they were. The most important reason was their utter helplessness before the American air war and the shock of the atomic bombs, combined with the entry of Russia into the Asian war. Most scholars have correctly recognized both the motive behind the American decision to use the bomb—to save American lives—and the effect it had on Japan's decision to surrender. We have direct testimony from Kodama Yoshio, writing immediately after the war:

> The dropping of the atomic bomb threw the Government and the clique within the Imperial Palace into the throes of fear. The leaders of the Japanese Army and Navy, however, were even more greatly taken aback at the appearance of this new weapon. Almost simultaneously with the dropping of the second atomic bomb on Nagasaki, the Soviet Union declared war on Japan. As a result, the Government and the clique within the Imperial Household completely lost all will to continue the war ... stunned by the appearance of the atomic bomb, and paralyzed of all feeling by the round-the-clock air raids of the enemy, the nation had no more ability left to feel consternated by the Soviet Union's attack on Japan.[112]

Nevertheless, in the decades following the war, the historical revisionists got busy. The fiftieth anniversary of the Hiroshima bombing saw renewed attacks on the conclusion that the bombs were dropped to save U.S. lives. Revisionists claimed that the decision to use the bomb was driven by ulterior motives—such as the desire to manipulate the Soviets, hatred of the Japanese, racism, and imperialism.

Early criticism of the bombs can be found in the memoirs of Acting Secretary of State Joseph C. Grew, as well as Secretary of War Henry Stimson.[113] But the most influential criticism with respect to American motives comes from Gar Alperovitz's *Atomic Diplomacy*, which claims that the bombs were used to shape postwar diplomacy with the Soviets. They were the first act of the Cold War, he claimed, not the last act of World War II. Alperovitz was anticipated by Blackett's *Fear, War and the Bomb* published in 1949. The revisionists' claims have been briefly summarized by historian Robert James Maddox, who correctly concluded that "revisionist allegations are based on pervasive misuses of the historical record and should not be taken seriously." In addition to his analysis of the effects of the bombs on Japanese decision making, Asada has delineated some of the historical controversies about the use of the bombs, from Japanese perspectives.[114]

Hasegawa Tsuyoshi has also created a revisionist history, based on extensive reading of Russian, Japanese, and American archival evidence: *Racing the Enemy: Stalin, Truman, and the Surrender of Japan*. Hasegawa agrees that the Japanese leadership had ample opportunity to end the war and did not, and that Hirohito was concerned primarily to preserve the imperial house, not to save his own people. Hasegawa brings forth a unique Japanese letter protesting the use of the bombs, which requires, he claims, serious consideration.[115] The letter, however, may illustrate not a desire to end the destruction and thus the war, but rather the ineffectual nature of the peace faction in Japan, and thus the need for a decisive American victory, achieved quickly. The letter may also have been propaganda intended to undercut the American will to win, a conclusion that is consistent with the repeated commitment of the Japanese to continue the war, to motivate the population for a national suicide charge, and to use a super-weapon if they had it. In the end, Hasegawa provides no evidence for his claims that Truman used Potsdam "to justify use of the bomb" or that Stalin saw "American duplicity" in the Potsdam declaration.[116] *Racing the Enemy* is full of fascinating archival material, but the arguments are often based on psychological conjectures that run contrary to the course and outcome of the war.

The postwar world was surely on Truman's mind at Potsdam in July 1945; he would have been shortsighted indeed if it were not. But the evidence overwhelmingly demonstrates that his main goal was to avoid the horrendous casualties that would have resulted from a land invasion of Japan. Americans knew that seventy-six thousand Japanese soldiers on Okinawa had killed or wounded some seventy thousand Americans.[117] All evidence, including the American actions during the occupation itself, shows that to defeat Japan while avoiding such vastly greater carnage was Truman's primary concern.

But did American leaders have knowledge of other options that would have eliminated the need for killing thousands at Hiroshima and Nagasaki? For instance, should the Americans have blockaded Japan and waited for surrender? Undoubtedly, a blockade would have worked eventually—perhaps four months later, in December. Would that have been reason to forgo use of the bombs?

First off, ten thousand American POWs would likely have died in those four months, adding to the many others worked to death while starving in subhuman prison camps. They had not started World War II; there is no rational moral basis for sacrificing them in order to save the lives of those who did start the war. But there is more. Hundreds of thousands of people were dying every month in the Asian war started by the Japanese. A four-month delay could have cost more than a million Asian lives. As Richard Frank put it, the Japanese civilians at Hiroshima and Nagasaki "held no stronger right not to be slaughtered than did the vast numbers of Chinese and other Asian noncombatants, the Japanese noncombatants in Soviet captivity in Asia, or the Japanese noncombatants (not to mention Allied prisoners of war and civilian internees) who would have perished of starvation and disease in the final agony of the blockade."[118]

A four-month blockade could also have killed one to two million Japanese. They were already living below sustenance levels—a famine would have killed far more than did the bombs. Had the Americans wished to commit genocide, a blockade and a news blackout would have been the most effective way to do so. But such a slow squeeze would have given the Japanese government time to react, to adjust its policies and its propaganda, and to put down domestic opposition. Further, to end the war the U.S. Army Air Force planned to drop up to 100,000 tons of conventional bombs every month on Japan—the equivalent, over four months, of more than twenty Hiroshima bombs. All told, the cost of a four-month delay in ending the war could have reached three million lives.

There is more. With four additional months to position their troops, the Soviets would have taken control of some northern islands—Sakhalin Island for certain, probably Hokkaido, and perhaps northern Honshu. Japan might have been divided, and two generations of northern Japanese forced to live under the Soviets. The bombs likely prevented the Soviet enslavement of twenty million Japanese.[119]

Some revisionists claim that America dropped the atom bombs not out of military necessity but from ulterior motives, such as vengeance or racial hatred. If the Americans had acted on such motives there would have been clear evidence from the occupation, when they had a free hand. There was no "rape of Tokyo" akin to Nanking. But such claims are also inconsistent with the nature of the Japanese war effort. The Japanese military machine was dispersed throughout civilian areas and was never limited to troops. It was impossible to bomb factories without hitting civilians. Hiroshima was the headquarters of the Japanese Second Army, which commanded the suicidal defense of southern Japan. Civilians did not protest the thousands of troops in their city—many joined in the effort and prepared to die. The survivors of the atomic bombs became advocates for peace only *after* the bombs fell—when it was safe to hold such opinions and to act on them.

George Weller, a news correspondent who defied MacArthur's orders against travel to Nagasaki, sent back dispatches with what he had seen. They have only recently been published. As Weller wrote on September 8, 1945, "All around the Mitsubishi plant are ruins which one would gladly have spared." But to spare those plants—and the prison camp placed next to the armor plant—would have spared "the war plants of death," which he details for his readers: the ammunition factory (with 1,740 employees), the ship parts plant (1,016 employees), the Mitsubishi torpedo plant (7,500 employees), and three steel foundries (3,400 employees).[120] In Weller's evaluation, Americans needed to know that Nagasaki had been mass-producing weapons with the active assistance of thousands of civilians.

Another revisionist claim is that that U.S. officials invented high projections of casualties in order to justify use of the bombs. Certainly U.S. officials disagreed about casualty estimates, and thanks to the victory, all such estimates remained conjectural, and the men they represented remained alive. In June 1945 Americans estimated that more than 300,000 Japanese defenders would greet them on the southern island of Kyushu. By July this estimate was revised to 600,000. After the war, Americans discovered that 900,000 trained defenders had been ready on the southern island of Kyushu alone.[121] Americans knew that the ratio of defenders to American casualties on Iwo Jima and Okinawa approached one to one. The lowest computations for an invasion of southern Japan put American casualties, in the first ninety days at 132,385, with 25,741 killed and missing. The 1945 estimate of William Shockley (who later invented the transistor) put American casualties at 1.7 to 4 million, with 400,000 to 800,000 dead. Later apologists did *not* invent the higher figures. It was the revisionists who invented the fiction that the war would have ended quickly without the bombs or an invasion.[122] Truman had every reason to think otherwise.

As a parallel to these claims, the estimates of how many were killed in Hiroshima and Nagasaki have risen in the decades since 1945. Weller was aware of this, and it contradicted what he had seen close up. In 1966 he wrote a memoir of his experiences, and he disputed the figures that had arisen in the previous two decades. "Nagasaki's casualties have been rising in multiples of five and ten thousand. At the most recent ban-the-bomb meetings the dead tripled to 65,000."[123] This also contradicted what he had seen in the city itself: "Nagasaki was never, strictly speaking, 'destroyed'. Nagasaki had about 300,000 people, about the size of Worcester, Peoria or Tacoma. About 20,000 died right away, the majority by concussion from falling buildings or by burning in ruins, not by concussion of air or direct singeing. I was told that about 35,000 had been hurt, mostly by burns. Harrison's [Sergeant Gilbert Harrison] figures were 25,000 and 40,000. About 18,000 homes, mostly two-room bungalows, were destroyed, for perhaps $20 million worth of total replacement."[124] Weller's figures may not be precise, but his reports about the city do not support claims to the total and gratuitous destruction of a purely civilian target.

Some critics have claimed that the Americans could have demonstrated the frightening power of the bomb in an uninhabited area. This ignores the many obvious problems with this strategy, including the possibility of a dud that would have proved a point opposite of that intended. A strike against Tokyo harbor was ruled out because the wind would likely blow the bomb off course—as it did at Nagasaki. But the decisive issue in this regard was that the Japanese leadership needed to know that Americans had not only the *capacity* but also the *will* to act. American weakness of will was central to Japanese thinking—the *Ketsu-Go* plan depended on breaking the American will to sustain massive casualties. A demonstration of the bomb would have shown to Japanese military fanatics that the Americans did not have the will to use it. In their minds, it would have been a demonstration of weakness.

Yet another claim of the revisionists is that Japan was already defeated and that the bombs were unnecessary. Although it is true that the Japanese were materially defeated by 1944, they had not reversed their decision and commitment to fight; they had not accepted the fact of defeat and surrendered. Even after the bombing of Nagasaki, the Big Six could not agree to sue for peace, and many Japanese military officers retained their commitment to fight. The emperor's decision was vital to securing the surrender of millions of troops, and that happened only after Hiroshima and Nagasaki.

But the objection to the use of the bomb that deserves the most emphatic repudiation is the claim that it is an inherently immoral weapon. All weapons—from bowie knives to hydrogen bombs—are designed to kill, and there is a scale of destructiveness on which they fall. Atom bombs are at the supreme end of that scale; they can quickly kill a lot of people. To break the Japanese leadership out of their ideological blinders and end the war, American leaders needed to kill a lot of Japanese in a visibly shocking way. The resulting shock led to an immediate end to the war.

It is worth repeating the well-known fact that those Americans fighting the war—the soldiers who prepared to invade Japan—had very different views of the bombs than do the revisionists. Paul Fussell, an American soldier in Europe, was ordered to the Pacific not long before the anticipated American land invasion. His "Thank God for the Atomic Bomb" emphasized the "experience, sheer vulgar experience" of combat that separated those who praised the use of the bomb from its critics.[125] He and his fellow soldiers thought they were going to die in Japan, and they were right to think so—until the bombs were dropped. Also typical of thousands was an American army medic in Europe, Technical Sergeant Arnold Taylor, who was ordered after Germany surrendered to join the 82nd Airborne Division for the pending invasion of Japan. He shipped out on an aircraft carrier—but was spared the agony of seeing thousands of fellow soldiers die.[126] The bombs literally saved the lives of these men—and tens of thousands of others like them.

For those Japanese who wished for an end to the bloodbath, what fell out of the sky on those two days in 1945 were, in the words of Japanese navy minister Yonai Mitsumasa, "gifts from heaven." Sakomizu Hisatsune, chief cabinet secretary of Japan, said after the war: "The atomic bomb was a golden opportunity given by Heaven for Japan to end the war." Okura Kimmochi, president of the Technological Research Mobilization Office, wrote before the surrender: "I think it is better for our country to suffer a total defeat than to win total victory … in the case of Japan's total defeat, the armed forces would be abolished, but the Japanese people will rise to the occasion during the next several decades to reform themselves into a truly splendid people.… the great humiliation [the bomb] is nothing but an admonition administered by Heaven to our country."[127]

The American victory smashed the infrastructure of Japan's sacrificial indoctrination machine; this empowered the Japanese to break the chains of their suicidal regime and to drown out the discredited exhortations of the past.

Conclusion

The Lessons of the Victories

The six wars examined in these seven chapters varied substantially in the pretexts by which they began, the political decisions that brought the opponents to war, the tactics and technology of battle, and the nature and effectiveness of the postwar settlements. But each exhibits a certain underlying cause that led to the initial attacks, followed by a period of stalemate and protracted carnage that was ended only when the side under attack launched a motivated, forthright offense against the center of its opponent's power. These wars were fought by commanders who were oriented toward solid objectives and who used flexible strategies to pursue firm goals with an inflexible will. Each struck to the center of his enemy's strength, and achieved a physical victory that extinguished the moral and ideological fire behind the fight.

These events affirm the adage of Sun-tzu, that no one ever profited from protracted warfare. But Sun-tzu saw more than the strategic implications of this; he understood something of motivations, noting that "the army values being victorious; it does not value prolonged warfare."[1] To "the army" in this statement of values may be added "and the civilian population," for both are impoverished by an indecisive war of attrition without a defined end, and both profit by progress toward success. In a conflict between moral purposes, both sides can be demoralized by the approach of a powerful, confident army that demonstrates its ability to bring a fast end to the war. In these events, the first wave of the counteroffense was psychological: a shockwave of rumors and propaganda that moved ahead of the army and prepared the ground for the moral and psychological collapse of resistance. With the open and visible collapse of the will to fight, the nature of the defeat was inescapable, and issues from the old war did not become pretexts for a new war.

This moral perspective reveals an issue of cognitive import: that the fact of defeat in these cases led to lasting peace only when the fact of defeat was openly recognized and the legitimacy of the victor's terms was accepted by the vanquished. The layers of evasion and public obfuscation that followed World War I must count as important causes of the war that followed. The veils of illusion were ripped aside when Xerxes, Hannibal, Jodl, Hirohito, and their supporters saw the war end in failure. Agesilaos and Hitler could not admit defeat; they had to leave or die. The assertion of the "fact of defeat" in the victories of 1945 was vital to achieving the real, and deep, reforms needed to prevent the causes of the previous war from returning with a literal vengeance.

Deception and propaganda can contribute to the defeat of an enemy—but they did not define the basic goals of these wars. The power of the victories over the Persians, the Spartans, Rome's enemies, the American Confederacy, and Japan and Germany emanated from the fact that the goals of the victors were openly stated; there was an integration of word and deed. With the goals of the war made publicly explicit, every action could be designed and evaluated against the progress needed to attain those goals. Visible movement toward the objectives required to attain these goals motivated the victors and brought a sense of impending defeat to the losers, precisely because these movements were understood in relation to the goals. This does not mean that there were no ambiguities in the details of those goals—there was much that was unclear at the start of the occupation of Japan, for instance—but the fact that the existing regime had been defeated was clear to all. The details were handled at the discretion of the victors—of that there was no doubt.

The point is not only that strategy must serve policy goals but also that rational policy goals are derived from a proper understanding of self and others. The wars in this book were launched by leaders who were bent on loot, conquest, or slaves and who grossly misjudged their enemy as well as their own people. The failure of these leaders to formulate proper policy goals elevated the consequences of these conflicts far beyond what their own people could support, or their enemies could accept. This put their regimes at risk of total defeat should the military expedition fail. They lost sight of the nature of their enemy, of the depth of the enemy's motivations, and of the moral status of their own policies in the eyes of their enemies. The failure of these leaders to challenge the ideas that drove them and their people forward led them to take actions that roused the righteous anger of their enemies and placed their own regimes in mortal danger. Their foes abetted the slaughter as long as they failed to confront the true source of the attacks. Once they did, the battle and the conflict turned quickly in their favor.

The overall lesson here is to *take ideas seriously* and, most of all, to *recognize the power of moral ideas*. With such respect for ideas, we can approach questions that "realist" considerations of power relations alone cannot answer. In the clash between moral purposes at the heart of each of these conflicts, strength of commitment to the purpose empowered those who held it—as Chamberlain found out at Munich. The desire of the German people for their nation to rise again, anchored by claims to unjust treatment under the Versailles Treaty, was far more energetic than the desire of the British people to maintain a tranquil status quo. The British turned irrevocably against appeasement only when they saw Hitler take Prague, and the nature of the threat became clear. On this level, it was the relative commitment of each side to its moral cause, not the truth of that cause, that affected the outcome of the conflict.

But truth matters. On a deeper level, the examples in this book also show that the strongest power belonged to those who were, in fact, right, if those who were right knew it. This may be unfashionable to say today—in an intellectual climate that sunders fact and value, and understands moral claims as inherently contested matters of opinion —but it remains a demonstrable fact that the Spartan and Confederate slave systems were morally debased and that the freedom upheld by the Thebans and the Union was good. The political autonomy upheld by the Greeks, as well as the political relationships between

Rome and its Italian allies, was superior to the alternatives presented by Persia and Carthage. Certainly, the war between America and Japan in 1945 was not fought over morally equivalent options—not if peace and prosperity for millions of people are valued. The tragedy of Munich is in the failure of the British to recognize that their own moral norms could become weapons when manipulated by a vicious dictator. The British and the Americans—like the Greeks—became truly unbeatable when they grasped how right they really were. As the war progressed, public exposure of the enemy's actions strengthened the victor's knowledge of its own moral rectitude and discredited their former enemies' failed policies in their own eyes.

Readers should question how this approach would apply to the many wars that have not been considered here. How, for instance, did moral ideals condition the armies of Napoleon, which reformed even after the defeat in Russia and his first exile—and why did those armies never again gain such strength after his final removal from the European scene? The defeats of the British by the American colonies, and of the Americans by the North Vietnamese, are also among the notable omissions, for which a few words must suffice here. The inquiry, in each case, would focus on the facts and issues at stake, on the moral terms by which each side fought as it did, on the way they understood those moral concepts and applied them to their situation, and on the reasons why those who lost were able to accept defeat. In each case, the stronger power decided that it was in its interests to battle the weaker. Yet in each case the stronger power accepted defeat without using its full force.

Why did the British not expend their full force against the colonists, which would surely have ended the rebellion? The answer is found in part in the wider context of Britain's involvement with the world. Tied down in a struggle to protect its interests in Europe and across the globe, a full-blown military effort to bring the distant colonists to heel could have jeopardized Britain's international position to an unacceptable degree. The colonies were simply not worth it. This much is similar to America's position with respect to Vietnam; fear of a conflagration far wider than Vietnam limited the amount of force the Americans were willing to use. For all of the bomb tonnage dropped on Vietnam, Hanoi was not obliterated, the docks at Haiphong remained operational, and nuclear weapons were not used.

In each case there was also a moral issue at stake. In the century before the American Revolution, England had created a new relationship between Parliament and the crown. The meaning of this development appeared in John Locke, whose Second Treatise was published in 1689, the year that brought William and Mary to the throne in the Glorious Revolution. Locke's ideas upholding the rights of man were respected in England and were nearly quoted in American founding documents. As the cultural source of these ideas, there was no deep philosophical problem in England with accepting the basic desires of the Americans to exist without tyranny—the problem was political. The Americans—led by a deeply committed group of revolutionaries, with strong but not unanimous support in the population—could accept nothing less than what was stated in their Declaration of Independence, a document that connected their moral ideals with political realities and a firm objective. British politicians and their king wanted to maintain the status quo: the colonial status of the Americans. But once the British recognized their inability to maintain their political hold on the colonies, they reconciled themselves to the fact of American independence, and they became, over time, lasting allies of the Americans. The "special relationship" between England and America is the result—a relationship that has lasted, despite particular disagreements between leaders, precisely because of their shared values.

Vietnam has been said to have parallels to the American Revolution, given propaganda inside America that presented communist expansion as a "people's liberation." In evaluating this situation, one would have to know the facts. Was the situation in South Vietnam an internal uprising, or had the Viet Cong, supported by North Vietnam, created terror by murdering some seventy-five hundred local officials between 1960 and 1965?[2] The latter was true—and it had no precedent in the American Revolution. Many Americans knew that "people's liberation" was a smokescreen for the violent expansion of communism, which was pursued with revolutionary fervor. Retired General Maxwell Taylor, in a 1966 statement to the Senate Foreign Relations Committee, provided the most succinct statement of this fact: "The term 'war of national liberation' is merely Communist jargon for the use of terrorism and guerrilla warfare to subvert a non-Communist government while disguising the aggression as a civil revolt."[3]

Given this conclusion, the next question would be whether the communist government of the North was ideologically capable of ending such terrorism, and accepting a noncommunist South. Could they reverse the doctrine of communist expansion? Were the North Vietnamese similar to the British in 1780? Or was violent revolution and communist dictatorship so deeply ingrained in their political ideology that they could never accept a noncommunist South? The answer to this last question was yes—they could not accept an independent South Vietnam, and they never did. History shows that everything they did—including every negotiation—was designed to subordinate the South to communism. Given this unshakable motivation, and given their proximity to the South, the result was a horrific jungle war, fed by a constant flow of materials, and fought by highly motivated warriors who never gave up.

In deciding what to do about this, American leaders were driven by the memory of Munich 1938, and they vowed "No more appeasement!" But better questions should have been asked about the moral principles on display in 1938 and their parallels to America and the communists. Should the Americans have granted to the government of North Vietnam the status of a legitimate state, with "equal rights" among others? Should Americans have defended the principle of national self-determination in South Vietnam, if it included the election of a communist government? If the answers to these questions were no, and given the nature of the government in the North, then the Americans should have had one major objective—the fall of the North Vietnamese government—or they should not have fought at all.

But the Americans disavowed up front the goal of ending the government of North Vietnam. General Taylor said, in 1966, "There is nothing in this definition of requirements, however, which in itself requires an unconditional surrender of the Communist forces or the destruction of the Communist state of North Vietnam." The means that American leaders were willing to use—a limited war—defined

American policy. Means and ends were reversed, and strategy dictated policy. American leaders evaded the fact that revolutionary expansion was inextricably part of communist ideology, because doing so would have required them to act in ways they were not willing to do. Yet they were not willing to withdraw from the conflict either. As a result, American leaders walked a middle road: they accepted the legitimacy of the government in North Vietnam, all the while that government was killing American soldiers in order to establish the same government in South Vietnam.

To show that peace was their goal, American leaders engaged in peace talks while American soldiers were still dying. The result was to emphasize the legitimacy of the enemy government and to demonstrate the Americans' weakness. The North Vietnamese used the peace talks as Hitler had used his negotiations with the British: to further their own aims. But Munich could have been used to understand that by seeking national self-determination for South Vietnam, American leaders actually sought the same goal as that pursued by the North Vietnamese: the national self-determination of South Vietnam's future. North Vietnam had determined its own national destiny through a communist takeover, which American leaders accepted as legitimate by negotiating with its government. How could the Americans avoid seeing the same takeover of South Vietnam as legitimate? The lessons of Munich would have shown that, given this agreement on basic goals, the North had to win. The question was only one of time, and body counts.

The Americans had only two courses of action open to them: to accept the existence of the North Vietnamese government and therefore the fall of the South, or to destroy the government in the North as a necessary condition to an independent South. In either case, Sun-tzu should have been consulted, for the protracted campaign that followed was more damaging than either a fast destruction of the northern capital or the swift fall of the South without a fight would have been.

In describing the objectives of the war, General Taylor cited the historian Polybius, who, Taylor claims, opposed the final destruction of Carthage when he wrote that the proper object of war is not "annihilation" but rather to lead opponents "to mend their ways." "Sooner or later," said Taylor, "it is to be hoped, the leadership in Hanoi will become convinced that they have no choice but to mend their ways." A sharper concern for history would have shown that the precedent of Carthage did not apply to Vietnam in this way. First off, Polybius wrote this passage with respect to the tyrant Philip V of Macedonia, who aggressively brutalized the Greeks.[4] Beyond this erroneous citation, Carthage had accepted the peace ending the Second Punic War and had upheld the peace for two generations. Their greatest general renounced the war and rebuked those who disagreed. The Carthaginians engaged in no aggression, asked the Romans to mediate border disputes, crucified a leader who may have urged war, sent a peace envoy to Rome, and surrendered their arms to the Romans. For two generations the prosperous, trade-oriented Carthaginians made clear in word and deed that there was no aggression left in them.

North Vietnam displayed none of this. Revolutionary communist expansion (not prosperity) was at the center of the regime's ideology, and communist revolution was openly stated to be the prime objective of the regime. This objective motivated the regime's supporters, especially the guerrillas in the South. The "ways" of North Vietnam could not have been "mended" other than by the visible fall of the communist regime. The proper historical example was not Carthage, but Sparta, Nazi Germany, and imperial Japan. If American leaders were not willing to undertake the task of ending the regime—and to make that goal public—then, once again, they should not have fought. Without clearly stated, rational objectives, the war became a circular process of bloodshed rather than a goal-directed offense, and military victories on the battlefield became irrelevant to the outcome of the war.[5]

This conclusion about Vietnam differs from what had happened with Carthage, because the character of the Carthaginian regime was not founded upon such an aggressive ideology. The Roman destruction of Carthage, far from demonstrating some universal rule about the need to destroy an enemy totally, rather demonstrates how a militaristic approach to a foreign policy—and the needless destruction of a population center—can ruin a proper peace.

Those interested in studying the establishment and preservation of peace would do well to examine where peace has actually been established and preserved. This is not in dictatorships that keep power by internal wars against their own people, but in free societies that protect their citizens' rights. The relationships between the United States and both Canada and Mexico—the longest undefended borders in history—as well between Germany and France, and Japan and Taiwan, are living examples of peace that emerged from war. The opposite is in the hatred and ongoing commitment to war expressed against Israel, in calls for holy war against unbelievers in whatever century they have occurred, and in a slew of indecisive tribal and ideological conflicts across time and space. Ideas matter. Visible success bolsters the legitimacy of a cause, gives hope to its proponents, and deepens their motivations to pursue the cause with violence. With the legitimacy of a cause unchallenged, the prime cause of war remains intact, smoldering for the moment but lashing out when it is able.

Matters of war and peace cannot be fully grasped without concern for moral ideas, which can lead one population into sacrificial slaughter for a leader, a cause, or a deity and can motivate another to defend their liberties with even greater violence. It is ideas that move people to act—and only widely held ideas of a fundamental moral nature can throw an entire continent into the chaos of war. The lives of soldiers and civilians depend upon clear statements of the objectives to be achieved and a commitment to create the resources necessary to prevail. *Sic vis pacem, para bellum*, in matter and in mind.

NOTES

INTRODUCTION

1. The *National Strategy for Combating Terrorism* (February 2003), 12, has four mentions of "victory"; the word is missing in the 2006 version. The army Field Manual for the "surge" in Iraq, "Counterinsurgency," FM 3-24, MCWP 3-33.5– (2006), discusses victory as gaining the support of the populace; for instance, "Victory is achieved when the populace consents to the government's legitimacy and stops actively and passively supporting the insurgency" (1.14).

2. Summers, *Strategy*, 58, citing army Field Manual 100-5– (1939).

3. Summers, *Strategy*, 61–63, citing the army Field Manual 100-5– (1954), 6. The 1962 manual went further: "The essential objective of United States military forces will be to terminate the conflict rapidly and decisively in a manner best calculated to prevent its spread to general (nuclear) war"; see Field Manual 100-5 –(1962), 9.

4. Aristotle, *Nicomachean Ethics* 1.1.3. In the 1970s American military commanders acknowledged a decline in the teaching of military history; the result was a volume to affirm the importance of history to military education, Jessup and Coakley, *Guide*.

5. Greek and Roman handbooks on waging war usually focus on tactics—for example, Aineias, *How to Survive;* Polyaenus, *Stratagems;* and Maurice, *Strategikon*. The "causes" of war, including motivations, are generally the province of historians—for example, Thucydides, *History;* Herodotus, *Histories,* and Polybius, *Rise*.

6. Freud "Why War?" letter to Albert Einstein, September 1932, in Maple and Matheson, *Aggression,* 21–23. Codevilla and Seabury, *War,* 40, find no cases of truly "accidental" war.

7. The role of moral ideas as motivations to fight deserves more attention by students of international relations. Welch, *Justice and the Genesis of War*, attempts to describe a desire for justice as a motive in war. Codevilla and Seabury, *War*, 33–46, reject deterministic causes, but fail to acknowledge the central role of ideas in decision making. Of course, errors in judgment about the intentions of an adversary can have a powerful effect on one's moral conclusions: Jervis, "War and Misperception."

8. Stout, *Upon the Altar*, to the American Civil War, equates morality in war with just-war theory; see my discussion in chapter 5. For a summary of a massive literature on basic just-war positions, see Johnson, "The Just War Idea." The definitive full-length treatment of a modern manifestation is Walzer, *Wars Just and Unjust*.

9. Liddell Hart, *Strategy,* 4–5.

10. Sun-tzu, *Art of War,* 191, for a similar thought: "One who excels at warfare compels men and is not compelled by them."

11. Clausewitz, *On War,* II.3, p. 149.

12. Jomini, *The Art of War*, 63, for example: "If every theatre of war forms a figure presenting four faces more or less regular, one of the armies, at the opening of the campaign, may hold one of these faces."

13. A definitive article in a growing literature is Beyerchen, "Clausewitz." Chaos theory is in the *Marine Corps Warfighting Publication* 5-4, chap. 1, "Complexity, Clausewitz, and the Edge of Chaos."

14. Wright, *Study of War*, 103–108, reduces the commonly accepted causes of war to (1) necessity, for survival; (2) a desire, for wealth, power, or social solidarity; (3) ideology, "which requires fighting in the presence of certain stimuli"; and (4) "Men and governments feel like fighting" from pugnaciousness, boredom, or frustrations. In the end, Wright concludes, "The object of a war, whether economic, political, religious, or dynastic, must rest on a systematization of ideas, or law in the broadest sense, which gives that object a value. Values do not grow out of events, but ideas" (110).

15. Douhet, *Command of the Air* (originally published in 1921).

16. Hansell, *Air Plan*, 28, is an insider's view of debates over U.S. air doctrine prior to World War II.

17. The "wrestlers" analogy breaks down, given that a wrestler follows rules and accepts the decisions of referees; he does not decide the match with a bayonet.

18. Clausewitz, *On War*, VIII.4, p. 595.

19. Clausewitz, *On War*, VIII.4, p. 596.

20. Liddell Hart, *Strategy;* Lynn, *Battle*. Lynn's omissions of later classical and Hellenistic Greece, as well as Rome, are worthy of special criticism, given that he is explicitly attempting to refute V. D. Hanson's view in *Carnage and Culture* that there is a continuity of war-fighting goals and methods from the Greeks through the Romans into the modern day.

Chapter 1: The Greco-Persian Wars

1. Balcer, *Persian Conquest*, 31–32, for Athenians commemorations of the victory.

2. Thucydides 1.23: "The decision was reached quickly as a result of two naval battles and two battles on land."

3. Balcer, *Persian Conquest*, 48, for Cyrus's relationship to Astyages; Cyrus's rule could be seen as "a dynastic change within the Median 'royal house.'"

4. Teixidor, *Pagan God*, 29–30. From Pritchard, *Ancient Near Eastern Texts*, 291. The stele of Esarhaddon: Pritchard, *The Ancient Near East*, pl. 121. Esarhaddon captured Tyre in 677 BC.

5. The Cyrus cylinder asserts Cyrus's claims: Pritchard, *Ancient Near East*, 206–208 and pl. 66.

6. Van De Mieroop, *Near East*, 113–114.

7. Van De Mieroop, *Near East*, 248, for Assyria's "creation of an Imperial structure."

8. Horne, *Arda Viraf;* Boyce, *Letter of Tosar;* Kelsay and Johnson, *Just War*, 40–41, for discussion and references.

9. Young, in *Cambridge Ancient History*, 2nd ed., IV, 99–203.

10. Van De Mieroop, *Near East*, 290–291, for the Behistun inscription. Connections to Zoroastrianism are disputed; Balcer, *Persian Conquest*, 126–128, sees Zoroastrianism as central to Darius's appeal to Ahura Mazda; Frye, *History*, 120–124, thinks that the lies-versus-truth view of the world is not exclusive to Zoroaster. On the

kings' lineage, Frye, *History,* chap. 5; Young, in *Cambridge Ancient History,* 2nd ed., IV.I.iv.

11. Wallinga, *Xerxes,* disagrees that universal domination was a Persian aspiration; he sees a series of episodes leading to reactions and new policies. Balcer, *Persian Conquest,* 19, begins with a "larger, rapid, and ever expanding growth of imperialism as the fundamental element to successful Achaemenid kingship." Wallinga concludes that the king attacked to counter Greek sea power, and Balcer sees an overextension of Persian power as responsible for his failure.

12. Diodorus 10.3.

13. Strauss and Ober, *Anatomy of Error,* chap. 1, has a brief but elegant treatment of how ideological motives shaped the actions of Xerxes; chap. 4 connects similar influences to the defeat of Darius II by Alexander.

14. Balcer, *Persian Conquest,* 40.

15. The account of the Persian invasion of Scythia is in Herodotus book 4, esp. 4.83–144.

16. Herodotus 5.30–31 claims that some "substantial citizens" of Naxos had fled to Miletus and called upon Aristagoras to regain their position; Aristagoras enlisted the Persian king's help by promising the king money and territory. Herodotus 4.137–138 for the decision against revolt.

17. The Ionian revolt begins at Herodotus 5.28; the decision to revolt is at 5.36.

18. Miltiades as "tyrant": Herodotus 6.34.

19. Herodotus 5.49–50 for the entreaty to Sparta; 5.55, 96–97 for Athens and the dispatch of aid; 5.105 for the burning of Sardis.

20. Herodotus 6.32. Miletus sacked: Herodotus 6.18.

21. Green, *Greco-Persian Wars,* 22.

22. Phrynichus's play: Herodotus 6.21. Badian, "Phrynichus," argues that the play was performed after 480 BC and represented the sackings of Athens in 480 and 479.

23. On the expedition of 492, Herodotus 6.43–45. Marathon is at 6.111–117; Diodorus 11.6–12 preserves an encomiastic fragment by the poet Simonides.

24. Herodotus 7.144 on the silver decision.

25. Diodorus 11.37 for this as "the end of the Median War." Green, *Greco-Persian Wars,* 275: "Though the war with Persia was not over yet—peace would only be ratified thirty years later, after innumerable crises and at least one major naval engagement—its shadow had, at last, receded from Greek soil."

26. Herodotus 8.140–144; 9.9 for Persian entreaties to Athens; Tod, *Historical Inscriptions,* no. 204, for the oath; Krentz, "Oath," for several Greek oaths and interpretations.

27. Diodorus 11.37; Herodotus 9.114–118; Plutarch, *Cimon* 9. Gomme's commentary to Thucydides 1.89.2 shows the revolt to be in response to the actions against Sestos.

28. Herodotus 8.133; Diodorus 11.34.

29. Herodotus 7.33; the siege is at 9.114; also Diodorus 11.37.

30. Herodotus 6.25–33 describes how the Persians took control of Samos, Caria, Chios, Lesbos, Tenebos, Perinthus, Byzantium, and other city-states. Care must be taken for confusions in Diodorus's *History*: Andros, Carystos, and Paros (Herodotus 8.111; Diodorus 11.37); first battle at Cyprus (Diodorus 11.44); Byzantium (Diodorus

11.44; Thucydides 1.128; Plutarch, *Cimon* 9); Eion, Scyros, the Carian and Lycian cities (Diodorus 11.60; Plutarch, *Cimon* 7, 8); Phaselis (Plutarch, *Cimon* 12.3); Cyprus second battle, the fall of Citium and Marium, and Artaxerxes' entreaty for peace (Diodorus 12.3–4; Plutarch, *Cimon* 18–19). Persian garrisons were attacked, for instance, at Cyprus, and renowned Persians taken at Byzantium: Diodorus 11.44.1–4.

31. Eurymedon: Plutarch, *Cimon* 12–13; Diodorus 11.100. Diodorus is confused, as noted in Gomme, *Historical Commentary*, comment to 100.1. Diodorus 11.61: the Persians feared the growing power of Athens.

32. Diodorus 12.4. The cities on Cyprus were Citium and Marium, and the king responded to a siege at Salamis, off of Asia Minor (a different Salamis than the island off the Greek mainland).

33. Herodotus 9.82.

34. Herodotus 7.48 has Xerxes say that another army can be raised; Herodotus 8.115 describes the Persian forces as eating grass and tree bark during the retreat.

35. The Persians tried to divide the Ionian Greeks, Herodotus 6.11–12; they wanted a separate peace with Athens, Diodorus 11.28; Sparta broke from the Athenians at Sestos, Diodorus 11.37. After Salamis, the Persians had to guard the Ionian areas (Diodorus 11.27) while the Phoenicians sailed home (Herodotus 9.96 and Diodorus 9.19; Green, *Greco-Persian Wars*, 277).

36. Preserved in a Greek inscription; see Meiggs and Lewis no. 12: *Gadatai doulōi* (to my slave Gadatas); Fornara, *Archaic*, no. 35.

37. Herodotus 7.27–29, 38–39.

38. Ancient military disasters caused by ideological blindness were explored by Strauss and Ober, *Anatomy of Error*. Xerxes went forth "for personal ideological legitimacy, rather than by any rational assessment of the long-term strategic needs of the empire" (41).

39. Herodotus 7.7 for Xerxes' suppression of the revolt. Kelly, "Persian Propaganda," 195: years of preparation were not needed for Xerxes' mission.

40. Aristotle, *Politics* 5.9.4–5, on tyrants and the pyramids of Egypt.

41. Herodotus 7.146–147; *The Three Hundred Spartans* (Twentieth-Century Fox, 1961).

42. Herodotus 6.112.

43. Wallinga, *Xerxes*, 21–26, sees the invasion of 480 as a response to growing Greek power. Thucydides 1.69.5 for the mistaken policy of the Persians as the chief cause of the failure.

44. Herodotus on Hecataeus: 5.36. The Spartan warning: Herodotus 1.152; Diodorus 9.36. Lazenby, *Defense*, 38: the small size of the forces sent by the Greeks to Tempe and Thermopylae suggests little knowledge of the Persians. Cartledge, *Agesilaos*, 217: in a Spartan campaign against Persia in 395, the commander Agesilaos "treated the Persian Empire as if it were a Peloponnesian country town that could be brought to its knees by the traditional ... strategy of ravaging."

45. Strauss, *Salamis*, 183–184, for lies told to the Great King.

46. Herodotus 5.78, translated by Strauss, *Salamis*, 13.

47. Herodotus 6.9–11.

48. Herodotus 6.31 recounts the taking of Chios, Lesbos, and other city-states.

49. Herodotus 7.135.

50. Plutarch, *Themistokles* 18.

51. Herodotus 8.57–64 for the debate, and the decision to fight at Salamis. Herodotus 8.75–76 for the spy Sicinnus; Plutarch, *Aristides* 10, for Themistokles' deception, and Athens' rebuff of a Persian peace offer; Diodorus 11.28.

52. Herodotus 6.103, 106 on the guidance of Hippias during the expedition to Marathon.

53. Aeschylus, *Persians* 337–343, sings of 1,207 Persian ships, versus 310 of the Greeks. Plutarch, *Themistokles*, sees 180 Greek ships.

54. Herodotus 8.143.

55. Herodotus 9.98.

56. Diodorus 11.60–61.

57. Young, in *Cambridge Ancient History*, 2nd ed., IV, 102.

58. Frye, *History*, 127. Xenophon, *Cyropaideia*, sees a moral decline since Cyrus, a view similar to Xerxes' *hubris* as portrayed by Aeschylus and Herodotus. Balcer, *Persian Conquest*, 21–22.

Chapter 2: The Theban Wars

1. Hanson, *Soul*, 51: Epaminondas "turned a fierce group of farmers into ideological warriors." Liddell Hart, *Strategy*, 13: what ended Sparta's ascendancy was "a man, and his contribution to the science and art of warfare." Cartledge, *Spartans*, 225, sees a "brilliant general and philosopher." Xenophon, a Spartan sympathizer and friend of the Spartan king Agesilaos, who lost his property in Sparta's battles, first omitted Epaminondas from his *Hellenica* but finally condescended to recognize that "for planning and audacity this man could not be criticized" (*Hellenica* 8.5.8).

2. Major ancient sources on the Theban wars are Diodorus, book 15; Xenophon, *Hellenica*, books 3–4., and *Agesilaos*; Plutarch, *Agesilaos* and *Pelopidas*.

3. The historical "mirage" began as "le mirage spartiate" by François Ollier; noted by Cartledge, "Socratic's Sparta," 312.

4. Tyrtaeus 12 (excerpted). Translations by the author; compare to Gerber, *Greek Elegiac Poetry*.

5. Tyrtaeus 10 (excerpted).

6. Primarily agricultural in their importance, the helots had a range of functions; for categories, Ducat, "Obligations," 196–199.

7. Thucydides 4.80. The "tall poppies" applies to grain in Herodotus 5.92–93; Aristotle, *Politics* 3.13 and 5.10. Aristotle and Herodotus reverse the figures of Thrasybulus and Periander. Aristotle, *Politics* 5.11: lopping off the eminent men is an "old way" of preserving tyranny.

8. *Oxyrhynchus Papyrus*, xi.38–xii.31, in Moore, *Aristotle*, 127–133; also Cartledge, "Boiotian Swine." On the Theban League: Buckler, *Theban Hegemony*, 15–45.

9. Hanson, *Soul*, 27, sees a "common hatred of Sparta" as a "catalyst" for the agrarian democracy.

10. Andrewes, "The Government of Classical Sparta," 49.

11. Mises, *Omnipotent Government*, 44. Mises prefers "etatism" to "statism" because the former "expresses the fact that etatism did not originate in the Anglo-Saxon countries, and has only lately got hold of the Anglo-Saxon mind" (footnote to p. 5).

12. Rand, "New Fascism," 202, citing *The American College Dictionary* (New York: Random House, 1957).

13. Hodkinson, "Classical Sparta," considers the evidence for Sparta as "militaristic."

14. Herodotus 9.28.2; Plutarch, *Cimon* 16.4–5; Xenophon, *Hellenica* 6.4.15–17. The statistics are in Rhodes, *Greek City States*, 91.

15. Plutarch, *Agis*: attempts, in the late third century BC, to reform laws and relieve debtors of their burdens; Rhodes, *Greek City States*, 79–80, 83.

16. Xenophon, *Hellenica* 3.3.4–6. Xenophon is our only source for the Cinadon case.

17. Xenophon, *Hellenica* 3.3.3–4; cf. Plutarch, *Lysander* 22.6, 10–13. Cartledge, *Agesilaos*, 77: the "intervention" of Lysander was the "decisive factor" in Agesilaos's elevation.

18. Plutarch, *Agesilaos* 35.

19. Diodorus 15.20.3.

20. Cartledge, *Spartans*, 222–223, on Phlius.

21. Xenophon, *Hellenica* 5.3.10–16.

22. Diodorus 15.23.4–5.

23. Hamilton, *Agesilaus*, 152: "the liberators of Thebes" first challenged Spartan hegemony.

24. Diodorus 15.30.2.

25. Xenophon, *Hellenica* 5.4.56–57; Buckler, *Agesilaus*, 178.

26. Xenophon, *Hellenica* 6.2.1. Xenophon claims that Athens asked for peace because the Thebans had failed to pay for Athenian naval aid.

27. Xenophon, *Hellenica* 6.3.18–20.

28. Plutarch, *Agesilaos* 28.

29. Cartledge, *Agesilaos* 206–208, on the qualities of a commander, and of Agesilaos.

30. Diodorus 15.51.1–2.

31. The battle of Leuktra is at Xenophon, *Hellenica* 6.4.1–15; Diodorus 15.51–56; Plutarch, *Pelopidas* 20–23. Xenophon pays Epaminondas the ultimate compliment of not mentioning his name.

32. Plutarch, *Pelopidas* 18, on the Sacred Band. Xenophon may reveal his own inability to understand the tactical innovations of the Thebans, for instance at *Hellenica* 4.2.18, all the while he derides their bravery.

33. Diodorus 15.56.2–3.

34. Diodorus 15.62.4–5.

35. Xenophon, *Hellenica* 6.5.23.

36. Xenophon, *Hellenica* 6.5.32.

37. Xenophon, *Hellenica* 2.2.3, at the end of the Peloponnesian war with Sparta.

38. Xenophon, *Hellenica* 6.5.27–28.

39. Diodorus 15.66.1.

40. Clausewitz, *On War*, VIII.4, p. 596.

41. Diodorus 15.66, on the refounding of Messene.

42. Polyaenus, *Strategems* 2.3.5: Epaminondas wanted such a balance, recognizing that Theban dominance of the area would create enmity. Hanson, *Soul*, 428, n. 58.

43. Epaminondas as the founder of Megalopolis is disputed by Larsen, *Greek Federal States*, 186, n. 1.

44. Xenophon, *Hellenika* 7.1.28–32, on the Tearless Battle—not a single Spartan was killed.

45. Sun-tzu, *Art of War*, 177.

46. Diodorus 15.33.1–3.

47. Roebuck, *Messenia*, 47, and n. 93, citing Demosthenes 16.9.

48. Diodorus 15.39.2, following Ephorus.

49. Diodorus 15.88.3.

50. Vidal-Naquet, "Epaminondas the Pythagorean," chap. 3, on Epaminondas, who did not follow the "natural" desire to strengthen his own right. One should not push this too far; after the Peloponnesian War, Sparta did not destroy Athens at least in part to maintain a balance against Theban power.

51. Diodorus 15.52.4, quoting Homer, *Iliad* 12.243.

52. Diodorus 15.52.6–7.

53. Herodotus 6.106–107 (Marathon) and 7.206 (Thermopylae). Herodotus 8.72: many Peloponnesian cities remained indifferent even after the festivals ended.

54. Herodotus 4.3.

55. Hornblower, "Sticks, Stones."

56. Thucydides 1.101–103 on the Mount Ithome revolt; 5.23, on the Athenian agreement to assist.

57. Xenophon, *Hellenika* 6.2.15–19.

58. Plutarch, *Lysander* 15.7.

59. Plutarch, *Agesilaos* 35–36.

60. Alcidamas is preserved in Aristotle, *Rhetoric*, 1.13 and 2.23. Isocrates, "Archidamas," for the Spartan side. The speech may be a rhetorical exercise. Roebuck, *Messenia*, 43–44, stresses the Spartan "shame and anger" at having to treat with the Messenians.

61. Hanson, *Soul*, 372.

62. Thucydides 4.40; the surrender is at 4.38.

63. Plutarch, *Agesilaos* 40, and *Agis*.

Chapter 3: The Second Punic War

1. Herodotus 7.165–166; Diodorus 11.21–24. In Aristotle's *Poetics* 23.3, the battles occurred "about the same time, but for a different end."

2. Major sources are Polybius, Livy, and Diodorus. Ancient historians had Carthaginian sources lost to us, but fragments of Sosylus of Sparta and Silenus of Caleacte, who were in Hannibal's camp, have been preserved in Jacoby, *Fragmente*, nos. 175–180. Nepos, *Hannibal* 13.3, on Sosylus and Silenus. Silius Italicus, *Punica*, for anecdotal verses. The magisterial *Commentary on Polybius* by Frank Walbank is essential, as are Lazenby, *First Punic War*, and *Hannibal's War*; also, Scullard, *Scipio Africanus*; Rich, "Origins"; Kagan, *Origins*; Gabriel, *Scipio Africanus*; and for overflowing praise of Scipio, Liddell Hart, *Scipio Africanus*. Some of the many books on Hannibal are in the bibliography.

3. Polybius 3.22–27 on the treaties.

4. On Pyrrhus, see Plutarch, *Pyrrhus*; also Polybius 1.6.

5. Polybius 1.9 tells us how Hiero of Syracuse solved his mercenary problem: he allowed the mercenaries to run forth ahead of his army into battle and then abandoned them to slaughter.

6. Polybius 1.7–12 on Rhegium, Messana, and the Roman crossing into Sicily.

7. Polybius 1.11.

8. Polybius 1.62–63 on the end of the first war.

9. Discussed in Polybius 1.65–88. Lancel, *Carthage*, 372–376, for connections to Carthaginian political developments.

10. Polybius 3.9–10.

11. Polybius 3.39.6 for a *hormē* that is an attack; 1.78 for a mercenary who defected to Hasdrubal, as "full of a war-like spirit."

12. The later Roman writer Justin, *Epitome* 18.7, leaves us the myth, perhaps constructed as a background for ritual sacrifice in Carthage. He treats Queen Dido as a historic person, who sacrificed herself by leaping into the flames. Lancel, *Carthage*, 112; Picard, *Life*, 56–57.

13. Picard, *Life*, 226; Lancel, *Carthage*, 210–211.

14. Lancel, *Carthage*, 194. Hasdrubal means something like "he who has Baal's help."

15. Pritchard, *Ancient Near Eastern Texts*, 533–534.

16. Livy 21.21.

17. Polybius 7.9 for the treaty and oath, discussed in Picard, *Life*, 214–216, and Lancel, *Carthage*, 208–209.

18. Lancel, *Carthage*, takes this position.

19. Picard, *Life*, accepts this conclusion.

20. Such pragmatism continued into Hannibal's later years. In 195 BC he courted the Seleucid king Antiochus III by recalling his oath against the Romans nearly fifty years earlier. Livy 21.1; Polybius 3.11. Polybius writes that he spoke in self-defense, "being at a loss for further arguments."

21. Polybius 3.8.

22. In 411 BC opponents of the democracy in Athens established an oligarchic government while thousands of rowers were away from Athens: Thucydides 8.45–98.

23. Polybius 3.8.

24. Polybius 3.10.

25. Polybius 2.13 for the treaty.

26. Polybius 3.14–17, 20–21, 30; Livy 21.6–7.

27. Polybius 3.30.

28. Cicero, *De Officiis* (On Duties), I.23; the divine sanctity of oaths is at III.104. In the quote I have used "constancy" rather than "fidelity" for *constantia*, in order to stress consistency and avoid circularity.

29. Zonaras 8.22, cited in Lazenby, *Hannibal's War*, 26. Polybius may have omitted it, given his support for the Scipio clan. On the debate, Scullard, *Politics*, 40–41.

30. Polybius 3.15.

31. Livy 21.9.

32. Livy 21.18; 30.22. The relationship of the Ebro River Treaty to all of this remains perplexing.

33. Lazenby, *Hannibal's War,* 26.

34. Polybius 3.35; Livy 21.23.

35. Gabriel, *Scipio Africanus,* 35, interprets the strategy as the Hellenic ideal of defeating the enemy in a single big battle. Rome's strategic endurance thus surprised Hannibal and negated the strategy. For a more positive view of Hannibal's strategy, see Hallward, in *Cambridge Ancient History* VIII, 33–36.

36. Livy 22.24.

37. Polybius 3.107 states eight legions.

38. Gabriel, *Scipio Africanus*, 47. Supporting figures are in Livy 22.49.

39. Livy 30.26.

40. Livy 27.9–10; 27.36 for the census; Polybius 2.24.16. Lazenby, "Was Maharbal Right?" 44: 40 percent of Rome's allies were not available in 212, including nearly all Campanians. Gabriel, *Scipio Africanus*, 50–51, sees Rome as "on the verge of losing the political war" in 210 BC. For discussion, see Reid "Problems."

41. Lazenby, "Was Maharbal Right?" for discussion of Hannibal's refusal to attack Rome itself.

42. Polybius 10.7 and Livy 26.20 disagree about the locations of the Carthaginian forces in the winter of 210—but agree that they were divided.

43. Gabriel, *Scipio Africanus* 85–102, reconstructs the attack, and shows that the historians Livy and Polybius have overstated Scipio's speed. If Gabriel is right, the march took at least twenty days, and Scipio gained tactical surprise without the benefit of secrecy. Polybius 10.7–9 on Scipio's careful planning.

44. On the legend of Scipio Africanus as a sage, see Macrobius, *Dream*, derived from Cicero's *On the Republic,* book 3.

45. Three Hasdrubals are involved here: the elder Hasdrubal, son-in-law of Hamilcar who succeeded Hamilcar in Spain and died in 221; Hannibal's younger brother Hasdrubal, who set out for Italy; and Hasdrubal son of Gisgo, no relation. Mago, Hannibal, and Hasdrubal were brothers.

46. For the uprising of Spanish allies after Scipio left Spain, Livy 29.1.14– 2.3.6.

47. Livy book 28 for Scipio's return to Italy, unconstitutional election as consul, and debate with Fabius. Book 29 for the passage to Africa.

48. On politics and family relationships, see Scullard, *Politics*.

49. Such relationships affected the written histories. Polybius was attached to the Scipio *gens* through Scipio Aemilianus, and he repeatedly condemned the historian Fabius Pictor, cousin to Fabius Maximus.

50. Livy 28.42.

51. Livy 28.41.13.

52. Livy 28.41.

53. Livy 28.44.

54. Gabriel, *Scipio Afrticanus,* 155–156.

55. Livy 30.16.

56. Livy 29.3.

57. Livy 30.16.

58. Livy 30.21. Or Hannibal had trouble leaving; Gabriel, *Scipio Africanus*, 175.

59. Gabriel, *Scipio Africanus*, 175–176, brings to life the intense activity immediately before Zama.

60. Livy 22.13.

61. Polybius 3.15.

62. Polybius 3.15.

63. Kagan, *Origins*, 255.

64. Scullard, *Scipio*, 65.

65. Livy 23.11–13.

66. Livy 30.2.

67. Livy 30.37.

68. Polybius 36.67.

69. A generation after the massacre of Carthage, the Romans were fighting the same African tribes that Carthage had sought relief from: Sallust, *Jugurthine War*.

70. Harris, *War and Imperialism*; Mattern, *Rome and the Enemy*. Dillon and Welch, *Representations of War*.

Chapter 4: The Campaigns of Aurelian

1. Studies of Roman strategy were energized after Luttwak, *Grand Strategy*. For criticism, Kagan, "Redefining Roman Strategy," citing Isaac, *Limits*. Whittaker, "Where Are the Frontiers Now?" responds to Luttwak, with bibliography. Mattern, *Rome and the Enemy*, discusses Roman conceptions of space and strategy.

2. *Scriptores Historiae Augustae*, "Aurelian" 37.1; Eutropius, *Breviarium* 9.13–14; Aurelius Victor, *De Caesaribus* 35.12; Ammianus Marcellinus, *History* 35.1.

3. Meaning, outside of customary, "constitutional" procedures—and in a barracks outside the city proper. Tacitus, *Histories* 1.4.

4. The literature is voluminous. Classic studies used here include Rostovtzeff, *History;* Mommsen, *Provinces;* Potter, *Roman Empire;* Millar, *Roman Near East,* for inscriptions; and Millar, "Government," 347, for early military enrollments of Goths. On the Goths, Heather, *Fall,* 17–18.

5. De Blois, *Policy*, 37–44, on the upward mobility of soldiers. Macrinus, who usurped in 217, was the first emperor of equestrian rank. The strength of Roman elite ideology is shown by Whittaker, *Frontiers*, 195–196. For the political and social role of the senatorial aristocracy, Arnheim, *Senatorial Aristocracy*.

6. Noted by Potter, *Roman Empire,* 264–265.

7. Millar, "Government," 352–353, 363, 369. *Scriptores Historiae Augustae,* "Aurelian" 30.4–5, treats the Carpi tribe as enemies; Mommsen, *Provinces,* I.249, notes their client status after Aurelian's victories.

8. In Greek and Roman accounts, Philip killed Gordion in 244, after his retreat from the Persians. In an inscription at Naqsh-e Rustam, the Persian king Shapur took credit for killing Gordion. If the latter is correct, Gordion was the first emperor killed by barbarians in battle. Frye, *History*, appendix 4 for the inscription; Millar, *Roman Near East*, 152–154.

9. Heather, *Goths*, 38–50, summarizes the third-century attacks. Drinkwater, in *Cambridge Ancient History*, 2nd ed., XII, 42, distinguishes the Goths of the Danube

River areas from the more hostile Goths of the Black Sea area. The two groups cooperated in 267.

10. Millar, "Government," 347, discusses the divide.

11. Only "Vopiscus of Syracuse" in the unreliable *Scriptores Historiae Augustae,* "Aurelian," tells us about his early life. Aurelian's campaigns: White, *Restorer,* 60–137; Saunders, *Biography,* 137–280. Every step of this narrative is inferential.

12. Jacoby, *Fragmente,* 100; Millar, "Herennius Dexippus," 19–28.

13. Mattern, *Rome and the Enemy,* 182–183. Barton, "Price of Peace," 248, distinguishes *pacem dare* "to make peace" from *pacem petere* "to petition (or seek) peace."

14. Zosimus, *History* 1.89, whitewashes the defeat. *Scriptores Historiae Augustae,* "Aurelian" 18.3–4, 21.1–4; Aurelius Victor, *De Caesaribus* 35.2.

15. *Scriptores Historiae Augustae,* "Aurelian" 18.4–20.8, for consultation of the Sibylline Oracles.

16. Heather, *Goths,* 41.

17. The withdrawal is unclear. Dacia Mediterranea is not attested prior to AD 283. Coins stamped Dacia Felix (Dacia Fortunate) and Dacia Maximus may have been minted *before* his victories. White, *Restorer,* 95–96; Saunders, *Biography,* 199–204.

18. Criticism of Hadrian in Aurelius Victor, *De Caesaribus* 14.1; Eutropius, *Breviarium* 8.6; and Festus, *Breviarium* 14.4; 20.3. Cassius Dio 69.9 praised Hadrian, and criticized expeditions in his own day. Isaac, *Limits* 24–26.

19. Festus, *Breviarium* 8.2: "Two Dacias were made in the regions of Moesia and Dardania." Eutropius, *Breviarium* 9.13: that Aurelian "extended the Roman empire, by various successes in the field, to its former limits."

20. Ammianus Marcellinus, *History* 31.5.17: people here "were kept quiet for many years, except that occasionally bands of robbers raided the parts nearest to them and brought destruction upon themselves."

21. Aurelius Victor, *De Caesaribus* 35.1.

22. Millar, *Roman Near East,* 164–169, for Roman rule in the Near East in the 250s and 260s. Isaac, *Limits,* 134–140, for inscriptions that suggest Roman police activity in the late second century, but "the army perhaps did not continue these activities far into the third century."

23. II Chronicles 8.4; I Kings 9.18; Josephus, *Jewish Antiquities* 8.6.1. Solomon is said to have built it. Starcky and Gawlikowski, *Palmyre,* for an overview in French, with photos.

24. Pliny, *Natural History* V.21.88, in Isaac, *Limits,* 143.

25. Millar, *Roman Near East,* 165; Isaac, *Limits,* 226.

26. Isaac, *Limits,* 225.

27. Isaac, *Limits,* 143: Severus Alexander used it as an invasion route. Herodian, *History* 6.5.2.

28. For several interpretations of Palmyra's extraordinary rise, Isaac, *Limits,* 224–248. Saunders, *Biography,* 150, sees Palmyrene influence, not conquest. Festus, *Breviarium* 23, calls it "shameful" that Rome allowed Palmyra to defend Rome's border. Aurelius Victor, *De Caesaribus* 33.34, blames the emperor Gallienus.

29. Sartre, in *Cambridge Ancient History,* 2nd ed., XII, 513, citing Peter the Patrician, fragment 10.

30. White, *Restorer*, 48–49, for *dux orientis* as his title, but "Corrector" on the inscription.

31. Millar, *Roman Near East*, 170, for two inscriptions. The commentators are Georgius Syncellus, *Chronography* I.716, and Zonaras XII.24, cited in Isaac, *Limits*, 220–223. Mattern, *Enemy*, 194–202, on language, symbols, and imperial prestige.

32. White, *Restorer*, 89.

33. For example, Mattingly et al., *RIC* no. 2, has ZENOBIA AUG on the obverse, with IUNO REGINA on the reverse. White, *Restorer*, pls. 14–16b, for coin images. Mattingly, in *Cambridge Ancient History*, 1st ed., XII, 301–302, for coinage dates.

34. Which Isaac, *Limits*, 31, denies.

35. Saunders, *Biography*, 160. Drinkwater, in *Cambridge Ancient History*, 2nd ed., XII, 50, disagrees; Zenobia needed a "power base."

36. Saunders, *Biography*, 208–209.

37. Potter, *Roman Empire*, 267.

38. An inscription in Aramaic and Greek grants multiple titles to young Vaballathus: Saunders, *Biography*, 149.

39. White, *Restorer*, 60. Zosimus, *History*, reports support for Palmyra in Egypt (1.44) and Antioch (1.51). Isaac, *Limits*, 227, infers support among troops in Judea and Arabia, given milestone inscriptions.

40. Isaac, *Limits*, 226, sees the title "King" as an innovation in Palmyra.

41. Zosimus, *History* 1.50–56, is the most complete account of the war with Zenobia. *Scriptores Historiae Augustae*, "Aurelian," alternates between Aurelian's greatness and his cruelty; 22.3 for the quotation. White, *Restorer*, 48–52, on Odenathus and Rome, esp. Gallienus.

42. Zosimus, *History*, extends Zenobia's ambitions to Chalcedon, "had the Bithynians not heard of Aurelian's elevation and shaken off Palmyrene control."

43. *Scriptores Historiae Augustae*, "Aurelian" 23.4, substituting "trust" for "faith."

44. Saunders *Biography*, 208, attributed to Peter the Patrician fragment 10. *Scriptores Historiae Augustae*, "Aurelian" 24, has Apollonius of Tyrana tell Aurelian to spare his fellow citizens.

45. Barton, "Price of Peace," discusses a range of related terms, during the republican period.

46. *Scriptores Historiae Augustae*, "Aurelian" 31.1.

47. Zosimus, *History* 1.52.

48. Mommsen, *Provinces*, II.108–109, on "Occidentals against Orientals."

49. Eusebius, *History* 7.27–30, for Paul's heresy, and Aurelian's support for Roman Christians. Eusebius then changes his attitude, accusing Aurelian of mounting persecutions. White, *Restorer*, 104.

50. Saunders, *Biography*, 223, 226, citing Starcky and Gawlikowski, *Palmyre*, 65–66.

51. Vopiscus criticizes the execution of Longinus and claims that Zenobia wrongly blamed him for the war: *Scriptores Historiae Augustae*, "Aurelian" 30.3.

52. Her fate and her son's cannot be verified. Zosimus, *History* 1.59, says Zenobia died en route to Rome.

53. *Scriptores Historiae Augustae*, "Probus" 9.5.

54. Zosimus, *History* 1.50.2.

55. Frye, *History*, 305, notes the expedition of Carus into Ctesiphon in 283, and the Persians' gain in territory in Mesopotamia. Isaac, *Limits*, 32.

56. Zosimus, *History* 1.60–61; *Scriptores Historiae Augustae*, "Aurelian" 31. Saunders, *Biography*, 245–248, cites inscriptions, including one restored to make Antiochus into the son of Zenobia. Saunders notes that Aurelian would not have left Zenobia's son alive in Palmyra.

57. Zosimus, *History* 1.61; *Scriptores Historiae Augustae*, "Aurelian" 32.2, is surely wrong to have him in Thrace between his second trip to Palmyra and Alexandria.

58. Aurelius Victor, *De Caesaribus* 35.3–4, and Eutropius, *Breviarium* 9.13.1. White, *Restorer*, 113–115.

59. Tetricus minted coins into 274; see Mattingly et al., *RIC* nos. 72, 100, 220; Saunders, *Biography*, 255–258 reassesses.

60. Gregory of Tours, *History* 3.19, credits Aurelian with Dijon. Potter, *Roman Empire*, 270, infers Aurelian's trust of the people of Rome.

61. Three inscriptions reproduced and discussed in White, *Restorer*, 119–120.

62. White, *Restorer*, 124–128, for a few of Aurelian's laws in Justinian's *Digest*.

63. Beard, North, and Price, *Religions*, I.254, cited in Fowden, in *Cambridge Ancient History*, 2nd ed., XII, 546.

64. Baal Shamim was associated with the sky, something like "Lord of the Heaven": Teixidor, *Pagan God*, 27–29; Rives, *Roman Religion*, 64. Gershevitch traces the development of the Indian Mithra, syncretized with the Iranian Mithra, "contract" through third-century Manichaeism: Gershevitch, "Sonne," 68–89.

65. Halsberghe, *Cult*, 26–37, for evidence of Sol in Rome prior to the first century AD.

66. Halsberghe, *Cult*, 112–117.

67. Halsberghe, *Cult*, 11, 116.

68. Tacitus has soldiers salute the sun "as is the custom in Syria," *Histories* 3.24.

69. *Scriptores Historiae Augustae*, "Elagabalus" 7.4, for the idea of hierarchy. Rives, *Roman Religion*, 205.

70. MacMullen, *Paganism* 84–85, maintains that a coin with Sol giving a globe to Jupiter is a sign of friendliness, not a claim of supremacy.

71. Halsberghe, *Cult*, 29. The other location was the Campus Martius.

72. Halsberghe, *Cult*, 153; Aurelius Victor, *De Caesaribus* 35.5.

73. Fowden, in *Cambridge Ancient History*, 2nd ed., XII, 557.

74. *Scriptores Historiae Augustae*, "Aurelian" 38.2–3; 21.5–9, has "Aurelian, over-violent by nature, and now filled with rage, advance to Rome eager for revenge." Eutropius, *Breviarium* 9.14, for the rebellion. The sources have certainly overstated the casualties, claiming seven thousand killed, for example, *Scriptores Historiae Augustae*, "Aurelian" 38.2. Aurelius Victor, *De Caesaribus* 35, and Bird's note 6 for sources and issues. Ammianus Marcellinus, *History* 30.8, for Aurelian's falling on the rich "like a tidal wave."

75. White, *Restorer*, 118–124, for currency reforms. Eutropius, *Breviarium* 9.14, implies that the mint workers adulterated the coinage. Potter, *Roman Empire*, 273, claims Aurelian made currency problems worse by disturbing the balance between gold and silver.

76. *Scriptores Historiae Augustae*, "Aurelian" 35.3, 39.2; Zosimus, *History* 1.61.

77. White, *Restorer,* 92–94; Potter, *Roman Empire,* 270, 645, n. 46, on the walls. *Scriptores Historiae Augustae,* "Aurelian" 21.9–11; Aurelius Victor, *De Caesaribus* 35.7; Eutropius, *Breviarium* 9.15.

78. *Scriptores Historiae Augustae,* "Aurelian" 36–37; Zosimus, *History* 1.62.2; Zonaras 12.27. Discussed in White, *Restorer*, 137–138, and in more detail, Saunders, *Aurelian*, 267–278.

79. Bird's comment to Aurelius Victor, *De Caesaribus* 36, disputes the claims from Victor that Tacitus was chosen by the senate, as in *Scriptores Historiae Augustae,* "Tacitus" 3.1–2, 12.1.

Chapter 5: Sherman's March through the American South

1. McPherson, *Battle Cry,* 742–750; 760–762; Catton, *Grant Takes Command,* 338–341, on northern moves toward a negotiated peace; Castel, *Decision in the West*.

2. Schott, *Alexander H. Stephens,* 334, also in Eicher, *Longest Night*, 49.

3. Thompson, *Antislavery Writings*.

4. Higginson, "The Ordeal of Battle," *Atlantic Monthly*, July 1861, in Stout, *Upon the Altar*, 36–37.

5. Hanson, *Soul,* 433, n. 44, quoting McPherson, *Cause and Comrades,* 20–22, 108–110, on slavery in the letters of Union and Confederate soldiers; for Confederates, slavery was "not controversial."

6. Commager, *Documents*, 26–28.

7. McPherson, *Cause and Comrades*, researches thousands of letters and hundreds of diaries to find a deep commitment among soldiers on both sides to the moral aspects of their respective causes.

8. For a short description of nullification, Hansen, *Civil War,* 14–18.

9. Stout, *Upon the Altar*, 9, quoting John Townsend, a southern planter.

10. Welles, *Diary,* I.107.

11. McClellan, letter to Simon Cameron, Secretary of War, September 13, 1861, in McClellan, *Civil War Papers*, 69. Also, Letter to Winfield Scott, August 8, 1861, in ibid., 79–80.

12. Grant, *Personal Memoirs,* II.127–132, in Commager, *Documents,* 151–53. Mosier, *Grant*, 49–50, shows the basis of this fragmentation in the earliest organization of the Union armies and calls on Grant's acknowledgment of the advantage this offered to the South: Grant, *Memoirs*, 144.

13. Sherman, "Grand Strategy," in *Battles and Leaders* IV.249.

14. Mosier, *Grant*, 32–33, notes that medical records show fifty-six deaths during the war from edged weapons and that rifling was the decisive technical innovation. Mosier concludes that the key was "in knowing when frontal assaults were necessary and when they were delusional" (163).

15. Sherman, "Grand Strategy," in *Battles and Leaders* IV.253. Sherman notes Hood was no. 44 in his West Point class; although rank bore little correlation to effectiveness as a general, Sherman knew his foes.

16. Sherman, *Memoirs,* II.39.

17. *Albany Patriot,* June 9, 1864, in Risley, "*Albany Patriot,*" 251.

18. Sherman, "Grand Strategy," in *Battles and Leaders* IV.257.

19. Letter from Grant of September 12, 1864, in Sherman, *Memoirs,* II.113.

20. Letter to Grant, September 20, 1864, in Sherman, *Memoirs,* II.115.

21. Clausewitz, *On War,* VII.4, p. 527.

22. Sherman, *Memoirs,* II.30. On logistics, provisions, and foraging, Sherman, *Memoirs,* II.175–176, 182–185; Liddell Hart, *Sherman,* 330.

23. Sherman, *Memoirs,* II.32.

24. Sherman, *Memoirs,* II.28.

25. Bull, *Soldiering,* 173.

26. Thucydides 1.11.

27. Osborn, *Fiery Trail,* 51.

28. Liddell Hart, *Sherman,* 323. One newspaper wrote that Sherman, "An Artful Dodger," was trying to "mystify our military authorities." See van Tuyll, "Two Men," 283; *Augusta Chronicle,* November 29, 1864. Osborn wrote that when approaching Clinton, the Confederates were led to think they were heading for Macon. Sherman's men then cut the railroad line to prevent them from moving. Osborn, *Fiery Trail,* 51, 57.

29. Sherman, *Memoirs,* II.42.

30. Grant, Letter of April 4, 1864, in Sherman, *Memoirs,* II.26.

31. Bull, *Soldiering,* 184.

32. On the correspondence between Lincoln, Halleck, Grant, and Sherman, see Grimsley, *Hard Hand,* 187–188.

33. Letter to Grant, November 6, 1864, in Sherman, *Sherman's Civil War,* 751; McDonough and Jones, *War So Terrible,* 319–320, for Sherman as assuring Lincoln's reelection.

34. Cooke, *Wearing of the Gray,* 511.

35. Clemson, *Rebel Came Home,* 54, in Eicher, *Longest Night,* 719.

36. Buck, *Shadows,* 57.

37. Buck, *Shadows,* 206, 230.

38. Hanson, *Soul,* 190, for Sherman's letter of September 17, 1863, to Halleck on the social structure of the South.

39. *Early County News* (Georgia), April 5, 1865, in Williams, *Johnny Reb's War,* 43.

40. Watkins, *Company Aytch,* 6–7.

41. Hanson, *Soul,* 176, citing Lewis, *Sherman,* 453.

42. Osborn, *Fiery Trail,* 110.

43. Nichols, *Great March,* 149, cited in Osborn, *Fiery Trail,* 111, n.1.

44. Sherman, Letter to brother John Sherman, January 28, 1864, in Sherman, *Sherman's Civil War,* 596.

45. Grant, *Memoirs,* 191.

46. Hanson, "Dilemmas."

47. Letter to George H. Thomas, October 2, 1864, in Sherman, *Sherman's Civil War,* 730.

48. Letter to Halleck, October 19, 1864, in Sherman, *Sherman's Civil War,* 734.

49. Letter to Halleck, September 17, 1863, in Sherman, *Sherman's Civil War,* 543–550.

50. Letter to Roswell M. Sawyer, Asst. Adj. General of Volunteers, January 31, 1864, in Sherman, *Sherman's Civil War,* 600, 602.

51. Sherman, *Memoirs,* II.186; Hitchcock, *Marching,* 83–85, on Cobb.

52. Bull, *Soldiering,* 187; Osborn, *Fiery Trail,* 57.

53. Osborn, *Fiery Trail,* 58, on arming the convicts in Milledgeville.

54. Sherman, *Memoirs,* II.189, in Eicher, *Longest Night,* 765; Lewis, *Sherman,* 449–450, for predictions of Sherman's failure.

55. Sherman, *Memoirs,* II.188.

56. *Albany Patriot,* September 22, 1864, in Risley, "*Albany Patriot,*" 253.

57. *Macon Telegraph,* "The Situation," November 21, 1864, in van Tuyll, "Two Men," 280.

58. Sherman, *Memoirs,* II.191.

59. Grimsley, *Hard Hand,* 199. Also, Caudill and Ashdown, *Sherman's March,* 25–26.

60. Letter to Gen. Wm. J. Hardee, December 17, 1864, in Sherman, *Memoirs,* II.210–211, 228.

61. Secessionists were never in the majority; see Williams, *Johnny Reb's War,* 45, n. 3.

62. Osborn, *Fiery Trail,* 70, relates the ineffectiveness of the defensive guns at Fort McAlister.

63. Sherman, *Sherman's Civil War,* 772, in McPherson *Ordeal,* 463.

64. Major Osborn recorded on December 13 that they had been "cut off 33 days from all communication with the known world"; Osborn, *Fiery Trail,* 46.

65. Letter to Halleck, December 24, 1864, in Sherman, *Memoirs,* II.227.

66. Letter of December 13, 1864, in Sherman, *Memoirs,* II.201.

67. Letter to Grant, December 24, 1864, in Sherman, *Memoirs,* II.224.

68. Grant, letter to Sherman, December 27, 1864, in Sherman, *Memoirs,* II.238.

69. Caudill and Ashdown, *Sherman's March,* 30.

70. Sherman, "Grand Strategy," in *Battles and Leaders* IV.258. The cause of the burning remains unresolved; like Richmond, it is possible that southern soldiers had begun it. It was not Sherman's fault, and he did not apologize. Caudill and Ashdown, *Sherman's March,* 27–28.

71. Sherman, *Memoirs,* II.255.

72. Sherman, *Memoirs,* II.255.

73. Letter to Grant of January 29, 1865,in Sherman, *Memoirs,* II.260–261; Lewis, *Sherman,* 435, for epithets and imagery attached to the march.

74. Caudill and Ashdown, *Sherman's March,* 38.

75. Letter to Halleck, September 4, 1864, in Sherman, *Sherman's Civil War,* 697.

76. Letter to Halleck, September 20, 1864, in Sherman, *Memoirs,* II.117–118. (emphasis in original).

77. Hood, Letter to Sherman, September 9, 1864, in Sherman, *Memoirs,* II.119.

78. Letter to Hood, September 10, 1864, in Sherman, *Sherman's Civil War,* 706.

79. Letter to Hood, September 14, 1864, in Sherman, *Memoirs,* II.128.

80. Sherman, *Memoirs,* II.236.

81. Grimsley, *Hard Hand,* 188, for evacuation figures. Sherman was called "Attila of the North" in the *Macon Telegraph,* December 13, 1864, editorial entitled "Sherman the Lunatic," cited in van Tuyll, "Two Men," 284.

82. Letter to Hood, September 10, 1864, in Sherman, *Memoirs*, II.120.

83. Letter to James M. Calhoun, Mayor of Atlanta et al., September 12, 1864, in Sherman, *Memoirs*, II.125–126.

84. Stout, *Upon the Altar*, 370–372. Stout must distort the issue factually in order to make his case; he writes of Sherman "destroying everything in his path" and describes 700 adults and 860 children, moved under a truce, as "The sight was pathetic, the event dangerous" (394, 370). Sherman had a better grasp of events when he wrote of the millions outside of Atlanta, torn by the war.

85. Sherman, *Memoirs*, II.249.

86. Caudill and Ashdown, *Sherman's March*, 16. Contrast Confederate president Jefferson Davis, who saw it as affirming northern abolitionist intentions and as an unconscionable incitement to insurrection by the slaves; Stout, *Upon the Altar*, 171.

87. Hansen, *Civil War*, 581.

88. Wheeler, *We Knew Sherman*, 93.

89. Paine, *Common Sense*, 43.

90. Related by Grant to Brevet Brigadier General Horace Porter, USA; see Porter, "Surrender," in Johnson and Buell, *Battles and Leaders*, IV.745.

91. "Battle for Missouri," in Lindeman, *Conflict*, 24–42. The one maneuver they quickly learned was retreat.

92. Sherman, Letter to U. S. Grant, April 25, 1864, in Sherman, *Sherman's Civil War*, 876.

93. Sherman, *Sherman's Civil War*, 883.

94. Letter to Major Gen. John A. Logan, May 12, 1865, in Sherman, *Sherman's Civil War*, 898–899.

95. Letter to the Mayor of Atlanta, September 12, 1864, in Sherman, *Memoirs*, II.127.

96. Caudill and Ashdown, *Sherman's March*, 38, observe two competing narratives: the Yankee view of "opportunism and efficiency" in meeting his goals, and the southern view of a machine "in which values and traditions were sacrificed for mere results." But, again, which results, which values, and which traditions? This "mere result" set four million people free.

97. Hanson, *Soul*, 442, n. 121, citing Royster, *Destructive War*, 329.

98. Described in Caudill and Ashdown, *Sherman's March*, 38.

99. Hanson, *Soul*, 438–439, n.105, citing Walters, *Merchant of Terror*, 182–183, 205.

100. Hansen, *Civil War*, 270–271.

101. Clinton, November 7, 2001, http://www.islamfortoday.com/clinton01.htm.

Chapter 6: The Prelude to World War II

1. Murray, *Change*, 184.

2. Goldhagen, *Hitler's Willing Executioners*. For analysis, Peikoff, *Ominous Parallels*. Waite, *Psychopathic God*, fails to explain why the Germans followed such a psychopath.

3. Kagan, *Origins*, 282–283.

4. Bell, *Origins,* 16.

5. Versailles Treaty, June 28, 1919, http://www.firstworldwar.com/source/versailles.htm. On the conference, Keylor, *Legacy*.

6. Wilson, "Fourteen Points" Speech, delivered in Joint Session of Congress, January 8, 1918, http://www.firstworldwar.com/source/fourteenpoints.htm.

7. Kant, *Perpetual Peace,* II.II, subtitled *The Law of Nations Shall be Founded on a Federation of Free States*.

8. Kant, *Perpetual Peace*, II.I.

9. "Democracy" reflects a deeper aspect of Kant's philosophy: that perceptions of reality are created by actions of the mind and that, in a social context, this subjectivism is collective. Despotic democracy—and appeals to the vote as the standard of self-determination—are in this deeper sense consistent with Kant's philosophy.

10. Poincaré, "Welcoming Address at the Paris Peace Conference," January 18, 1919, http://www.firstworldwar.com/source/parispeaceconf_poincare.htm.

11. The Commission on the Responsibility of the Authors of the War and on Enforcement of Penalties, May 6, 1919, http://www.firstworldwar.com/source/commission warguilt.htm.

12. Article 232 limited responsibility to civilian damages.

13. Overy, *Road to War,* 122.

14. Several works at the time maintained that Germany would return to aggression in pursuit of empire, in terms that were remarkably prescient: Veblen, *Inquiry*; Bainville, *Consequences*; Schuman, *Politcs*; discussed in Hughes, "Origins."

15. The *Kriegsschuldreferat,* or War-Guilt Section; Kagan, *Origins,* 290, citing Geiss, "Outbreak," 71–74; Baylen and Evans, "History."

16. Marks, "1918," 23–24, contrasts her view, that the lack of evidence in German cities for defeat undercut the peace, with the view of Taylor, *Origins,* e.g., 24–25, that failure to divide Germany destroyed the peace. But dividing Germany was not possible, and had it been tried, the motivations among Germans to correct the injustice would have been strong. The cause of the new war would have remained in place.

17. For the *Dolchstoss,* see, for example, Watt, *Kings,* chap. 16.

18. Hitler's speech to the Reichstag, reported in the *Times,* February 21, 1938, 9.

19. Lamb, *Drift,* 71.

20. Lamb, *Drift,* 69.

21. Kagan, *Origins,* 398.

22. Shotwell, *Germany,* 82.

23. Marks, "Myths," 233–234.

24. Keynes, *Consequences,* 5, 35–36.

25. Evans, *Coming,* 66; Kagan *Origins,* 292. Theobald von Bethmann-Hollweg was chancellor from 1909 to 1917. Craig, *Germany,* 33, 106. The occupation army left eighteen months ahead of schedule, because the French quickly paid off the indemnities rather than use them as political grievances.

26. Gilbert, *Roots,* 62.

27. Keynes, *Consequences,* 225.

28. Gannon, *Press,* 12.

29. Gilbert, *Roots,* 52.

30. Marks, "Myths," is the essential article. Kagan, *Origins,* 303–307, condenses this history.

31. Marks, "Myths," 237, describes three types of bonds: A, B, and C. C bonds were virtually worthless and accounted for more than 80 billion of the 132 billion marks assessed. Keynes, *Revision,* 24, for the figure of 132 milliards.

32. In addition to Marks, "Myths," see "Kagan, *Origins,* 292, and nn. 19 and 20. In particular, Schuker, "End."

33. Marks, "Myths," 254. Franck, in Germany immediately after the peace, notes that "the assertion that Germany had been 'starved to her knees' was scarcely borne out by observation in the occupied area," although Berlin was hungry "in a superlative degree" (*Vagabonding,* 61, 137).

34. Peikoff, *Parallels,* 216.

35. Marks, "Myths," 255: the Germans "from start to finish deemed reparations a gratuitous insult."

36. Occupation of the valley was discussed at the London Conference, March 1920; Marks, "Myths," 236.

37. Marks, "Myths," 238, n. 28.

38. Rand, "Roots," 37.

39. Van Evera, "Cult," 67–69, citing Otto Richard Tennenburg in 1911, Crown Prince Wilhelm in 1913, and Hermann Vietinghoff-Scheel in 1912.

40. The source is a memorandum from Hitler's adjutant, Colonel Friedrich Hossbach. Taylor, *Origins,* 131–134, doubts the memorandum, but the criticism fails; Trevor-Roper, "A.J.P. Taylor," 93–95; Kagan, *Origins,* 382–384; Koch, "Hitler," 168–171.

41. Marks, "Myths," 240–241.

42. Marks, "Myths," 243, and n. 47.

43. Marks, "Myths," 240–241, 248.

44. Marks, "1918," 29, for post-1918 events as a continuation of World War I, which lasted until Locarno in 1925.

45. Treaty of Mutual Guarantee between Germany, Belgium, France, Great Britain, and Italy; October 16, 1925 (Locarno Pact), at the *Avalon Project,* http://www.yale.edu/lawweb/avalon/intdip/formulti/locarno_001.htm.

46. Vittorio Scialoja, Italian foreign minister, and president of the league in 1925 and 1929, on Locarno: "If the dream has taken us into an ideal world, if it has enabled us to give shape to lofty moral and legal aspirations, if it has been the image of a more human and more divine future, let us not regret that we dreamt." Spoken at the Sixth Assembly of the League of Nations, December 12, 1925, http://www.unog.ch/80256EDD006B8954/(httpAssets)/A2BDB2541806916CC1256F33002DD5F8/$file/locarno_eng.pdf.

47. The Peace of Toruń in 1466, following the Thirteen Years War, awarded the area to Poland against the Teutonic Knights. John Stuart Mill wrote in 1861, "The German colony of East Prussia ... must ... be either under a non-German government, or the intervening Polish territory must be under a German one" ("Considerations," 549). Marks, "1918," 25, notes that the Germans resented any transfers of land and that a more lenient treaty would not have alleviated this resentment.

48. The Rapallo Agreement, from the *Avalon Project*, from Article 1, http://www.yale.edu/lawweb/avalon/intdip/formulti/rapallo_001.htm; Treaty of Berlin: http://www.yale.edu/lawweb/avalon/intdip/formulti/berlin_001.htm.

49. Voigt, *Unto Caesar*, published in 1938, recognized this point.

50. Marks, "Myths," 252.

51. Some historians claim that a "cult of the offensive" before World War I was followed by a "cult of the defensive" before World War II. This led to reliance on the Maginot Line and deepened the defensive mentality. Kagan, *Origins*, 356, citing Van Evera, "Cult," 58–107.

52. On Stresemann, Shirer, *Rise and Fall*, 88, 192, 295.

53. Kagan, *Origins*, 324.

54. The controversy is exemplified most plainly in reactions to Taylor, *Origins*.

55. Shirer, *Rise and Fall*, 292.

56. Rock, *Appeasement*, 86.

57. "'Abyssinia's Sad Fate' by Scrutator," *Sunday Times*, May 10, 1936, in Morris, *Roots*, 36.

58. "Democracies and Dictatorships," *Tablet*, January 8, 1938, in Morris, *Roots*, 39 (emphasis added).

59. Gannon, *Press*, 70.

60. Letter to George Ferguson, assistant editor of the *Winnipeg Free Press*, April 27, 1939, in Caputi, *Appeasement*, 156. Pages 149–161 deal with the contentious issue of Dawson's and Barrington-Ward's roles in allegedly shaping the news to promote the policy of appeasement, including the assessment of Gannon, *Press*, that the news was not censored.

61. Letter to Master of Balliol, October 7, 1938, in Gannon, *Press*, 11.

62. For instance, Gannon, *Press*, 8–10, 32–33, for the *Daily Mail*'s claim that a rearmed Britain could be friends with a rearmed Germany.

63. Geoffrey Dawson, writing to correspondent H. G. Daniels, May 23, 1937, in Shirer, *Rise and Fall*, 396; Gannon, *Press*, 114.

64. Goering testified at Nuremberg, March 14, 1946: "When the Civil War broke out in Spain, Franco sent a call for help to Germany.... I urged him [Hitler] to give support under all circumstances, firstly, in order to prevent the further spread of communism in that theater and, secondly, to test my young Luftwaffe at this opportunity in this or that technical respect." *Avalon Project*, http://www.yale.edu/lawweb/avalon/imt/proc/03-14-46.htm#Goering2.

65. Adams *Politics*, appendix 1.

66. Murray, *Change*, 72–73; his policies as chancellor were continued by John Simon in July 1937. For a view of Chamberlain as deeply concerned for British security, opposed to disarmament, and as cognizant that Germany was the "ultimate potential enemy," Hughes, "Origins," 286.

67. Kagan, *Origins*, 367.

68. Lamb, *Drift*, 89.

69. Murray, *Change*, 66.

70. Murray, *Change*, 193, recounts the visit of General Joseph Vuillemin, chief of staff of the French air force, to the Luftwaffe. The Germans painted a frightening

picture of their air capabilities, with figures "all out of proportion to reality." His report weakened the French commitment to defend Czechoslovakia.

71. *Manchester Guardian*, October 29, 1938, in Gannon, *Press*, 9.

72. Shirer, *Rise and Fall*, 386.

73. Keylor, *Legacy*, 53.

74. Kagan, *Origins*, 344.

75. Hitler's speech is excerpted in Shirer, *Rise and Fall*, 393–397, along with the response of the *Times* and the British government to the proposed arms limits. Rowse, *Appeasement*, 6–7, on events of 1935.

76. Earlier, Goering told the British that they had an air force; the British proceeded with talks. Shirer, *Rise and Fall*, 391. Hitler publicly revealed the air force in March; Kagan, *Origins*, 344.

77. Kagan, *Origins*, 360, n. 225, citing Craig, *Germany*, 691.

78. Kagan, *Origins*, 360, on Jodl, citing Schmidt, *Hitler's Interpreter*, 320, for Hitler, and Bullock, *Hitler*, 342–343, on international promises.

79. Shirer, *Rise and Fall*, 400, reports learning of Blomberg's decision six days later. The *Times* reported, on February 7, 1938, that German generals had argued for withdrawal should the French mobilize—and that Hitler knew that the French government had been told that Germany would withdraw, should the French mobilize. Watt, "German Plans," concludes that the Germans would have fought while withdrawing to the Rhine.

80. Klemperer, *Witness*, 155–156.

81. March 26, 1936, *Parliamentary Debates* 310:1435–1436.

82. Kagan, *Origins*, 348.

83. March 26, 1936, *Parliamentary Debates* 310:1443.

84. *Parliamentary Debates* 310:1446.

85. *Times*, February 25, 1938, 8. Philip Henry Kerr, Marquess of Lothian (1882–1940).

86. Lothian soon read *Mein Kampf* and changed his opinions about the Third Reich.

87. Lord Cecil said that the prime minister's statement disavowing league action to protect the Czechs was "almost an invitation to Hitler to go on." First Viscount Cecil of Chelwood Edgar Algernon Robert Cecil (1864–1958) advocated British participation in the League of Nations.

88. Churchill, *Gathering Storm*, 263, from Schuschnigg's record, *Ein Requiem in Rot-Weiss-Rot*, 37–38.

89. Austro-German Agreement of July 11, 1936, *Avalon Project*, http://www.yale.edu/lawweb/avalon/wwii/yellow/ylbk001.htm.

90. Churchill, *Gathering Storm*, 269. Hitler told his envoy to Italy to thank Mussolini: "I will never forget it, whatever may happen. If he should ever need any help, or be in any danger, he can be convinced that I shall stick to him whatever might happen even if the whole world were against him" (Kagan, *Origins*, 385).

91. Gannon, *Press*, 145.

92. *Times*, April 11, 1938, in Gannon, *Press*, 151.

93. *Times*, March 15, 1938.

94. Kagan, *Origins*, 386; Fuchser, *Chamberlain*, 112, at a cabinet meeting on March 12, 1938.

95. Churchill, *Gathering Storm*, 271–272; Murray, *Change*, 141–149, on German weakness.

96. Murray, *Change*, 148, citing *Hitler's Secret Conversations*, 207.

97. Guderian, *Panzer Leader*, 49–56, for the entry into Austria. Perhaps, as Guderian notes here, Churchill "was anxious to prove that the political leaders of Great Britain and France could have gone to war in 1938 with good prospects of achieving victory."

98. Cabinet meeting of September 13, 1938, in Murray, *Change*, 199. Similarly for a September 14 meeting, Fuchser, *Chamberlain*, 141.

99. Kagan, *Origins*, 379; Murray, *Change*, 58, 62–64, 177.

100. *Times*, September 7, 1938, in Gannon, *Press*, 177–178.

101. Overy, *Road to War*, 101.

102. Murray, *Change*, 61. On the moralizing views of British statesmen, Overy, *Road to War*, chap. 2. For Chamberlain as concerned with "cold realism" and raison d'etat, Feiling, *Life*, 253.

103. Murray, *Change*, 63; discussion of the air staff report is at 156–162. Kagan, *Origins*, 408: Chamberlain did not say he was delaying to build up British military strength, and he opposed substantial increases in armament spending even after Munich.

104. Kagan, *Origins*, 394; Fuchser, *Chamberlain*, 136.

105. Fuchser, *Chamberlain*, 116; this was Churchill's advice to Parliament on March 14 and 24. Fuchser notes that the Foreign Policy Committee, over the advice of the Foreign Office, agreed a week after the Austrian absorption not to fight for Czechoslovakia.

106. Fuchser, *Chamberlain*, 142.

107. Fuchser, *Chamberlain*, 143.

108. Konrad Henlein (1898–1945), leader of the Sudetan German Party.

109. Fuchser, *Chamberlain*, 145.

110. Murray, *Change*, 195.

111. Kagan, *Origins*, 411, for a Gallup poll in the summer of 1939, in which three-quarters of the British public thought that a war to stop Hitler would be worth it.

112. Churchill, *Gathering Storm*, 301.

113. Churchill, *Gathering Storm*, 315; Fuchser, *Chamberlain*, 156.

114. In response to a note from Hitler, who stated his willingness to give guarantees to the Czechs: Fuchser, *Chamberlain*, 158.

115. Churchill, *Gathering Storm*, 315; Fuchser, *Chamberlain*, 158.

116. From Hitler's Obersalzberg Speech, August 22, 1939, in Adamthwaite, *Making*, 220.

117. *Avalon Project*, April 4, 1946, http://www.yale.edu/lawweb/avalon/imt/proc/04-04-46.htm.

118. Murray, *Change*, 240.

119. Murray, *Change*, 3–49, analyzes Germany's position in raw materials and military capacities.

120. Murray, *Change*, 210, 212.

121. Kagan, *Origins*, 409–410, citing Murray, *Change*, 262–263.

122. Shirer, *Rise and Fall*, 290–291, describes Germany's desperate state when Hitler took power.

123. Peikoff, *Ominous Parallels*, 19.

124. Gannon, *Press*, 31.

125. Kennedy and Imlay, "Appeasement," 117, note that "appeasement originally was a positive concept—as in the appeasing of one's appetite."

126. Kagan, *Origins*, 322.

127. Rand, "Anatomy," discusses compromise in terms of basic principles.

128. *Parliamentary Debates* 310:1462, Sir Archibald Sinclair, March 26, 1936.

Chapter 7: The American Victory over Japan

1. Takemae, *Allied Occupation*, xxx–xxxi.

2. Dower, *Embracing Defeat*, 87 referencing Tsurumi, *Social Change*.

3. Takemae, *Allied Occupation*, 372.

4. Woodard, *Allied Occupation*, 10 (emphasis added). On Shinto and related terms, see pp. 9–13.

5. On the idea of kokutai in relation to the mythology of Shinto, see Brownlee, "Four Stages," http://www.iar.ubc.ca/centres/cjr/seminars/semi2000/jsac2000/brownlee .pdf. Woodard's conclusion that the "The Kokutai Cult was not a form of Shinto" does not follow, to the extent that the cult "derived from Shinto mythology," raised "a traditional religious concept ... to the status of a religio-political absolute," and was practiced using the shrines and priests of Shinto; Woodard, *Allied Occupation*, 9–13.

6. Meiji Constitution, articles 1, 2, 3, 11, 13, 20, *Hanover Historical Texts Project*, http://history.hanover.edu/texts/1889con.html.

7. *The Way of the Subject*, August 1941, in Dower, *Embracing*, 277.

8. For the Educational Rescript and educational reforms, Takemae, *Allied Occupation*, 347–371; Woodard, *Allied Occupation*, 164–175; Dower, *Embracing*, 244–250.

9. Woodard, *Allied Occupation*, 165. See note 1 for the suicide of Jiro Ishiroku.

10. Woodard, *Allied Occupation*, 164; Takemae, *Allied Occupation*, 347.

11. Yoshida, *Requiem*, 107; Dower, *Embracing*, 415–416.

12. Takemae, *Allied Occupation*, 357.

13. Dower, *Embracing*, 33–34.

14. Dower, "Japanese Cinema," 33–54, examines Japanese wartime cinema for clues to how such ideals were popularly received.

15. Dower, *War without Mercy*, 264.

16. Toland, *Rising Sun*, 7.

17. Toland, *Rising Sun*, chap. 1, summarizes these events, including the army in Manchuria, which acted "to the dismay of not only the world but Tokyo itself" (8).

18. Toland, *Rising Sun*, 9.

19. Toland, *Rising Sun*, 26.

20. Dower, *Embracing*, 277.

21. Dower, *Embracing*, 277.

22. Iris Chang, *The Rape of Nanking,* has documented this gruesome story. A group of foreigners established a "safe zone" and saved thousands. The precise figures for such atrocities are usually contentious, and Nanking is no exception. Takemae, *Allied Occupation,* 561, and n. 8, for several views. The International Military Tribunal for the Far East, for instance, accepted 200,000. In Takemae's words, Japanese troops "rampaged, engaging in an orgy of rape, pillage, arson and murder," all part of a "wider campaign of annihilation" in Manchuria and China. Dower, *Embracing,* 22, sees a "blood-soaked monster" and the "rape of Nanking." Hasegawa, *Racing,* 13, 299: the event was "brutal" and "cruel." Lynn, *Battle,* 227: Chang's "basic narrative is accurate."

23. Reproduced as a plate in Chang, *The Rape of Nanking.*

24. One reason often cited for the war was the American-British-Dutch oil embargo, in July 1941. But the embargo did not cause the attacks of 1931 or 1937. The Japanese need for thousands of tons of oil was due to the decision to wage war. Sagan, "Origins," examines the lead-up to Pearl Harbor as "a mutual failure of deterrence" but notes that the Americans knew of the Japanese intentions to attack Indochina before the embargo. The embargo was a contributing, but not a primary, cause of Japan's war with the United States. After 1945, the Japanese acquired all the oil they needed through nonviolent, productive trade.

25. *E-Speeches.com,* http://www.espeeches.com/fdr.htm.

26. Rosenberg, *Date,* 15. Rosenberg examines how the "infamy" phrase passed down through history.

27. Campbell, "Roosevelt," examines Roosevelt's central role in this policy; he had "the deepest commitment to it" of the major leaders (219).

28. Armstrong, *Unconditional Surrender,* 12 (emphasis added).

29. Frank, *Downfall,* 27. Frank notes that documents show that this policy was oft discussed, although its effect on Japan was not considered.

30. *Birth of the Constitution of Japan,* http://www.ndl.go.jp/constitution/e/shiryo/01/002_46/002_46tx.html.

31. Maddox, *Weapons,* 11.

32. Campbell, "Roosevelt," 222, sees a "shift in attitudes" that is "surprising."

33. Walker, *Prompt and Utter Destruction,* 46.

34. Frank, *Downfall,* 27.

35. Burrell, *Ghosts,* presents many acts that could be described as "suicidal." One may shudder to think of what it took to defeat the Japanese on Iwo Jima. But Americans did not go to war *intending* to die. They conducted no death rituals, and no *banzai* charges where death was the goal.

36. Dockrill and Freedman, "Hiroshima," 195.

37. For the *Ketsu-Go* plan, in English, see MacArthur, *Reports,* II.II, 601–607.

38. Yoshida, *Requiem,* 109, for the saying.

39. Inoguchi, *Divine Wind,* 58.

40. Kodama, *I Was Defeated,* 174. His life involved plots against the Japanese government, cooperation with American authorities, and organized crime.

41. Frank, *Downfall,* 188–189.

42. Frank, *Downfall,* 189.

43. The Berlin (Potsdam) Conference, July 17 to August 2, 1945, Annex II, at the *Avalon Project*, http://www.yale.edu/lawweb/avalon/decade/decade17.htm.

44. Butow, *Decision*, 146, n. 14.

45. Frank, *Downfall*, 235. The emperor confirmed the order in his *Showa Tenno Doluhekuroku*, dictated in March and April 1946.

46. Frank, *Downfall*, 113.

47. Frank, *Downfall*, 225, with intercepted communications at 221–239. Butow, *Japan's Decision*, 56–57.

48. Takemae, *Allied Occupation*, 220–221; Dower, *Empire*, 227–272.

49. Takemae, *Allied Occupation*, 24–25.

50. For the Konoe Memorial, Dower, *Empire*, 259–265; text at 260–264. Butow, *Japan's Decision*, 47–50.

51. Dower, *Empire*, 271. His arrest is at 265–272.

52. All quotes from Butow, *Japan's Decision*, 66, 68.

53. Frank, *Downfall*, 91–92, 96; Bix, "Japan's Delayed Surrender," 80–115.

54. Discussed in Butow, *Japan's Decision*, 70–75.

55. Frank, *Downfall*, 113. Despite this record, many commentators continue to assert that Japan was trying to surrender before the summer of 1945, for example, Codevilla and Seabury, *War*, 132. Campbell, "Roosevelt," 239.

56. Butow, *Japan's Decision*, 151.

57. Asada, "Shock," 505.

58. In Kido's recollections, which were probably intended to protect the emperor. Asada, "Shock," 487.

59. Asada, "Shock," 494. The three August 9–10 meetings are reconstructed at 490–496.

60. Toland, *Rising Sun*, 833.

61. Discussed by Dower, *Embracing*, 88–89.

62. Wetzler, *Imperial Tradition*, 50–57.

63. Woodard, *Allied Occupation*, 8, citing SCAP document *Political Reorientation of Japan*, September 1945–September 1948, Report of the Government Section.

64. Toland, *Rising Sun*, 847. Dower, *Embracing*, 492 for Tōjō's transformation into "prime culprit," abetted by his failure to commit suicide with a sword.

65. An overview of the occupation is in Bailey, *Postwar Japan*, 21–66. The definitive treatments used here are Takemae, *Allied Occupation*, and Dower, *Embracing*.

66. Takemae, *Allied Occupation*, 56–57.

67. The Potsdam principles were issued in longer form in the "U.S. Initial Post-Surrender Policy for Japan," (SWNCC-150/4/A) on September 22. The Joint Chiefs of Staff directive cited here, JCS1380/15, was the third of the three major control documents for the occupation. Takemae, *Allied Occupation*, 226–227; Dower, *Embracing*, 73–75. Documents may be read at *Birth of the Constitution of Japan*, http://www.ndl.go.jp/constitution/e/shiryo/01/022shoshi.html.

68. Cohen, *Remaking Japan*, 57, 59.

69. Cohen, *Remaking Japan*, 57, 59.

70. Cohen, *Remaking Japan*, 8–9. The meeting was of the State-War-Navy Coordinating Committee Subcommittee on the Far East (SWNCCFE), August 29, 1945,

which, Cohen writes, brought the only question ever raised about limits to occupation authority. Eugene Dooman, special assistant to Undersecretary of State Joseph C. Grew, raised the matter; his replacement was John Vincent Carter, who sent the Shinto telegram cited in this chapter.

71. Joint Chiefs of Staff, JCS 1380/15, II.13.

72. Dower, *Embracing,* 115, 420.

73. Passim, "Occupation," 111; Hata, "Occupation," 99, for Japanese machinery rusting on Shanghai docks.

74. Takemae, *Allied Occupation,* 347. Of course, scholars disagree about the degree of change, or continuity, between prewar and postwar periods. Gluck, "Idea," poses two periods in Hirohito's reign (*showa*); Dower, "Useful War," focuses on the continuities. But it is beyond doubt that Japanese policies changed radically.

75. Dower, *Embracing,* 215–216. The movie was *Our Job in Japan,* by the War Department, November 1945.

76. Dower, *Embracing,* 217; *Saturday Evening Post,* December 15, 1945.

77. Dower, *Embracing,* 25.

78. Hata, "Occupation," 96–97, for MacArthur's eleven categories of early occupation objectives. The October 4, 1945, Civil Liberties Directive (SCAPIN-93) was the start.

79. Takemae, *Allied Occupation,* 300, for a photo of a 1947 graduating class.

80. Reproduced and discussed in Woodard, *Allied Occupation,* 54–56.

81. Woodard, *Allied Occupation,* 66–68.

82. Joint Chiefs of Staff, JCS 1380/15, I.10.

83. The "Administration of the Educational System of Japan" directive, SCAPIN-178. Takemae, *Allied Occupation,* 349–351, for "dangerous" courses that were eliminated.

84. "The Suspension of Courses in Morals (*Shushin*)," SCAPIN-519, December 31.

85. Woodard, *Allied Occupation,* 171, quoting from a Ministry of Education notice of June 3, 1947.

86. Takemae, *Allied Occupation,* 347.

87. Woodard, *Allied Occupation,* 168. The notice was dated May 13, 1946.

88. Dower, *Embracing,* 242.

89. Takemae, *Allied Occupation,* 362.

90. Dower, *Embracing,* 247, 157; Takemae, *Allied Occupation,* 361, on the student's activities.

91. Manchester, *American Caesar,* 603, from *Life Magazine,* August 22, 1955.

92. Dower, *Embracing,* 171.

93. Dower, *Embracing,* 236.

94. Dower, *Embracing,* 80–84, on the first reforms; 29, on contempt for returning soldiers.

95. Dower, *Embracing,* 170–171.

96. Dower, *Embracing,* 181.

97. Dower, *Embracing,* 176–177, for a school lesson by Prince Akahito, who succeeded Hirohito.

98. Cartoon by Katō Etsurō, from his booklet *Okurareta Kukumei*, roughly "The Revolution We Have Been Given," published in August 1946 by Kobarutosha. The context of the cartoon is MacArthur's "Civil Liberties Directive" directive of October 4, 1945. Reproduced and discussed in Dower, *Embracing*, 65–71.

99. Dower, *Embracing*, 374.

100. Clausewitz, *On War*, I.1.2, p. 75: war is "an act of force to compel the enemy to do our will." Clausewitz, influenced by German philosophy, may have accepted *voluntarism*, the idea that the "will" is an independent faculty. But the "will" to war is not a separate faculty in a person's psychology. Similarly, "voluntaristic" theories of war in Codevilla and Seabury, *War*, 43–46.

101. Noted in Asada, "Shock," 479.

102. Frank, *Downfall*, 337, citing Schaffer, *Wings of Judgment*, 88–89.

103. Sherman, *Sherman's Civil War*, 543–550.

104. Armstrong, *Unconditional Surrender*, 143; Jodl's statement is in Holles, *Unconditional Surrender*, 13. Campbell, "Roosevelt," 239–240, concludes that unconditional surrender "neither lengthened the war nor determined its outcome." But, I respond, it defined the end of the war.

105. Frank, *Downfall*, 337.

106. Armstrong, *Unconditional Surrender*, 155.

107. Dower, *Empire*, 277–278; Dower, *Embracing*, 77.

108. Dower, *Embracing*, 185.

109. Sakaguchi Ango, "On Decadence"; discussed in Dower, *Embracing*, 155–156.

110. Dower, *Embracing*, 187–188.

111. Dower, *Embracing*, 183.

112. Kodama, *I Was Defeated*, 175.

113. Asada, "Shock," 500–501.

114. Alperovitz, *Diplomacy*; Blackett, *Fear, War and the Bomb*. Responses: Asada, "Shock"; Maddox, *Weapons*; Dockrill and Freedman, "Hiroshima"; Kagan, "Why America Dropped the Bomb"; and Asada, "Japanese Perceptions."

115. Hasegawa, *Racing the Enemy*, 299–301.

116. Hasegawa, *Racing the Enemy*, 5, 165.

117. Frank, *Downfall*, 71.

118. Frank, *Downfall*, 359–360.

119. Frank, *Downfall*, 324, notes that Truman's firm refusal to Soviet demands on August 18 for the occupation of Hokkaido saved hundreds of thousands of Japanese from death.

120. Weller, *First into Nagasaki*, 30–31.

121. Frank, *Downfall*, 191 for the 900,000 figure.

122. Frank, *Downfall*, 194, 340–342, and 342n.

123. Weller, *First into Nagasaki*, 17. First published as "Back to Nagasaki," in Brown and Bruner, *How I Got That Story*.

124. Weller, *First into Nagasaki*, 16.

125. Fussell, *Atomic Bomb*, 14.

126. Private correspondence with Mr. Taylor's daughter, Hannah Krening.

127. Asada, "Shock," 507, 509.

Conclusion

1. Sun-tzu, *Art of War*, 174.

2. Noted in Walzer, *Wars*, 201, who nevertheless maintained that the National Liberation Front was legitimate.

3. Retired general Maxwell Taylor, statement to the Senate Foreign Relations Committee, 1966, published as "Peace and Stability for Vietnam Is Constant U.S. Objective."

4. Polybius 5.11.5.

5. Summers, *Strategy*, 1, for a North Vietnamese colonel's reply to Colonel Summers, that American success on battlefields was "irrelevant."

BIBLIOGRAPHY

Dates are for the edition I used and may not indicate the date of first publication. Citations of classical writers are given in the format standard to classical scholarship, using book, section, and line numbers as appropriate. This is an abbreviated list; no attempt has been made to be comprehensive.

Introduction and General Readings

Aineias the Tactician. *How to Survive under Siege*, tr. D. Whitehead (Oxford: Clarendon, 1990).

Aristotle. *Nicomachean Ethics*, tr. T. Irwin (Indianapolis: Hackett, 1999).

Beyerchen, A. "Clausewitz, Nonlinearity, and the Unpredictability of War," in *Coping with the Bounds: Speculations on Nonlinearity in Military Affairs*, ed. T. Czerwinski (Washington, DC: Dept of Defense, Command and Control Research Program, by the Institute for National Strategic Studies, National Defense University, 1998), 151–197. Originally in *International Security* 17.3 (1992): 59–90.

Clausewitz, Carl von. *On War*, ed. and tr. M. Howard and P. Paret (New York: Routledge, 2004).

Codevilla, A., and P. Seabury. *War: Ends and Means*, 2nd ed. (Washington, DC: Potomac Books, 2006).

Douhet, G. *Command of the Air*, tr. D. Ferrari (New York: Coward-McCann, 1942).

Freud, S. "Why War?" in *Aggression, Hostility and Violence: Nature or Nurture?* ed. T. Maple and D. W. Matheson (New York: Holt, Rineheart and Winston, 1973), 16–27.

Hanson, V. D. *Carnage and Culture: Landmark Battles in the Rise of Western Power* (New York: Anchor, 2001).

———. "The Dilemmas of the Contemporary Military Historian," in *Reconstructing History: The Emergence of a New Historical Society*, ed. E. Fox-Genovese and E. Lasch-Quinn (New York: Routledge, 1999), 189–201.

Jervis, R. "War and Misperception," in *The Origin and Prevention of Major Wars*, ed. R. I. Rotberg and T. K. Rabb (Cambridge: Cambridge University Press, 1988), 101–126.

Jessup, J. E., and R. W. Coakley, eds., *A Guide to the Study and Use of Military History* (Washington, DC: U. S. Government Printing Office, 1979).

Johnson, J. T. "The Just War Idea: The State of the Question," *Social Philosophy and Policy* 23.1 (2006): 167–195.

Jomini, A. H. de. *The Art of War*, special edition, tr. G. H. Mendell and W. P. Craighill (El Paso: El Paso Norte, 2005).

Liddell Hart, B. H. *Strategy*, 2nd ed. (New York: Meridien, 1991).

Lynn, J. *Battle: A History of Combat and Culture* (Boulder, CO: Westview, 2003).

Maurice. *Strategikon: Handbook of Byzantine Military Strategy*, tr. G. T. Dennis (Philadelphia: University of Pennsylvania, 1984).

National Strategy for Combating Terrorism (February 2003). https://www.cia.gov/news-information/cia-the-war-on-terrorism/Counter_Terrorism_Strategy.pdf.

———. (September 2006). http://www.whitehouse.gov/nsc/nsct/2006.

Strauss, B. S., and J. Ober. *The Anatomy of Error* (Boston: St. Martin's, 1992).

Polyaenus. *Stratagems of War*, ed. and tr. P. Krentz and E. L. Wheeler (Chicago: Ares, 1994).

Stout, H. S. *Upon the Altar of a Nation: A Moral History of the American Civil War* (New York: Viking, 2006).

Summers, H. *On Strategy: A Critical Analysis of the Vietnam War* (Novato, CA: Presidio, 1982).

Sun-tzu. *The Art of War*, tr. R. D. Sawyer (Boulder, CO: Westview, 1994).

Taylor, M. "Peace and Stability for Vietnam Is Constant U.S. Objective," statement to the Senate Foreign Relations Committee, 1966, http://www.ndu.edu/library/taylor/mdt-0312.pdf.

Thucydides. *History of the Peloponnesian War*, tr. R. Warner (New York: Penguin, 1972).

U.S. Department of the Army. "Counterinsurgency," Army Field Manual 3-24; Marine Corps Warfighting Publication 3-33.5 (Washington, DC: Department of the Army, 2006).

———. Field Manual 100-5, *Field Service Regulations: Operations* (Washington, DC: Government Printing Office, September 27, 1954).

———. Field Manual 100-5, *Field Service Regulations: Operations* (Washington, DC: Government Printing Office, February 19, 1962).

U.S. Department of the Navy. *Marine Corps Warfighting Publication 5-4: Small Wars* (Washington, DC: CSC, 1996).

U.S. Department of War. Field Manual 100-5, *Tentative Field Service Regulations: Operations* (Washington, DC: Government Printing Office, October 1, 1939).

Walzer, M. *Wars Just and Unjust: A Moral Argument with Historical Illustrations*, 3rd ed. (New York: Basic Books, 2006).

Welch, D. *Justice and the Genesis of War* (Cambridge: Cambridge University Press, 1995).

Wright, Q. A *Study of War*. Abridged ed. (Chicago: University of Chicago Press, 1964).

Chapter 1: The Greco-Persian Wars

Primary Sources

Aeschylus. *The Persians,* tr. H. W. Smith. (Cambridge, MA: Harvard University Press, 1988).

Aristotle. *Aristotle: The Complete Works,* ed. J. Barnes (Princeton: Princeton University Press, 1984).

———. *Politics,* tr. T. A. Sinclair, rev. and re-presented by T. J. Saunders (New York: Penguin, 1981).

Boyce, M. *The Letter of Tosar* (Rome: Instituto Italiano per il Medio ed Estremo Oriente, 1968).

Diodorus Sikulus 11–12.37.1: Greek History, 480–431—The Alternative Version, tr. P. Green (Austin: University of Texas, 2006).

Diodorus the Sicilian. *Library of History,* vols. III and IV, tr. C. H. Oldfather (Cambridge, MA: Harvard University Press, 1961).

Fornara, C. W., ed. *Archaic Times to the End of the Peloponnesian War* (Cambridge: Cambridge University Press, 1988).

Herodotus. *The Histories,* tr. A. de Sélincourt (New York: Penguin, 1996).

Horne, C. F. *The Book of Arda Viraf: A Dantesque Vision of Heaven and Hell* (Whitefish, MT: Kessinger, 2005).

Meiggs, R., and D. Lewis, eds. *A Selection of Greek Historical Inscriptions to the End of the Fifth Century B.C.* (Oxford: Clarendon, 1980).

Plutarch. *The Lives of the Noble Greeks and Romans,* tr. J. Dryden (New York: Modern Library, 1992).

Pritchard, J. B., ed. *The Ancient Near East I: An Anthology of Texts and Pictures* (Princeton: Princeton University Press, 1958).

———, ed. *Ancient Near Eastern Texts Relating to the Old Testament,* 2nd ed. (Princeton: Princeton University Press, 1969).

Thucydides. *History of the Peloponnesian War,* tr. R. Warner (New York: Penguin, 1972).

Tod, M. N., ed. *A Selection of Greek Historical Inscriptions* (Oxford: Clarendon, 1948).

Xenophon. *The Persian Expedition,* tr. R. Warner (New York: Penguin, 1967).

Secondary Sources

Badian, E. "Phrynichus and Athens' *Oikeia Kaka,*" *Scripta Classica Israelica* 15 (1996): 55–60.

Balcer, J. M. *The Persian Conquest of the Greeks, 545–450 BC* (Konstanz: Universitätsverlag Konstanz, 1995).

Burn, A. R. *Persia and the Greeks: The Defense of the West, 546–478 B.C.,* 2nd ed. (Stanford, CA: Stanford University Press, 1968).

The Cambridge Ancient History, vol. IV, *Persia, Greece and the Western Mediterranean, c. 525 to 479 BC*, 2nd ed., ed. J. Boardman, N. G. L. Hammond, D. M. Lewis, and M. Ostwald. (Cambridge: Cambridge University Press, 1988).

Cartledge, P. *Agesilaos and the Crisis of Sparta* (London: Duckworth, 1987).

Frye, R. N. *The History of Ancient Iran* (Munich: Beck Verlagsbuchhandlung, 1984).

Gomme, A. W. *A Historical Commentary on Thucydides*, vol. 1, (Oxford: Oxford University Press, 1945).

Green, P. *The Greco-Persian Wars*, repr. ed. (Berkeley: University of California Press, 1998).

Hanson, V. D. *The Western Way of War: Infantry Battle in Classical Greece* (Berkeley: University of California Press, 1989).

Hignett, C. *Xerxes' Invasion of Greece* (Oxford: Clarendon, 1963).

Kelly, T. "Persian Propaganda—a Neglected Factor in Xerxes' Invasion of Greece and Herodotus," *Iranica Antiqua* 38 (2003): 173–219.

Kelsay, J., and J. T. Johnson. *Just War and Jihad: Historical and Theoretical Perspectives on War and Peace in Western and Islamic Traditions* (New York: Greenwood, 1991).

Krentz, P. M. "The Oath of Marathon, Not Plataia?" *Hesperia* 76 (2007): 731–742.

Lazenby, J. *The Defense of Greece, 490–479 B.C.* (Warminster: Aris and Philips, 1993).

Olmstead, A. T. *History of the Persian Empire* (Chicago: University of Chicago Press, 1959).

Souza, P. de. *Piracy in the Graeco-Roman World* (Cambridge: Cambridge University Press, 2000).

Strauss, B. *The Battle of Salamis: The Naval Encounter That Saved Greece—and Western Civilization* (New York: Simon and Schuster, 2004).

Strauss, B. S., and J. Ober. *The Anatomy of Error* (Boston: St. Martin's, 1992).

Teixidor, J. *The Pagan God: Popular Religion in the Greco-Roman Near East* (Princeton: Princeton University Press, 1977).

Van De Mieroop, M. *A History of the Ancient Near East ca. 300–323 BC*, 2nd ed. (Malden: Blackwell, 2007).

Wallinga, R. T. *Xerxes' Greek Adventure: The Naval Perspective* (Leiden: Brill, 2005).

Warry, J. *Warfare in the Classical World* (New York: Barnes and Noble, 2000).

Chapter 2: The Theban Wars

Primary Sources

Aristotle. *Politics*, tr. T. A. Sinclair, rev. and re-presented by T. J. Saunders (New York: Penguin, 1981).

———. *The Rhetoric and Poetics of Aristotle*, tr. W. Rhys Roberts and I. Bywater (New York: McGraw Hill, 1984).

Diodorus the Sicilian. *Library of History*, vols. VI and VII, tr. C. H. Oldfather (Cambridge, MA: Harvard University, 1963, 1952).

Gerber, D. L., ed. *Greek Elegiac Poetry* (Cambridge, MA: Harvard University, 1999).

Herodotus. *The Histories*, tr. A. de Sélincourt (New York: Penguin, 1996).

Isocrates. "Archidamas" in *Isocrates I*, tr. G. Norlin (Cambridge, MA: Harvard University, 1928), 343–411.

Moore, J. M. *Aristotle and Xenophon on Democracy and Oligarchy*, 2nd rev. ed. (Berkeley: University of California Press, 1975).

Plutarch. *The Lives of the Noble Greeks and Romans*, tr. J. Dryden (New York: Modern Library, 1992).

Rhodes, P. J. *The Greek City States: A Sourcebook* (Norman: University of Oklahoma, 1986).

Xenophon. *History of My Times*, tr. R. Warner (New York: Penguin, 1979).

Secondary Sources

Anderson, J. K. *Military Theory and Practice in the Age of Xenophon* (Berkeley: University of California Press, 1970).

Andrewes, A. "The Government of Classical Sparta," in *Sparta*, ed. M. Whitby (New York: Routledge, 2002), 49–68.

Buck, R. J. *Boiotia and the Boiotian League, 432–371 BC* (Edmonton: University of Alberta Press, 1994).

Buckler, J. *The Theban Hegemony, 371–362 BC* (Cambridge, MA: Harvard University Press, 1980).

Cartledge, P. *Agesilaos and the Crisis of Sparta* (London: Duckworth, 1987).

———. "Boiotian Swine (For)ever? The Boiotian Superstate 395," in *Polis and Politics: Studies in Ancient Greek History; Presented to Mogens Herman Hansen on His Sixtieth Birthday, August 20, 2000*, ed. P. Flensted-Jensen, T. H. Nielsen, and L. Rubinstein (Copenhagen: Museum Tusculanum, University of Copenhagen, 2000), 397–418.

———. "The Effects of the Peloponnesian (Athenian) War on Athenian and Spartan Societies," in *War and Democracy: A Comparative Study of the Korean War and the Peloponnesian War*, ed. D. McCann and B. Strauss (New York: East Gate, 2001), 104–123.

———. "The Socratic's Sparta and Rousseau's," in *Sparta: New Perspectives*, ed. S. Hodkinson and A. Powell (London: Duckworth, 1999), 311–337.

———. *Sparta and Lakonia: A Regional History 1300–362 BC*, 2nd ed. (London: Routledge, 2002).

———. "A Spartan Education," in *Spartan Reflections* (London: Duckworth, 2001), 79–90.

———. *The Spartans: The World of Warrior-Heroes of Ancient Greece* (New York: Overlook, 2003).

Clausewitz, Carl von. *On War,* ed. and tr. M. Howard and P. Paret (New York: Routledge, 2004).

Demand, N. *Thebes in the Fifth Century: Heracles Resurgent* (London: Routledge and Kegan Paul, 1982).

Ducat, J. "The Obligations of Helots," tr. by S. Coombes, in *Sparta,* ed. M. Whitby (New York: Routledge, 2002), 196–214.

Finley, M. I., and B. Shaw *Economy and Society in Ancient Greece* (New York: Viking, 1983).

———. *The Use and Abuse of History* (New York: Penguin, 1987).

Forrest, W. G. *A History of Sparta* (New York: Norton, 1968).

Hamilton, C. D. *Agesilaus and the Failure of Spartan Hegemony* (Ithaca, NY: Cornell University Press, 1991).

Hanson, V. D. *The Other Greeks: The Family Farm and the Agrarian Roots of Western Civilization* (Berkeley: University of California Press, 1999).

———. *The Soul of Battle* (New York: Anchor, 1999).

Hodkinson, S. "Was Classical Sparta a Military Society?" in *Sparta and War,* ed. S. Hodkinson and A. Powell (Swansea: Classical Press of Wales, 2006), 111–162.

Hornblower, S. "Sticks, Stones, and Spartans: The Sociology of Spartan Violence," in *War and Violence in Ancient Greece,* ed. H. van Wees (London: Duckworth, 2000), 57–82.

Kagan, K. *The Eye of Command* (Ann Arbor: University of Michigan, 2006).

Keegan, J. *The Face of Battle* (New York: Penguin, 1976).

Larsen, J.A.O. *Greek Federal States: Their Institutions and History* (Oxford: Clarendon, 1968).

Lazenby, J. F. *The Spartan Army* (London: Aris and Philips, 1985).

Lewis, D. M. *Sparta and Persia* (Leiden: Brill, 1977).

Liddell Hart, B. H. *Strategy,* 2nd ed. (New York: Meridien, 1991).

Lloyd, A. B., ed. *Battle in Antiquity* (London: Duckworth, 1996).

Mises, Ludwig von. *Omnipotent Government* (Spray Hill, PA: Libertarian Press, 1985).

Rand, A. "The New Fascism: Rule by Consensus," in *The Ayn Rand Column,* 2nd ed., ed. M. Podritske and P. Schwartz (New Milford, CT: Second Renaissance, 1998), 95–111.

———. "The Roots of War," in *Capitalism the Unknown Ideal* (New York: Signet, 1967), 35–43.

Roberts, W. R. *The Ancient Boiotians* (Whitefish, MT: Kessinger Publishing, 2007).

Roebuck, C. *A History of Messenia from 369 to 146 B.C.* (Chicago: University of Chicago Press, 1941).

Sekunda, N. *The Spartan Army* (Oxford: Osprey Publishing, 2000).

Sun-tzu. *The Art of War,* tr. R. D. Sawyer (Boulder, CO: Westview, 1994).

Vidal-Naquet, P. "Epaminondas the Pythagorean, or the Tactical Problem of Right and Left," in *The Black Hunter: Forms of Thought and Forms of Society in the Greek*

World, tr. tr. A. Szegedy-Maszak, (Baltimore: Johns Hopkins University Press, 1986), 61–84.

CHAPTER 3: THE SECOND PUNIC WAR

Primary Sources

Aristotle. *The Rhetoric and Poetics of Aristotle,* tr. W. Rhys Roberts and I. Bywater (New York: McGraw Hill, 1984).

Cicero. *De Officiis,* tr. W. Miller (Cambridge, MA: Harvard University Press, 1997).

Cornelius Nepos. *On Great Generals, On Historians,* tr. J.C. Rolfe (Cambridge, MA: Harvard University Press, 1929).

Diodorus of Sicily. *Library of History,* vol. 12, tr. F. W. Walton (Cambridge, MA: Harvard University Press, 1967).

Jacoby, F., ed. *Die Fragmente der Griechischen Historiker* (Leiden: Brill, 1958).

Justin. *Epitome of the Philippic History of Pompeius Trogus,* tr. J. C. Yardley (Atlanta: Scholars Press, 1994).

Livy. *The Early History of Rome,* tr. A. de Sélincourt (New York: Penguin, 1971).

———. *The War with Hannibal,* tr. A. de Sélincourt (New York: Penguin, 1972).

Macrobius. *Commentary on the Dream of Scipio,* tr. A. H. Stahl (New York: Columbia University Press, 1952).

Polybius. *Rise of the Roman Empire,* tr. I. Scott-Kilvert (New York: Penguin, 1980).

Pritchard, J. B. *Ancient Near Eastern Texts Relating to the Old Testament,* 3rd ed. with suppl. (Princeton: Princeton University Press, 1969).

Sallust. *The Jugurthine War / The Conspiracy of Catiline,* tr. S.A. Handford (New York: Penguin, 1963).

Tiberius Catius Silius Italicus. *Punica,* tr. J. D. Duff (Cambridge, MA: Harvard University Press, 1934).

Secondary Works

Astin, A. E. *Scipio Aemilianus* (Oxford: Clarendon, 1967).

Bagnall, N. *Essential Histories: The Punic Wars* (Oxford: Osprey, 2002).

Baker, G. P. *Hannibal* (1930; New York: Barnes and Noble, 1967).

The Cambridge Ancient History, vol. VIII, *Rome and the Mediterranean, 218–133 BC,.* ed. S. A. Cook, F. E. Adcock, and M. P. Charlesworth (Cambridge: Cambridge University Press, 1965).

Cary, M., and H. H. Scullard. *A History of Rome,* 3rd ed. (London: Macmillan, 1975).

Caven, B. *The Punic Wars* (London: Weidenfeld and Nicolson, 1980).

Daly, G. *Cannae: The Experience of Battle in the Second Punic War* (London: Routledge, 2002).

De Beer, G. *Alps and Elephants: Hannibal's March* (London: Geoffrey Bles, 1955).

———. *Hannibal: Challenging Rome's Supremacy* (New York: Viking, 1969).

Dillon, S., and K. E. Welch. *Representations of War in Ancient Rome* (Cambridge: Cambridge University Press, 2006).

Dodge, T. A. *Great Captains: Hannibal*, 2 vols. (Boston: Houghton Mifflin, 1891).

Dorey, T. A., and D. R. Dudley. *Rome against Carthage* (New York: Doubleday, 1972).

Errington, R. M. *The Dawn of Empire: Rome's Rise into History* (Ithaca, NY: Cornell University Press, 1972).

Gabriel, R. *Scipio Africanus: Rome's Greatest General* (Dulles, VA: Potomac Books, 2008).

Goldsworthy, A. *Cannae* (London: Cassell, 2001).

Harris, W. V. *War and Imperialism in Republican Rome, 327–70 BC* (Oxford: Clarendon, 1979).

Healey, M. *The Battle of Cannae: Hannibal's Greatest Victory* (Oxford: Osprey, 1994).

Hoyos, D. *Hannibal's Dynasty: Power and Politics in the Western Mediterranean, 247–183 BC* (New York: Routledge, 2003).

Kagan, D. *On the Origins of War and the Preservation of Peace* (New York: Anchor, 1996).

Lancel, S. *Carthage: A History*, tr. A. Nevill (Oxford: Blackwell, 1997).

Lazenby, J. F. *The First Punic War: A Military History* (London: UCL Press, 1996).

———. *Hannibal's War: A Military History of the Second Punic War* (Warminster: Aris and Philips, 1978).

———. "Was Maharbal Right?" in *The Second Punic War: A Reappraisal*, ed. T. Cornell, B. Rankov, and P. Sabin. (London: Institute of Classical Studies, 1996), 39–48.

Liddell Hart, B. H. *Scipio Africanus: Greater than Napoleon* (New York: Da Capo, 1994).

Luttwak, E. *The Grand Strategy of the Roman Empire* (Baltimore: Johns Hopkins University Press, 1976).

Mattern, S. B. *Rome and the Enemy: Imperial Strategy in the Principate* (Berkeley: University of California Press, 2002).

Morris, W. O. *Hannibal and the Crisis of the Struggle between Carthage and Rome* (New York: Putnam's Sons, 1897).

Peddie, J. *Hannibal's War* (Phoenix Mill: Sutton, 1997).

Picard, G. C., and C. Picard. *The Life and Death of Carthage: A Survey of Punic History and Culture from Its Birth to the Final Tragedy*, tr. D. Collion (New York: Taplinger Publishing, 1969).

Prevas, J. *Hannibal Crosses the Alps: The Invasion of Italy and the Punic Wars* (Cambridge MA: Da Capo, 1998).

Proctor, D. *Hannibal's March into History* (Oxford: Oxford University Press, 1994).

Reid, J. S. "Problems of the Second Punic War: Rome and Her Italian Allies," *Journal of Roman Studies* 15 (1915): 87–124.

Rich, J. W. *Declaring War in the Roman Republic in the Period of Transmarine Expansion* (Brussels: Latomus, 1976).

———. "The Origins of the Second Punic War," in *The Second Punic War: A Re-appraisal*, ed. T. Cornell, B. Rankov, and P. Sabin (London: Institute of Classical Studies, 1996), 1–33.

Scullard, H. H. *Roman Politics: 220–150 B. C.* (Oxford: Clarendon, 1973).

———. *Scipio Africanus in the Second Punic War* (Cambridge: at the University Press, 1930).

———. *Scipio Africanus: Soldier and Politician* (Ithaca, NY: Cornell University, 1970).

Smith, R. B. *Carthage and the Carthaginians* (London: Longmans, Green, 1913).

Walbank, F. *Commentary on Polybius* (Oxford: Oxford University Press, 1999).

Warmington, B. H. *Carthage* (New York: Barnes and Noble, 1994).

Chapter 4: The Campaigns of Aurelian

Primary Sources

Ammianus Marcellinus. *The History*, tr. W. Hamilton and A. Wallace-Hadrill (New York: Penguin, 1986).

Aurelius Victor. *De Caesaribus*, tr. H. W. Bird (Liverpool: Liverpool University Press, 1994).

Blockley, R. C. *The Fragmentary Classicising Historians of the Later Roman Empire: Eunapius, Olympiodorus, Priscus and Malchus*, 2 vols. (Liverpool: Liverpool University Press, 1981–1983).

Chaffin, C. *Olympiodorus of Thebes and the Sack of Rome: A Study of the Historikoi Logoi, with Translated Fragments, Commentary and Additional Material* (Lampeter: Edwin Mellen, 1993).

Cicero. *On Duties*, tr. W. Miller (Cambridge, MA: Harvard University Press, 1997).

———. *Selected Political Speeches*, tr. M. Grant (New York: Penguin, 1989).

Dimaio, M. "Zonaras' Account of the Neo-Flavian Emperors: A Commentary" (Ph.D. diss., University of Missouri-Columbia, 1977).

Dio Cassius. *Roman History*, vol. VIII, *Books 61–70*, tr. E. Cary (Cambridge, MA: Harvard University Press, 1925).

Dodgeon, M. H., and S.N.C. Lieu, eds. *The Roman Eastern Frontier and the Persian Wars (AD 226–363)* (London: Routledge, 1991).

Eusebius. *The History of the Church from Christ to Constantine*, tr. A. Louth (New York: Penguin, 1989).

Eutropius. *Breviarium*, tr. H. W. Bird (Liverpool: Liverpool University Press, 1993).

Festus. *Breviarium*, tr. T. M. Banchich and J. A. Meka, Canisius College Translated Texts, no. 2 (Buffalo, NY, 2001), http://www.roman-emperors.org/festus.htm.

Gregory of Tours. *History of the Franks*, tr. L. Thorpe (New York: Penguin, 1974).

Herodian. *History*, in 2 vols. tr. C. R. Whittaker (Cambridge, MA: Harvard University Press, 1969).

———. *History of the Roman Empire* (Berkeley: University of California Press, 1961).

Jacoby, F., ed. *Die Fragmente der Griechischen Historiker* (Leiden: Brill, 1958).

Jordanes. *The Origin and Deeds of the Goths,* tr. Charles C. Mierow, http://www.ucalgary.ca/~vandersp/Courses/texts/jordgeti.html.

Josephus. *Jewish Antiquities* (London: Wordsworth, 2006).

Marcus Aurelius. *Meditations,* tr. G. Long (Mineola, NY: Dover, 1997).

Orosius. *The Seven Books of History against the Pagans,* tr. R. J. Deferrari (Washington, DC: Catholic University Press, 1964).

Pliny. *Natural History: A Selection*, tr. J. F. Healy (New York: Penguin, 1991).

Mattingly, H., et al, *Roman Imperial Coinage* (*RIC*) (London: Spink and Son, 1923). *Virtual Catalog of Roman Coins*, http://artemis.austincollege.edu/acad/cml/rcape/vcrc/.

Sallust. *The Jugurthine War / The Conspiracy of Catiline,* tr. S. A. Handford (New York: Penguin, 1963).

The Scriptores Historiae Augustae, tr. D. Magie (Cambridge, MA: Harvard University Press, 1998).

Tacitus. *The Complete Works,* ed. M. Hadas, tr. A. J. Church and W. J. Brodribb (New York: Modern Library, 2003).

Zonaras. *Ioannis Zonarae Annales*, in *Corpus Scriptorum Historiae Byzantinae*, vols. 29–31, ed. M. Pinder (Bonn: Weber, 1841–1897).

Zosimus. *New History*, tr. R. T. Ridley (Sydney: Australian Association for Byzantine Studies, 1982).

Zosimus: Hostoria Nova; The Decline of Rome, tr. J. J. Buchanan and H. T. Davis (San Antonio, TX: Trinity University, 1967).

Secondary Works

Arnheim, M.T.W. *The Senatorial Aristocracy in the Later Roman Empire* (Oxford: Clarendon, 1972).

Barnes, T. D. *Constantine and Eusebius* (Cambridge, MA: Harvard University Press, 1981).

Barton, C. A. "The Price of Peace in Ancient Rome," in *War and Peace in the Ancient World*, ed. K. Raaflaub (Oxford: Blackwell, 2007), 245–255.

Beard, M., J. North, and S. Price. *Religons of Rome,* vol. I, *A History* (Cambridge: Cambridge University Press, 1998).

De Blois, L. *The Policy of the Emperor Gallienus* (Leiden: Brill, 1976).

Bowder, D. *The Age of Constantine and Julian* (London: Paul Elek, 1978).

Bradley, H. *The Story of the Nations: The Goths* (London: T. Fisher Unwin, 1888).

Burns, T. *A History of the Ostrogoths* (Bloomington: Indiana University, 1984).

The Cambridge Ancient History, vol. XII, *Imperial Crisis and Recovery*, ed. S. A. Cook, F. E. Adcock, M. P. Charlesworth, and N. H. Baynes (Cambridge: Cambridge University Press, 1965).

The Cambridge Ancient History, vol. XII, *The Crisis of Empire, A.D. 193–337*, 2nd ed., ed. A. K. Bowman, P. Garnsey, and A. Cameron (Cambridge: Cambridge University Press, 2005).

Drinkwater, J. *The Gallic Empire: Separatism and Continuity in the North-Western Provinces of the Roman Empire, A.D. 260–274* (Stuttgart: Franz Steiner, 1987).

Ferguson, J. *The Religions of the Roman Empire* (London: Thames and Hudson, 1982).

Frye, R. N. *The History of Iran* (Munich: Beck'sche, 1984).

Gershevitch, I. "Die Sonne das Beste," in *Mithraic Studies: Proceedings of the First International Congress of Mithraic Studies*, vol. 1, ed. J. Hinnells (Manchester: Manchester University, 1975), 69–89.

Gibbon, E. *The Decline and Fall of the Roman Empire*, 3 vols. (New York: Modern Library).

Grant, M. *The Emperor Constantine* (London: Weidenfeld and Nicolson, 1993).

Halsberghe, G. H. *The Cult of Sol Invictus* (Leiden: Brill, 1972).

Heather, P. J. *The Fall of the Roman Empire: A New History of Rome and the Barbarians* (Oxford: Oxford University Press, 2006).

———. *The Goths* (Oxford: Blackwell, 1996).

Isaac, B. *The Limits of Empire: The Roman Army in the East*, rev. ed. (Oxford: Oxford University Press, 1992).

Jones, A.H.M. *A History of Rome through the Fifth Century*, vol. II, *The Empire* (New York: Harper and Row, 1970).

Kagan, K. *Eye of Command* (Ann Arbor: University of Michigan, 2006).

———. "Redefining Roman Grand Strategy," *Journal of Military History* 70.2 (2006): 333–362.

LeGlay, M., J.-L. Voisin, and Y. Le Bohec. *A History of Rome*, 2nd ed., tr. A. Nevill (Oxford: Blackwell, 2003).

Luttwak, E. N. *Grand Strategy of the Roman Empire: From the First Century A.D. to the Third* (Baltimore: Johns Hopkins University Press, 1976).

MacMullen, R. *Constantine* (New York: Dial, 1969).

———. *Paganism in the Roman Empire* (New Haven: Yale University Press, 1981).

Mattern, S. B. *Rome and the Enemy* (Berkeley: University of California Press, 1990).

Millar, F. "Government and Diplomacy in the Roman Empire during the First Three Centuries," *International History Review* 10.3(1988): 345–377.

———. "P. Herennius Dexippus: The Greek World and the Third-Century Invasions," *Journal of Roman Studies* 59 (1969): 12–29.

———. *The Roman Near East, 31BC–AD 337* (Cambridge, MA: Harvard University Press, 1993).

Mommsen, T. *The Provinces of the Roman Empire from Caesar to Diocletian*, 2 vols. in 1 (New York: Barnes and Noble, 1996).

Moorhead, J. *The Roman Empire Divided, 400–700* (New York: Pearson Education Limited, 2001).

Pennella, R. J. *Greek Philosophers and Sophists in the Fourth Century A.D.* (Leeds: University of Leeds Press, 1990).

Potter, D. S. *The Roman Empire at Bay, AD 180–395* (New York: Routledge, 2004).

Rich, J., and G. Shipley. *War and Society in the Roman World* (New York: Routledge, 1993).

Rives, J. B. *Roman Religion* (Oxford: Blackwell, 2007).

Rostovtzeff, M. I. *The Social and Economic History of the Roman Empire,* ed. P. M. Fraser (Oxford: Oxford University, 1957).

Saunders, R. T. "A Biography of the Emperor Aurelian" (Ph.D. diss., University of Cincinnati, 1991).

Starcky, J., and M. Gawlikowski. *Palmyre* (Paris: Libraire d'Amerique et d'Orient, 1985).

Teixidor, J. *The Pagan God* (Princeton: Princeton University Press, 1977).

Ward, A. M., et al. *A History of the Roman People,* 4th ed. (Upper Saddle River, NJ: Prentice Hall, 2003).

Ware, W. *Aurelian* (London: Estes and Lauriat, 1850).

Watson, A. *Aurelian and the Third Century* (New York: Routledge, 1999).

White, J. F. *Restorer of the World* (Staplehurst: Spellmount Publishers, 2004).

Whittaker, C. R. *Frontiers of the Roman Empire* (Baltimore: Johns Hopkins University Press, 1994).

———. "Where Are the Frontiers Now?" in *The Roman Army in the East,* ed. D. L. Kennedy and D. Braund (Ann Arbor: University of Michigan, 1996), 25–42).

Chapter 5: Sherman's March through the American South

Primary Sources

Aristotle. *Politics,* tr. T. Sinclair, rev. ed. T. J. Saunders (New York: Penguin, 1981).

Buck, L. R. *Shadows on My Heart: The Civil War Diary of Lucy Rebecca Buck of Virginia*, ed. E. R. Baer (Athens: University of Georgia Press, 1997).

Bull, Sgt. Rice C. *Soldiering: The Civil War Diaries of Rice C. Bull, 123rd New York Volunteer Infantry,* ed. K. J. Bauer (San Rafael: Presidio, 1977).

Clausewitz, Carl von. *On War,* ed. and tr. M. Howard and P. Paret (New York: Routledge, 2004).

Clemson, F. *A Rebel Came Home: The Diary and Letters of Floride Clemson, 1863–1866* (Columbia: University of South Carolina, 1989).

Commager, H. S. *Fifty Basic Civil War Documents* (Malabar, FL: Robert E. Krieger Publishing, 1982).

Cooke, J. E. *Wearing of the Gray: Being Personal Portraits, Scenes and Adventures of the War* (Bloomington: Indiana University Press, 1959).

Grant, U. S. *Personal Memoirs of U. S. Grant,* ed. E. B. Long (New York: Da Capo Press, 1982).

Jomini, Baron de. *Summary of the Art of War,* tr. Capt. G. H. Mendell (Philadelphia: J. B. Lippincott, 1873).

McClellan, G. *The Civil War Papers of George B. McClellan: Selected Correspondence,* 1860–1865, ed. S. W. Sears (New York: Da Capo Press, 1992).

Nichols, G. W. *The Story of the Great March from the Diary of a Staff Officer* (Williamstown, MA: Corner House Publishers, 1984).

Osborn, T. W. *The Fiery Trail: A Union Officer's Account of Sherman's Last Campaigns*, ed. R. Harwell and P. N. Racine (Knoxville: University of Tennessee, 1986).

Paine, T. *Common Sense, The Rights of Man, and Other Essential Writings* (New York: Meridian, 1969).

Porter, H. "The Surrender at Appomattox Court House," in *Battles and Leaders of the Civil War*, vol. IV, ed. R. U. Johnson and C. C. Buel (New York: Thomas Yoseloff, 1956), 729–746.

Sherman, W. T. *From Atlanta to the Sea* (London: Folio Society, 1961).

———. "The Grand Strategy of the Last Year of the War," in *Battles and Leaders of the Civil War*, vol. IV, ed. R. U. Johnson and C. C. Buel (New York: Thomas Yoseloff, 1956), 247–259.

———. *Memoirs of General William T. Sherman*, 2 vols. (Bloomington: Indiana University Press, 1957).

———. *Sherman's Civil War: Selected Correspondence of William T. Sherman, 1860–1865*, ed. B. D. Simpson and J. V. Berlin (Chapel Hill: University of North Carolina, 1999).

Thompson, C .B., ed. *Antislavery Political Writings, 1833–1860: A Reader* (Armonk: M. E. Sharpe, 2003).

Upson, F. T. *With Sherman to the Sea: The Civil War Letters, Diaries, and Reminiscences of Theodore F. Upson*, ed. O. O. Winther (Baton Rouge: Louisiana State University, 1943).

The War of the Rebellion: A Compilation of the Official Records of the Union and Confederate Armies (known as O.R.), 128 vols. (Washington, DC: Government Printing Office, 1880–1901).

Watkins, S. *"Company Aytch"; or, A Side Show of the Big Show and Other Sketches*, edited with an introduction by M. Thomas Inge (New York: Plume, 1999).

Welles, G. *Diary of Gideon Welles* (Boston: Houghton Mifflin, 1911).

Wheeler, R. *We Knew William Tecumseh Sherman* (New York: Thomas Y. Crowell Company, 1977).

Secondary Works

Castel, A. *Decision in the West: The Atlanta Campaign of 1864* (Lawrence: University Press of Kansas, 1995).

Catton, B. *Grant Takes Command* (New York: Little, Brown, 1968).

Caudill, E., and P. Ashdown. *Sherman's March in Myth and Memory* (Lanham: Rowman and Littlefield, 2008).

Coburn, M. *Terrible Innocence: General Sherman at War* (New York: Hippocrene Books, 1993).

Cox, J. D. *The March to the Sea, Franklin and Nashville* (New York: Charles Scribner's Sons, 1882).

Eicher, D. *The Longest Night: A Military History of the Civil War* (New York: Simon and Schuster, 2001).

Fredrickson, George M. *The Inner Civil War: Northern Intellectuals and the Crisis of the Union* (New York: Harper and Row, 1965).

Glatthaar, J. *The March to the Sea and Beyond: Sherman's Troops in the Savannah and Carolinas Campaigns* (Baton Rouge: Louisiana State University, 1995).

Grimsley, M. *The Hard Hand of War: Union Military Policy toward Southern Civilians, 1861–1865* (Cambridge: Cambridge University Press, 1995).

Hansen, H. *The Civil War* (New York: Signet, 2002).

Hanson, V. D. "The Dilemmas of the Contemporary Military Historian," in *Reconstructing History: The Emergence of a New Historical Society,* ed. E. Fox-Genovese and E. Lasch-Quinn (New York: Routledge, 1999), 189–201.

———. *The Soul of Battle from Ancient Times to the Present Day: How Three Great Liberators Vanquished Tyranny* (New York: Anchor, 1999).

Hitchcock, H. *Marching with Sherman* (Lincoln: University of Nebraska, 1995).

Hughes, N. *Bentonville: The Final Battle of Sherman and Johnston* (Chapel Hill: University of North Carolina, 1996).

Kennedy, F. H., ed. *The Civil War Battlefield Guide* (Boston: Houghton Mifflin, 1990).

Lewis, L. *Sherman: Fighting Prophet* (New York: Smithmark, 1994).

Liddell Hart, B. H. *Sherman: Soldier, Realist, American* (New York: Da Capo, 1993).

Lindeman, J., ed. *The Conflict of Convictions: American Writers Report the Civil War* (Philadelphia: Chilton, 1968).

McPherson, J. M. *Battle Cry of Freedom: The Civil War Era* (New York: Oxford University Press, 1998).

———. *For Cause and Comrades: Why Men Fought the Civil War* (New York: Oxford University Press, 1997).

———. *Ordeal by Fire: The Civil War and Reconstruction* (New York: McGraw-Hill, 2000).

Marszalek, J. F. *Sherman: A Soldier's Passion for Order* (New York: Free Press, 1993).

McCormick, E. L., E. McGehee, and M. Strahl. *Sherman in Georgia* (Boston: D. C. Heath, 1961).

McDonough, J. L., and J. P. Jones. *War So Terrible: Sherman and Atlanta* (New York: W. W. Norton, 1987).

Mitchell, J. B., and E. S. Creasy. *Twenty Decisive Battles of the World* (New York: Macmillan, 1964).

Risley, F. "*The Albany Patriot*, 1861–1865: Struggling to Publish and Struggling to Remain Optimistic," in *The Civil War and the Press,* ed. D. B. Sachsman, S. Kittrell Rushing, and D. R. van Tuyll (New Brunswick: Transaction Publishers, 2000), 245–256.

Royster, C. *The Destructive War: William Tecumseh Sherman, Stonewall Jackson, and the Americans* (New York: Vintage, 1991).

Schott, T. E. *Alexander H. Stephens of Georgia* (Baton Rouge: Louisiana State University, 1988).

Stout, H. S. *Upon the Altar of a Nation: A Moral History of the American Civil War* (New York: Viking, 2006).

van Tuyll, D. R. "Two Men, Two Minds: An Examination of the Editorial Commentary of Two Georgia Editors during Sherman's March to the Sea," in *The Civil War and the Press*, ed. D. B. Sachsman, S. Kittrell Rushing, and D. R. van Tuyll (New Brunswick: Transaction Publishers, 2000), 275–290.

Walters, J. B. "General William T. Sherman and Total War," *Journal of Southern History* 14.4 (1948): 447–480.

———. *Merchant of Terror: General Sherman and Total War* (New York: Bobbs-Merrill, 1973).

Williams, D. *Johnny Reb's War: Battlefield and Homefront* (Abilene, TX: McWhiney Foundation, 2000).

Chapter 6: The Prelude to World War II

Primary Sources

Avalon Project at Yale Law School, primary documents, http://www.yale.edu/lawweb/avalon (1996–2007).

Bainville, J. *Les consequences politiques de la paix* (Paris: Fayard, 1920).

"Cato" (pseudo.). *Guilty Men* (New York: Frederick A. Stokes, 1940).

Chamberlain, N. *In Search of Peace* (New York: G. P. Putnam's Sons, 1939).

EyeWitness to History, www.eyewitnesstohistory.com (2004).

First World War.com: The War to End All Wars, primary documents, http://www.firstworldwar.com/source/ (2000–2007).

Franck, H. A. *Vagabonding through Changing Germany* (New York: Harper and Brothers, 1920).

German Foreign Office. *Documents on the Events Preceding the Outbreak of the War* (New York: German Library of Information, 1940).

Hitler, A. *Hitler's Second Book: The Unpublished Sequel to Mein Kampf* (New York: Enigma Books, 2006).

———. *Hitler's Secret Conversations, 1941–1944*, ed. H. R. Trevor-Roper (New York: Signet, 1961).

———. *Mein Kampf* (New York: Houghton Mifflin, 1998).

Hobbes, T. *Leviathan: Authoritative Text; Backgrounds Interpretations*, ed. R. E. Flatman and D. Johnston (New York: Norton, 1996).

Kant, I. *Perpetual Peace: A Philosophical Essay*, tr. T. Humphrey (Indianapolis: Hackett Publishing, 1983).

Keynes, J. M. *The Economic Consequences of the Peace* (London: Harcourt Brace and Howe, 1920).

———. *A Revision of the Treaty* (London: Macmillan, 1971).

Klemperer, V. *I Will Bear Witness, 1933–1941: A Diary of the Nazi Years*, tr. M. Chalmers (New York: Modern Library, 1999).

Lloyd George, D. *The Truth about Reparations and War Debts* (New York: Fertig, 1970).

Parliamentary Debates, 5th ser. (London: HMSO, 1918–1939).

Schmidt, P.O. *Hitler's Interpreter* (London: Heinemann, 1951).

Schuman, F. L. *International Politics*, 2nd ed. (New York: McGraw Hill, 1937).

Shirer, W. *Berlin Diary* (New York: Alfred A. Knopf 1941).

Veblen, T. *An Inquiry into the Nature of the Peace* (New York: Macmillan, 1917).

Voigt, F. A. *Unto Caesar* (New York: G. P. Putnam's Sons, 1938).

World War I Document Archive, http://wwi.lib.byu.edu (2007).

Secondary Works

Adams, R. J. Q. *British Politics and Foreign Policy in the Age of Appeasement, 1935–39* (Stanford, CA: Stanford University, 1993).

Adamthwaite, A. P. *The Making of the Second World War* (New York: Routledge, 1989).

Baylen, J. O., and E. L. Evans. "History as Propaganda: The German Foreign Office and the 'Enlightenment' of American Historians on the War Guilt Question, 1930–1933," *Canadian Journal of History* 10.2 (1975): 185–207.

Bell, P.M.H. *The Origins of the Second World War in Europe*, 3rd ed. (London: Longman, 2007).

Boyce, R., and E. M. Robertson. *Paths to War: New Essays on the Origins of the Second World War* (Basingstoke: Macmillan Education, 1989).

Bullock, A. *Hitler: A Study in Tyranny* (New York: Harper and Row, 1962).

Caputi, R. J. *Neville Chamberlain and Appeasement* (Selinsgrove: Susquehanna University, 2001).

Churchill, W. *The Second World War*, vol. 1, *The Gathering Storm* (New York: Houghton Mifflin, 1948).

Craig, G. A. *Germany, 1866–1945* (New York: Oxford University, 1978).

Dear, I.C.B., gen. ed. *The Oxford Companion to World War II* (Oxford: Oxford University, 1995).

Dilks, D. "We Must Hope for the Best and Prepare for the Worst: The Prime Minister, the Cabinet, and Hitler's Germany," in *The Origins of the Second World War*, ed. P. Finney (London: Arnold, 1997), 43–62.

Evans, R. J. *The Coming of the Third Reich* (New York: Penguin, 2004).

Feiling, K. *The Life of Neville Chamberlain* (London: Macmillan, 1946).

Fuchser, L. W. *Neville Chamberlain and Appeasement* (New York: Norton, 1982).

Gannon, F. R. *The British Press and Germany, 1936–1939* (Oxford: Clarendon, 1971).

Geiss, I. "The Outbreak of the First World War and the German War Aims," in *1914, The Coming of the First World War*, ed. W. Laqueur and G. L. Mosse (New York: Harper and Row, 1966), 71–78.

Gilbert, M. *The Roots of Appeasement* (London: Weidenfeld & Nicolson, 1966).

Goldhagen, D. J. *Hitler's Willing Executioners* (New York: Vintage, 1997).

Guderian, H. *Panzer Leader: General Heinz Guderian* (New York: Da Capo, 2001).

Holles, E. *Unconditional Surrender* (New York: Howell Soskins Publishers, 1945).

Hughes, J. L. "The Origins of World War II in Europe: British Deterrence Failure and German Expansionism," in *The Origin and Prevention of Major Wars,* ed. R. I. Rotberg and T. K. Rabb (Cambridge: Cambridge University Press, 1993), 281–321.

Kagan, D. *On the Origins of War and the Preservation of Peace* (New York: Anchor, 1995).

Keegan, J. *The Second World War* (London: Pimlico, 1997).

Kennedy, P. M. *The Rise and Fall of British Naval Mastery* (New York: Penguin, 2004).

———. "The Tradition of Appeasement in British Foreign Policy, 1865–1939," *British Journal of International Studies* 2 (1976): 195–215.

Kennedy, P. M., and T. Imlay. "Appeasement," in *The Origins of the Second World War Reconsidered: A.J.P. Taylor and the Historians,* 2nd ed., ed. G. Martel (New York: Routledge, 1999), 116–134.

Keylor, W. R., ed. *The Legacy of the Great War: Peacemaking, 1919* (New York: Houghton Mifflin, 1998).

Koch, H. W. "Hitler and the Origins of the Second World War: Second Thoughts on the Status of Some of the Documents," in *The Origins of the Second World War: Historical Interpretations,* ed. E. M. Robertson (London: Macmillan, 1973), 158–188.

Lamb, R. *The Drift to War: 1922–1939* (London: W. H. Allen, 1989).

Liddell Hart, B. H. *The Liddell Hart Memoirs: The Later Years II* (New York: G. P. Putnam's Sons, 1965).

Marks, S. "The Myths of Reparations," *Central European History* 11(1978): 231–255.

———. "1918 and After: The Postwar Era," in *The Origins of the Second World War Reconsidered: A.J.P. Taylor and the Historians,* ed. G. Martel, 2nd ed. (New York: Routledge, 1999), 13–37.

McDonough, F. *Neville Chamberlain, Appeasement and the British Road to War* (Manchester: Manchester University Press, 1998).

Mill, J. S. "Considerations on Representative Government," from *The Collected Works of John Stuart Mill,* vol. XIX, ed. J. M. Robson (Toronto: University of Toronto, 1977).

Morris, B. *The Roots of Appeasement: The British Weekly Press and Nazi Germany during the 1930s* (London: Frank Cass, 1991).

Murray, W. *The Change in the European Balance of Power, 1938–1939: The Path to Ruin* (Princeton: Princeton University Press, 1984).

Overy, R. *The Road to War,* 2nd ed. (New York: Penguin, 1999).

Peikoff, L. *The Ominous Parallels: The End of Freedom in America* (New York: Meridian, 1993).

Rand, A. "The Anatomy of Compromise," in *Capitalism: The Unknown Ideal* (New York: Signet, 1967), 144–149.

———. "The Roots of War," in *Capitalism: The Unknown Ideal* (New York: Signet, 1967), 35–43.

Rock, W. R. *Appeasement on Trial: British Foreign Policy and Its Critics, 1938–1939* (North Haven, CT: Archon Books, 1966).

———. *British Appeasement in the 1930s* (London: Edward Arnold, 1977).

Rowse, A. L. *Appeasement: A Study in Political Decline* (New York: Norton, 1960).

Schaffer, R. *Wings of Judgment: American Bombing in World War II* (New York: Oxford University Press, 1985).

Schuker, S. A. "The End of Versailles," in *The Origins of the Second World War Reconsidered: A.J.P. Taylor and the Historians*, 2nd. ed., ed. G. Martel (New York: Routledge, 1999), 38–56.

Shirer, W. L. *The Rise and Fall of the Third Reich* (New York: Simon and Schuster, 1959).

Shotwell, J. T. *What Germany Forgot* (New York: Macmillan, 1940).

Taylor, A.J.P. *The Origins of the Second World War* (New York: Simon and Schuster, 1996).

Thompson, N. *The Anti-appeasers: Conservative Opposition to Appeasement in the 1930s* (Oxford: Clarendon, 1971).

Trevor-Roper, H. R. "A.J.P. Taylor, Hitler and the War," in *The Origins of the Second World War: Historical Interpretations*, ed. E. M. Robertson (London : Macmillan, 1973), 83–99.

Van Evera, S. "The Cult of the Offensive and the Origins of the First World War," in *Military Strategy and the Origins of the First World War: An International Security Reader*, ed. S. E. Miller, S. M. Lynn-Jones, and S. Van Evera (Princeton: Princeton University Press, 1991), 59–108.

Waite, R.G.L. *The Psychopathic God: Adolf Hitler* (New York: Da Capo, 1993).

Watt, D. C. "German Plans for the Rhineland: A Note," *Journal of Contemporary History* 1.4 (1966): 193–199.

Watt, R. M. *The Kings Depart: The Tragedy of Germany: Versailles and the German Revolution* (New York: Simon and Schuster, 1968).

Wrench, J. E. *Geoffrey Dawson and Our Times* (London: Hutchinson, 1955).

Chapter 7: The American Victory over Japan

Primary Sources

Birth of the Constitution of Japan, http://www.ndl.go.jp/constitution/e/shiryo/01shiryo.html.

Cohen, T. *Remaking Japan: The American Occupation as New Deal*, ed. H. Passim (New York: Free Press, 1987).

E-Speeches.com, http://www.espeeches.com/.

Hanover Historical Texts Project, Hanover College, Department of History, on-line resource, http://history.hanover.edu/project.html.

Holles, E. *Unconditional Surrender* (New York: Howell Soskins Publishers, 1945).

Inoguchi Rikihei, Nakajima Tadashi, and Pineau, R. *The Divine Wind* (Annapolis: Bantam, 1960).

Japan's Struggle to End the War: The United States Strategic Bombing Survey (Washington, DC: Government Printing Office, 1946).

Kodama Yoshio. *I Was Defeated* (Japan: Radiopress, 1959).

MacArthur, D. A. *Reports of General MacArthur: Japanese Operations in the Southwest Pacific Area*, vol. II, part II (Washington, DC: Government Printing Office, 1966).

Sakaguchi Ango. "Darakuron" (On Decadence), tr. I. Smith, http://mcel.pacificu.edu/aspac/papers/scholars/Smith/SAKAGUCHI.html.

Yoshida Mitsuru. *Requiem for the Battleship Yamato*, tr. Richard H. Minear (Seattle: University of Washington, 1985).

Weller, G. *First into Nagasaki: The Censored Eyewitness Dispatches on Post-Atomic Japan and Its Prisoners of War* (New York: Crown, 2006).

Secondary Works

Alperovitz, G. *Atomic Diplomacy: Hiroshima and Potsdam; The Use of the Atomic Bomb and the American Confrontation with Soviet Power* (New York: Vintage Books, 1967).

Armstrong, A. *Unconditional Surrender: The Impact of the Casablanca Policy upon World War II* (Westport, CT: Greenwood, 1974).

Asada Sadao "Japanese Perceptions of the A-Bomb Decision, 1945–1980," in *The American Military and the Far East: Proceedings of the Ninth Military History Symposium, United States Air Force Academy, 1–3 October 1980*, ed. J. C. Dixon (Washington, DC: Government Printing Office, 1980), 199–219.

———. "The Mushroom Cloud and National Psyches," in *Living with the Bomb: American and Japanese Cultural Conflicts in the Nuclear Age*, ed. L. Hein and M. Selden (Armonk: M. E. Sharpe, 1997), 173–201; originally in *Journal of American-East Asian Relations* 4 (1995): 95–116.

———. "The Shock of the Atomic Bomb and Japan's Decision to Surrender—A Reconsideration," *Pacific Historical Review* 67.4 (1998): 477–513.

Bailey, P. J. *Postwar Japan: 1945 to the Present* (Oxford: Blackwell, 1996).

Bernstein, B. J. "Understanding the Atomic Bomb and the Japanese Surrender: Missed Opportunities, Little-Known Disasters, and Modern Memory," in *Hiroshima in History and Memory*, ed. M. J. Hogan (Cambridge: Cambridge University, 1996), 38–79.

Bix, H. P. "Japan's Delayed Surrender: A Reinterpretation," in *Hiroshima in History and Memory*, ed. M.J. Hogan (Cambridge: Cambridge University, 1996), 80–115 ; originally in *Diplomatic History* 19.2 (1995): 197–225.

Blackett, P. M. S. *Fear, War and the Bomb* (New York: McGraw-Hill, 1949).

Brown, D., and R. Bruner, eds. *How I Got That Story*, by Members of the Overseas Press Club (New York: E. P. Dutton, 1967).

Brownlee, J. S. "Four Stages of the Japanese *Kokutai* (National Essence)." JSAC Conference, University of British Columbia, 2000, http://www.iar.ubc.ca/centres/cjr/seminars/semi2000/jsac2000/brownlee.pdf.

Burrell, R. S. *The Ghosts of Iwo Jima* (College Station: Texas A&M, 2006).

Butow, R.J.C. *Japan's Decision to Surrender* (Stanford, CA: Stanford University, 1954).

Campbell, A. E. "Franklin Roosevelt and Unconditional Surrender," in *Diplomacy and Intelligence during the Second World War: Essays in Honour of F. H. Hinsley*, ed. R. Langhorne (Cambridge: Cambridge University Press, 1985), 218–241.

Chang, I. *The Rape of Nanking: The Forgotten Holocaust of World War Two* (New York: Basic Books, 1997).

Clausewitz, Carl von. *On War*, ed. and tr. M. Howard and P. Paret (New York: Routledge, 2004).

Codevilla, A., and P. Seabury. *War: Ends and Means*, 2nd ed. (Washington, DC: Potomac Books, 2006).

Dixon, J. C., ed. *The American Military and the Far East: Proceedings of the Ninth Military History Symposium, United States Air Force Academy, 1–3 October 1980* (Washington, DC: Government Printing Office, 1980).

Dockrill, Saki, and L. Freedman. "Hiroshima: A Strategy of Shock," in *From Pearl Harbor to Hiroshima: The Second World War in the Pacific, 1941–1945*, ed. S. Dockrill (New York: St. Martin's, 1994), 191–212.

Dower, J. *Embracing Defeat: Japan in the Wake of World War II* (New York: W. W. Norton, 1999).

———. *Empire and Aftermath: Yoshida Shigeru and the Japanese Experience, 1878–1954* (Cambridge, MA: Harvard University Press, 1988).

———. "Japanese Cinema Goes to War," in *Japan in War and Peace: Selected Essays* (New York: New Press, 1993), 33–54.

———. "The Useful War," in *Japan in War and Peace: Selected Essays* (New York: New Press, 1993), 9–32.

———. *War without Mercy: Race and Power in the Pacific War* (New York: Random House, 1986).

Drea, E. "Previews of Hell: The End of the War with Japan," *Military History Quarterly* 7 (1995): 74–81.

Frank, R. B. *Downfall: The End of the Imperial Japanese Empire* (New York: Random House, 1999).

Fussell, P. *Thank God for the Atom Bomb* (New York: Summit Books, 1988).

Gentile, G. P. *How Effective Is Strategic Bombing? Lessons Learned from World War II to Kosovo* (New York: New York University Press, 2001).

Giangreco, D. M. "Casualty Projections for the US Invasions of Japan, 1945–1946: Planning and Policy Implications," *Journal of Military History* 61 (1997): 521–582.

Gluck, C. "The Idea of Showa," in *Showa: The Japan of Hirohito*, ed. C. Gluck and S. R. Graubard (New York: Norton, 1992), 1–26.

Hasegawa Tsuyoshi. *Racing the Enemy: Stalin, Truman, and the Surrender of Japan* (Cambridge, MA: Belknap Press, 2006).

Hata, Ikuhiko. "The Occupation of Japan, 1945–1952," in *The American Military and the Far East: Proceedings of the Ninth Military History Symposium, United States Air Force Academy, 1–3 October 1980,* ed. J. C. Dixon (Washington, DC: Government Printing Office, 1980), 92–108.

Kagan, D. "Why America Dropped the Bomb," *Commentary* 100.3 (September 1995): 17–23.

Lynn, J. *Battle: A History of Combat and Culture* (Boulder, CO: Westview, 2003).

Maddox, R. J. *Weapons for Victory: The Hiroshima Decision* (Columbia: University of Missouri Press, 1995).

Manchester, W. *American Caesar* (New York: Dell, 1983).

Passim, H. "The Occupation—Some Reflections," in *Showa: The Japan of Hirohito,* ed. C. Gluck and S. R. Graubard (New York: Norton, 1992), 107–129.

Pearlman, M. D. *Unconditional Surrender, Demobilization and the Atomic Bomb* (Fort Leavenworth, KS: Combat Studies Institute, 1996).

Rosenberg, E. *A Date Which Will Live in Infamy: Pearl Harbor in American Memory* (Durham, NC: Duke University, 2003).

Sagan, S. "The Origins of the Pacific War," in *The Origin and Prevention of Major Wars,* ed. R. I. Rotberg and T. K. Rabb (Cambridge: Cambridge University Press, 1988), 323–352.

Sills, D. L., ed. *The International Encyclopedia of the Social Sciences* (New York: Macmillan, 1968).

Takemae Eiji. *The Allied Occupation of Japan,* tr. R. Ricketts and S. Swann (New York: Continuum, 2003).

Toland, J. *The Rising Sun: The Decline and Fall of the Japanese Empire, 1936–1945* (New York: Modern Library, 1970).

Tsurumi Kazuko. *Social Change and the Individual: Japan before and after Defeat in World War II* (Princeton: Princeton University Press, 1996).

Walker, J. S. *Prompt and Utter Destruction: Truman and the Use of the Atomic Bombs against Japan* (Chapel Hill: University of North Carolina, 2004).

Weintraub, S. *The Last Great Victory: The End of World War II, July / August 1945* (New York: Dutton, 1995).

Wetzler, P. *Imperial Tradition and Military Decision Making in Prewar Japan* (Honolulu: University of Hawai'i, 1998).

Woodard, W. P. *The Allied Occupation of Japan, 1945–1952 and Japanese Religions* (Leiden: Brill, 1972).

Conclusion

Summers, H. *On Strategy: A Critical Analysis of the Vietnam War* (Novato, CA: Presidio, 1982).

Sun-tzu. *The Art of War,* tr. R. D. Sawyer (Boulder, CO: Westview, 1994).

Taylor, M. "Peace and Stability for Vietnam Is Constant U.S. Objective," http://www.ndu.edu/library/taylor/mdt-0312.pdf.

Walzer, M. *Wars Just and Unjust: A Moral Argument with Historical Illustrations,* 3rd ed. (New York: Basic Books, 2006).

INDEX